Handbook of Human Growth and Developmental Biology

Volume III
Developmental Biology of Organs and Systems

Part A:
Muscle, Blood, and Immunity

Editors
Esmail Meisami, Ph.D.
Paola S. Timiras, M.D., Ph.D.

CRC Press
Boca Raton   Ann Arbor   Boston

Library of Congress Cataloging-in-Publication Data

Developmental biology of organs and systems / editors, Esmail Meisami, Paola S. Timiras.
    p.   cm. -- (Handbook of human growth and developmental biology; v. 3)
    Includes bibliographical references.
    Contents: Pt. A. Muscle, blood, and immunity -- pt. B. Cardiovascular and respiratory development.
    ISBN 0-8493-3186-2 (pt. A). -- ISBN 0-8493-3187-0 (pt. B)
    1. Developmental biology. 2. Muscles--Differentiation. 3. Immune system--Differentiation. 4. Cardiovascular system--Differentiation. 5. Respiratory organs--Differentiation. 6. Hematopoietic system--Differentiation. I. Meisami, Esmail. II. Timiras, Paola S. III. Series.
    [DNLM: 1. Cardiovascular System--growth & development. 2. Hematopoiesis. 3. Human Development. 4. Immune System--growth & development. 5. Respiratory System--growth & development. 6. Muscles--growth & development.   WS 103 H236 1988 v. 3]
RJ131.H273   1988 vol. 3
[QP84]
612.6'5 s--dc20
[612.6]
DNLM/DLC
for Library of Congress
                                                  90-1522
                                                     CIP

    This book represents information obtained from authentic and highly regarded sources. Reprinted material is quoted with permission, and sources are indicated. A wide variety of references are listed. Every reasonable effort has been made to give reliable data and information, but the author and the publisher cannot assume responsibility for the validity of all materials or for the consequences of their use.

    All rights reserved. This book, or any parts thereof, may not be reproduced in any form without written consent from the publisher.

    Direct all inquiries to CRC Press, Inc., 2000 Corporate Blvd., N.W., Boca Raton, Florida, 33431.

© 1990 by CRC Press, Inc.

International Standard Book Number 0-8493-3186-2 (Volume III, Part A)
International Standard Book Number 0-8493-3187-0 (Volume III, Part B)

Library of Congress Card Number 90-1522
Printed in the United States

This handbook is dedicated to human progeny, with best wishes for optimal development!

PREFACE

The study of development, in particular human development, is unique and important. Not only is the unraveling of the complexities of development an exciting end in itself, but since development leads to the formation of the individual, it is a crucial means to better understand the structure and function of the adult. Indeed it would not be an exaggeration to say that the future of human society depends largely on the quality, welfare, and optimal biological development of its embryos, fetuses, neonates, infants, children, and adolescents. A proper knowledge of the biological foundation of human development and the genetic and environmental factors which regulate and/or influence its normal and abnormal course is necessary in order to optimize normal growth and development and to eradicate or minimize the harmful influences.

In the past, the study of human development has often been viewed either embryologically, stressing prenatal development, or behaviorally, focusing on postnatal psychological and social development. It is now time to consider human biological development as a continuum spanning the pre- and postnatal periods. Although the foundations of human biological development are rooted in, and share much with, animal development, there are important differences which warrant a separate and unique discipline of human growth and developmental biology. Therefore, on heuristic, altruistic, societal, scientific, and medical grounds, an urgent need exists to bring together, in one place, a comprehensive interdisciplinary treatise on human growth and development from conception to maturity, including particularly the cellular and functional aspects of developing organs and systems. It is to fulfill this need that the present handbook has been undertaken.

This handbook is intended to be useful for a wide readership, comprising professionals and researchers from diverse biological and biomedical fields, including advanced students, educators, and child development specialists. In order to facilitate its use by diverse readership, the handbook is organized along three major themes. The first theme is *neural, sensory, motor, and integrative development,* comprising the chapters in Volume I. Part A covers the basic aspects of neural development, while Part B reviews the sensory, motor, and integrative aspects of development; the genetic, hormonal, and environmental factors that regulate or influence brain development, normally and abnormally, are treated in Part C. The second theme, *endocrines, sexual development, growth, nutrition, and metabolism,* is presented in the chapters of Volume II. Part A covers the developmental aspects of endocrine glands, hormonal regulation, and reproductive and sexual development. The subjects of factors influencing growth, developmental digestion, nutrition, and metabolism are treated in Part B. The third and last theme, comprising Volume III, involves the *development biology of organs and systems.* The developmental aspects of muscle, blood, and immunity are presented in Part A and the development of cardiovascular and respiratory systems in Part B.

The authors of the individual chapters were asked to write overviews of their subject matter and not exhaustive lengthy reviews, as the latter may be found in more specialized journals, books, or series. Plus, detailed treatment may not be suitable for the readership of the handbook which is presumed to be diverse. The references provided at the end of the chapters are intended to lead the users to the appropriate original articles or review journals where the subject is discussed in more depth.

Although the authors were requested to focus on human development, this was not always possible, in which case the relevant materials from experimental animal studies are described. The inclusion of the latter materials was occasionally necessary as they elucidated mechanisms of development not known in the human or not accessible to investigation.

Although we intended the handbook to be truly comprehensive, we have not quite achieved this goal. For example, some topics, such as development of tactile senses, pain and nociception, normative growth studies, and development of the kidney, liver, and skin, are not included mainly because the original authors who undertook these parts were unable to complete their sections. As we wished to avoid further delays and were pressed for a less massive volume, we were forced to cope with these deficiencies and hope that they will be corrected in future editions.

The editors are indebted to the internationally known authors who kindly cooperated with the editors during the revisions and showed exceptional patience and forebearance over the usual delays associated with the publication of a work of this size. Special thanks are also due to the distinguished members of the Advisory Board who not only nominated prospective authors, but helped edit some of the manuscripts. For Volume III, Part A, we are particularly indebted to Profs. Mehdi Tavassoli and Rodrick Westerman, without whose invaluable contribution, the sections on blood and muscle development, respectively, would not have been possible. We are also grateful to Dr. Doherty B. Hudson for editorial assistance and to Laura Elliot, Darla Wigginton, Patricia LaForce, Sara Goering, Angela Keleher, and Angela Roberts for valuable secretarial assistance. We would like to also thank the editorial staff of CRC Press, in particular Ms. Amy Skallerup, Sandy Pearlman, Barbara Caras, Jocelyn Makepeace, and Carolyn Lea, for helping us complete this work.

E. Meisami and P. S. Timiras, Editors

THE EDITORS

Esmail (Essie) Meisami, Ph.D., was born and raised in Tehran, Iran. He attended the University of California at Berkeley where he received both the B.A. and Ph.D. degrees in Physiology. In 1971, he joined the Faculty of Science of the University of Tehran where he taught and carried out research in physiology and neuroscience and helped establish and direct the International Research Institute of Biochemistry and Biophysics and its Neurobiology Laboratory. In 1980, he returned to Berkeley where he worked until 1986 as a visiting professor and research physiologist. He is now an Associate Professor at the Department of Physiology and Biophysics and the Neuroscience Graduate Program of the University of Illinois at Urbana-Champaign. Dr. Meisami's research has been mainly concerned with the problems of brain development and the intrinsic and extrinsic factors, such as hormones and sensory stimulations, which influence it. He has pioneered the use of the olfactory system as a model for developmental neurobiology and has demonstrated the marked retarding effects of sensory deprivation on the developing olfactory bulb. In addition to authoring numerous research papers and monographs, Dr. Meisami has previously edited a book on Neural Growth and Differentiation (with Dr. M. A. B. Brazier, Raven Press, 1979). He is a member of several scientific societies and has served on the Councils of the International Brain Research Organization and International Society for Developmental Neuroscience.

Paola S. Timiras, M.D., Ph.D., is Professor of Physiology, Department of Molecular and Cell Biology, University of California, Berkeley (UCB). A native of Rome, Italy, Dr. Timiras is a graduate of the University of Grenoble, France (B.A.), University of Rome (B.A., M.D. Summa cum Laude), and the University of Montreal, Quebec, Canada (Ph.D.). Before joining the University of California (1956) she was a faculty member of the University of Montreal (1950 to 1953) and the University of Utah (1953 to 1955). She was chairman of the Department of Physiology-Anatomy at UCB from 1978 to 1984. Her interest in development and aging has led to her participation in several related societies. She has directed an innovative medical program (Health and Medical Sciences) (1973 to 1975) and a training program in developmental physiology and aging (1965 to 1976) at UCB. She has been one of the founders and president of the International Society of Developmental Neuroscience (1978 to 1981) and vice-president and president of the International Society of Psychoneuroendocrinology (1974 to 1982). Her research on the neuroendocrinology of development and aging has resulted in the publication of more than 350 papers and several books. Dr. Timiras is also a consultant for many government agencies concerned with biological research and is on the editorial board of several specialty journals. Her contributions to teaching and research have been recognized by several awards from the U.S. and abroad.

ADVISORY BOARD

Robert Balázs, M.D., Ph.D.
Netherlands Institute for Brain Research
Amsterdam, The Netherlands

Colin T. Jones, Ph.D.
Laboratory of Cellular and Developmental
  Physiology
Institute for Molecular Medicine
University of Oxford
Oxford, England

Dominick P. Purpura, M.D.
Professor and Dean
Albert Einstein College of Medicine
Bronx, New York

Mehdi Tavassoli, M.D.
Professor, Department of Medicine
University of Mississippi
Jackson, Mississippi

J. J. Van Der Werff Ten Bosch, M.D.
Professor
Department of Endocrinology, Growth
  and Reproduction
Erasmus University Medical School
Rotterdam, The Netherlands

R. A. Westerman, Ph.D.
Associate Professor
Department of Physiology
Monash University
Clayton, Victoria, Australia

CONTRIBUTORS

Edward J. Benz, Jr., M.D.
Department of Medicine and Human
  Genetics
Yale University School of Medicine
New Haven, Connecticut

Ilan Bleiberg, Ph.D.
Department of Histology and Cell
  Biology
Sackler School of Medicine
Tel-Aviv University
Tel-Aviv, Israel

Paul J. Leibson, M.D., Ph.D.
Department of Immunology
Mayo Clinic and Foundation
Rochester, Minnesota

Lee-Nien Lillian Chan, Ph.D.
Department of Human Biology,
  Chemistry, and Genetics
University of Texas Medical Branch
Galveston, Texas

W. B. Ershler, M.D.
Department of Medicine
School of Medicine
University of Wisconsin
Madison, Wisconsin

Dennis Hatcher, Ph.D.
Health Maintenance Australia
South Melbourne, Victoria, Australia

Anthony R. Hayward, M.D., Ph.D.
Department of Pediatrics
University of Colorado Health Sciences
  Center
Barbara Davis Center for Childhood
  Diabetes
National Jewish Hospital
Denver, Colorado

G. John Lang, Ph.D.
Health Australia
North Sydney, NSW, Australia

A. R. Luff, Ph.D.
Department of Physiology
Monash University
Clayton, Victoria, Australia

Ian S. McLennan, Ph.D.
Department of Anatomy
University of Otago
Dunedin, New Zealand

Esmail Meisami, Ph.D.
Department of Physiology and Biophysics
University of Illinois
Urbana, Illinois

Nancy A. Noble, Ph.D.
Department of Medicine
University of Utah
Salt Lake City, Utah

Marilyn S. Pollack, Ph.D.
Department of Microbiology and
  Immunology
Baylor College of Medicine
Histocompatibility and Clinical
  Immunology Laboratory
The Methodist Hospital
Houston, Texas

Sandra F. Schnall, M.D.
Department of Internal Medicine
Division of Hematology/Oncology
Temple University School of Medicine
Philadelphia, Pennsylvania

N. T. Shahidi, M.D.
Department of Pediatrics
School of Medicine
University of Wisconsin
Madison, Wisconsin

Myra Small, Ph.D.
Department of Histology and Cell
  Biology
Sackler School of Medicine
Tel-Aviv University
Tel-Aviv, Israel

Martin H. Steinberg, M.D.
Department of Medicine
University of Mississippi School of
  Medicine
Veterans Administration Medical Center
Jackson, Mississippi

Mehdi Tavassoli, M.D.
Department of Medicine
University of Mississippi School of
 Medicine
Jackson, Mississippi

Paola S. Timiras, M.D., Ph.D.
University of California at Berkeley
Berkeley, California
 and Universite de Bordeaux
Talence, France

CRC
Handbook of Human Growth and Developmental Biology

Volume I: Neural, Sensory, Motor, and Integrative Development
    Part A: Development Neurobiology
    Part B: Sensory, Motor, and Integrative Development
    Part C: Factors Influencing Brain Development

Volume II: Endocrines, Sexual Development, Growth, Nutrition, and Metabolism
    Part A: Endocrines and Sexual Development
    Part B: Growth, Nutrition, and Metabolism

Volume III: Developmental Biology of Organs and Systems
    Part A: Muscle, Blood, and Immunity
    Part B: Cardiovascular and Respiratory Development

TABLE OF CONTENTS

MUSCLE DEVELOPMENT

Early Development and Fusion of Muscle Cells 3
I. S. McLennan

Differentiation of Muscle Fibers: Myofibrils and Contractility 25
I. S. McLennan

Differentiation of Muscle Fibers: Biochemistry and Energetics 37
D. D. Hatcher and A. R. Luff

Skeletal Muscle Growth, Hypertrophy, Repair, and Regeneration 57
G. J. Lang and A. R. Luff

DEVELOPMENT OF BLOOD AND HEMATOPOIESIS

Developmental Changes in Blood as a Whole 77
N. T. Shahidi and W. B. Ershler

Ontogeny of Hemopoiesis 101
M. Tavassoli

Aging of the Red Cell: Metabolic Changes During Development
and Senescence 113
N. A. Noble

Developmental Patterns of Human Hemoglobin Synthesis 135
S. F. Schnall and E. J. Benz, Jr.

Genetic Abnormalities of Hemoglobin 149
M. H. Steinberg

Developmental Pattern of Red Cell Membrane 165
L.-N. L. Chan

DEVELOPMENT OF IMMUNITY AND IMMUNE SYSTEM

Development of Lymphoid Tissues 179
I. Bleiberg and M. Small

Immunity Development 197
P. Leibson and A. Hayward

The Histocompatibility System in Health, Disease, and Aging 211
M. S. Pollack

Index 231

MUSCLE DEVELOPMENT

EARLY DEVELOPMENT AND FUSION OF MUSCLE CELLS

Ian S. McLennan

EMBRYOLOGICAL ORIGIN

The basic embryology of skeletal muscle formation has been extensively studied in many species and will only be briefly described here. For more extensive information, the reader is referred to works on the embryological origin of muscles[1-3] and the mechanism of pattern formation of muscles.[3-6]

The myogenic precursors of muscles are derived from the myotomal component of somitic mesoderm, whereas the connective tissue and tendons originate from the somatopleural mesoderm.[2,7,8] The musculature of the body wall was originally thought to be formed from lateral plate mesoderm,[9] but more recent studies indicate that it is probably also derived entirely from somitic mesoderm.[10,11] Somites contribute to various groups of muscle, depending on their position in the rostral caudal axis. For instance, human extraocular muscles are derived from the mesoderm within the orbit,[12] and somites 16 to 20, in the chicken, give rise to the wing musculature.[13] Previous studies have indicated that the myogenic precursors from all limb level somites become mixed in the limb and are not selectively incorporated into particular limb muscles.[14] This view has, however, been challenged by Beresford and Lance-Jones, who contend that each muscle in a limb is derived from a particular group of somites.[15,16]

The precursors of limb muscles migrate into the limbs at the earliest stage of limb development and condense into dorsal and ventral premuscle masses within each limb segment. The division of the premuscle masses to give rise to individual muscles is temporally correlated with the arrival of motor axons and the onset of myotube formation, but is probably not causally related to either of these events.[17,18]

HISTOGENESIS

Introduction

There are at least two modes of production of myotubes in vertebrate muscles. Myotubes are initially formed by the end-to-end fusion of myoblasts within the premuscle masses. This primary generation of myotubes then acts as a template for the formation of a second generation of myotubes, from which the bulk of muscle is formed.[19,20] Some authors have suggested that the earliest formed secondary myotubes act as a template for the formation of a third generation of myotubes (tertiary myotubes). A recent immunological study of developing human muscle has provided strong evidence for the existence of primary, secondary, and tertiary myotubes.[21] In this respect, human muscle development may differ from that occurring in small laboratory animals. Studies of developing mouse,[22,23] rat,[24,25] and *Xenopus*[26] muscles have shown that myogenesis occurs in two temporally distinct phases which are separated by a period in which myotube formation does not occur. Myogenesis in the second phase cannot be further subdivided on temporal grounds into a secondary and tertiary generation.

The uncertainty about the number of myotube generations has lead to some ambiguity in the literature. For instance, Kelly and Zacks[19] refer to the existence of a primary, secondary, and tertiary generation of myotubes in rat hindlimb muscles with the latter two generations corresponding to the secondary myotubes of Ontell and Dunn.[20,24] In this chapter, I have designated all myotubes formed in the initial phase of myogenesis as being primary and have designated those in subsequent phases as being secondary, as this is the terminology that has been most widely used in the literature. Furthermore, there are distinct physiological

Table 1
DIFFERENCES BETWEEN PRIMARY AND SECONDARY MYOTUBES

1. They are formed from distinct subpopulations of myoblasts. (See section entitled Myoblast Diversity.)
2. Their morphology is different during the early stages of development. (See section entitled Secondary Myotubes.)
3. The production of secondary but not primary myotubes is dependent upon functional innervation of the developing muscle. (See section entitled Neural Influences.)
4. The generation of a myotube is one of the factors which determines whether it is becomes innervated by a fast or slow motoneuron.[29]
5. The acquisition of their ATPase staining characteristics has a different developmental sequence.[29]

and anatomical differences between the myotubes which I classify as primary and secondary (Table 1), the most demonstrative of these differences being the dependence of secondary, but not primary, myotubes on functional innervation for their formation.[27,28] This is not to assert that all primary, or secondary, myotubes are identical. Indeed, primary and secondary myotubes can both be separated into various subtypes by ATPase histochemistry[29] or myosin immunohistochemistry.[30,31]

Primary Myotubes

The primary generation of myotubes is formed from a mass of myoblasts around the time the premuscle masses are beginning to split into individual muscles. Myotube production at this stage occurs in a short burst and then ceases until the second generation of myoblasts is formed. In the rat IV lumbrical, for instance, 107 primary myotubes are formed between the 16th and 17th days of gestation. There is then a hiatus until the 19th day, when secondary myogenesis occurs and a further 800 myotubes are formed.[25] The proportion of myotubes belonging to the primary generation varies from muscle to muscle, with slow muscles having a higher proportion than fast muscles. Even in slow muscles most fibers are of secondary myotubal origin.[23]

During primary myogenesis, muscles consist of independent myotubes or small numbers of cell aggregates, depending on the muscle. Cell aggregates typically contain between one and several myotubes, as well as some undifferentiated cells. The myotubes within these clusters are in close association with each other and there are occasional tight-junctions between their plasma membranes. In most muscles there is no basal lamina or collagen between the myotubes. The myotubes are often of irregular shape, contain high levels of glycogen, and have occasional long thin cytoplasmic processes. Most of the myotubes are of similar size and state of development. During the hiatus in myotube production between primary and secondary myogenesis, the primary myotubes elongate and enlarge and there is a rapid disaggregation of the clusters of myotubes (human,[32] bat,[33] chicken,[27,34] cow,[35] mouse,[36-38] pig,[39] rat,[19,20,25] and Xenopus[26]).

Secondary Myotubes

The mode of production of secondary myotubes differs from that of the first-formed myotubes. Secondary myotubes use primary myotubes as a template for their formation. Myoblasts proliferate on the surface of the primary myotubes and then fuse to form small secondary myotubes. This process results in the formation of small clusters of myotubes, consisting of a central primary myotube surrounded by undifferentiated cells and secondary myotubes in various stages of development. The secondary myotubes often do not stretch the entire length of the primary myotube (human,[32,40,41] bat,[33] chicken,[27,34] cow,[35] mouse,[36-38] pig,[39] rat,[19,20,24,25] and Xenopus[26]).

The cell clusters present during secondary myogenesis differ in several respects from the clusters which are present during primary myotube formation:

1. There is marked variation in the size and state of development of the myotubes within a cluster.
2. The entire cluster is surrounded by a common basal lamina.
3. The myotubes are of more regular shape and the smaller myotubes do not contain high levels of glycogen.
4. Pseudopodial processes from the secondary myotubes invaginate the walls of their associated primary myotube.

A further feature of these secondary clusters is that tight junctions occur between the primary myotubes and their associated myoblasts and secondary myotubes.[19,20,22,25] Such junctions allow depolarization to spread between the connected cells and also permit the movement of small molecules between the cytoplasm of the connected cells.[42,43] Thus, during secondary but not primary myotube development, both the proliferating myoblasts and immature myotubes may be exposed to small regulatory molecules, such as Ca^{2+} and cyclic nucleotides, from a more mature myotube.

Timing of Myogenesis

The timing of production of primary and secondary myotubes varies depending on the muscle being examined. In general, myotube production is complete in the human prior to six months of gestation. In quadriceps femoris, for instance, primary myotube production begins around the 7th week of gestation and secondary myotube formation begins at 10 weeks of gestation. Myotube production is complete and all fibers have differentiated to the myofiber stage by 25 weeks of gestation.[32] In the sartorius, myotube production is complete by 21 weeks of gestation.[44] Subsequent growth of muscle is due to hypertrophy of existing muscle fibers. Skeletal muscle fiber production only occurs in response to muscle injury or disease and is not part of the mechanism of normal muscle growth.[45]

In most laboratory and domestic animals, the production of myotubes is complete prior to the birth of the animal[22,46-48] (or, in the case of chickens, prior to hatching[29]). The notable exception to this generalization is the rat, where secondary myotube production occurs in some muscles during the first few postnatal days.[24,25]

MYOBLAST DIVERSITY

Laboratory Animals

Clonal analysis of cells from developing animal and human limbs has revealed that they contain a variety of myoblasts which can be identified according to their trophic requirements,[49-54] type of myosin that they produce,[55,56] or surface characteristics.[57] The initial studies in this area identified five subpopulations of chicken myoblasts on the basis of their trophic requirements and morphology and designated them FMS I and II and CMR I to III. FMS myoblasts will form myotubes in a basic medium containing horse serum and chicken embryo extract. FMS I and II myoblasts differ from each other in that the latter are dependent on functional innervation of the muscle for their proliferation (see below). CMR myoblasts, on the other hand, will only form myotubes when exposed to medium which has been conditioned by prior culture of muscle cells. When exposed to this type of medium, CMR I myoblasts form short myotubes with only a few nuclei. CMR II and III myoblasts form long myotubes with hundreds of nuclei, but differ from each other in that CMR II myoblasts, but not CMR III, lose their ability to respond to conditioned medium after a brief exposure to basic medium.[50-52] Each subpopulation of myoblast has a distinct temporal and spatial distribution in the developing limb. CMR I myoblasts are the first formed. CMR II myoblasts become common as the number of CMR I myoblasts begins to decline and the number of CMR II myoblasts declines in turn as FMS and CMR III myoblasts become common.[51]

These changes in myoblast type occur in the proximal before the distal regions of the limb.[52,53] The generation of this proximodistal gradient is dependent on the presence of the apical ectodermal ridge,[54] which is responsible for pattern organization in the limbs.

The temporal sequence of changes of myoblast type is suggestive that the later-appearing subpopulations of myoblasts are derived from the earlier-appearing subpopulations. This, however, does not appear to be the case. The subpopulations are stable over as many as 45 cell doublings[52] and experiments using chimeric birds indicate that CMR I myoblasts are not transformed in the limb to CMR II or III myoblasts.[58] Rather, it appears as if there are at least two distinct myogenic precursors which enter the limb at different stages of development.[59]

A consequence of the temporal pattern of appearance of the various types of myoblasts is that they will be differentially incorporated into either primary or secondary myotubes. The initial formation of primary myotubes will be exclusively from CMR I myoblasts, with CMR II myoblasts contributing to the growth and elaboration of primary myotube clusters. Both CMR I and II myoblasts cease to be found in the limb once secondary myotube formation occurs. Secondary myotubes must therefore be derived from CMR III or FMS myoblasts (compare References 27, 29, 50, 52, and 53). During secondary myogenesis in rats, new nuclei are not added to primary myotubes,[60] which raises the possibility that the various subpopulations of myoblasts are restricted in their capacity to fuse with each other.

Muscle growth and regeneration are due to myosatellite cells,[45] and it has often been assumed that the properties of myoblasts and myosatellite cells are similar. Recent studies have, however, indicated that myoblasts and myosatellite cells have at least some different characteristics.[61-64] Results obtained from the study of myoblasts and muscle development are thus not necessarily applicable to muscle growth or regeneration.

Humans

The results of clonal analysis of human embryos are analogous to those obtained with chicken embryos. Myoblasts are first detected in the human limb early in the sixth week of gestation. These myoblasts have similar properties to CMR I myoblasts. More mature myoblasts with properties similar to the late-forming myoblasts in chickens are detected during the ninth week of gestation and become more common by 100 d of gestation. The timing of the appearance of the early and late types of myoblasts has a proximodistal gradient, as occurs in the chicken.[51,65]

MYOBLAST PROLIFERATION AND FUSION

Neural Influences

The initial phase of myogenesis is independent of neural influences. The splitting of the premuscle masses to form individual muscles and the production of primary myotubes occur in the absence of functional innervation of the developing muscle.[17,18,27,28,66,67] The production of secondary myotubes, in contrast, does not occur if the muscle is exposed to a cholinergic receptor antagonist[27] or is denervated.[28,68] The muscles of paralyzed embryos do not contain large numbers of unfused myoblasts, indicating that the lack of secondary myotube production in paralyzed embryos is probably due to failure of myoblast proliferation or survival rather than inhibition of myoblast fusion.[27,69] There is direct evidence for this contention, which is summarized below.

Functional denervation of the developing hindlimb by either cauterizing the spinal cord or administering a cholinergic receptor antagonist reduces the number of myoblasts in the hindlimb. This reduction is due to a selective loss of CMR III and FMS myoblasts.[70,71] The critical interaction between the nervous system and the developing chicken limb is restricted to a period between the fifth and seventh day *in ovo*. Cholinergic receptor blockage before

Table 2
EFFECT OF VARIOUS FACTORS ON MYOBLASTS *IN VITRO*

Factor	Mitosis	Fusion	Chemotaxic
FGF	+ + + +	− −	?
IGF	+ + +	+	?
Insulin	+ +	+	?
ACTH	+ +	?	?
PGI$_2$	+?	?	?
T3	+	0	?
Glucocorticoid	+	0	?
Testosterone	+	0	?
PGE$_1$	0/−	+ + +	?
Catecholamines	0/−	+ +	?
Leukotrienes	?	+?	?
TGF	0	− − − −	?
PDGF	?	?	+ + +

Note: Positive symbol (+) indicates that the substance promotes the process, whereas negative symbol (−) indicates that it is inhibitory. The relative potency of each substance is indicated by the number of positive (+) or negative (−) symbols.

or after this period has no effect on the number of myoblasts which can be cloned from 10 to 18-d old hindlimbs.[71] The loss of CMR III clones from denervated hindlimbs can be reversed by inclusion of dorsal root ganglion or spinal cord cells in the culture. Thus, denervation is not preventing the migration of CMR III precursor cells into the hindlimb.[72] The inductive effect of neurons in culture is blocked by curare, indicating that acetylcholine is an important part of the effect.[72] It is unclear, however, if it is the sole agent, as for this induction to occur, the neuron must have neurites in close contact with the myoblasts,[72] which would not be expected if the neurons were acting solely by releasing an agent as diffusable as acetylcholine. Further, carbamylcholine, a cholinergic agonist, cannot by itself induce CMR III cells. These results have been interpreted to suggest that one needs a synapse between the neuron and myoblast in order for acetylcholine to be effective.[72] Alternatively, the neurons may additionally be releasing a nondiffusable factor which influences the induction of CMR III myoblasts.

Motor and sensory neurons have been postulated to release substances which control muscle fiber differentiation.[73] The involvement of such factors in the regulation of myoblast proliferation and fusion is uncertain. Extracts of peripheral nerves stimulate muscle colony formation *in vitro*.[74] The significance of this observation cannot be judged until the active ingredient has been purified. The importance of such caution has been illustrated by the recent demonstration that at least some of the effects of nerve extracts on muscle differentiation are due to traces of transferrin and ascorbic acid in the extract. Neither substance is synthesized by neurons, although it is possible that these substances are accumulated by neurons and released around their target muscles.[75-77]

Hormonal Influences
Introduction

The process of myogenesis may be influenced by a wide variety of hormones and autocrine factors (Table 2). Our knowledge in this area is very limited and based almost entirely on studies of cultured muscle. Although these studies are useful, it is unclear to what extent results obtained from *in vitro* investigations of skeletal muscle can be applied to *in vivo* myogenesis. The major rationale for using tissue culture is to avoid the complexity of *in vivo* myogenesis. This approach assumes, however, that the proliferation and fusion of myoblasts is normal in the absence of other cell types. In my mind, there is significant doubt about this asumption. As discussed above, there is evidence that the majority of myotubes will not form *in vivo* in the absence of functional innervation. There is also circumstantial evidence that preexisting myotubes and fibroblasts may influence myoblast proliferation *in vivo*. On the basis of these observations, skeletal myoblasts would not be expected to proliferate *in vitro*, due to the absence of neural influence. However, myoblasts do proliferate *in vitro*. There are several possible explanations of this paradox:

1. Myogenesis *in vitro* may be due to a minor subpopulation of myoblasts whose induction and proliferation are independent of motoneurons.
2. The effect of motoneurons on muscle formation *in vivo* may be mediated by growth factors which stimulate myoblast proliferation *in vitro*. This could arise as a result of:

    i. Motoneurons synthesizing and releasing growth factors
    ii. Motoneurons influencing the production of growth factors by primary myotubes or by proliferating myoblasts
    iii. Motoneurons controlling the expression of growth factor receptors by myoblasts

3. Myoblasts may be producing or responding to regulatory substances *in vitro* which they do not produce or respond to *in vivo*. Numerous examples of this type of phenomenon have been described for nonmyogenic cells[78-80] and the possibility that the factors which stimulate myoblast proliferation *in vitro* do not influence myogenesis *in vivo* needs to be seriously considered. This is particularly true when myoblast cell lines have been used instead of primary cell cultures.

The use of myoblast cell lines has been favored by many investigators as they are a homogenous cell population and are easy to prepare. Primary cell cultures, in contrast, contain a mixture of cell types — fibroblasts, various types of skeletal myoblasts, and possibly other cell types such as mast cells and smooth muscle myoblasts (precursors of the vasculature of the muscles). Thus, when studies are done with primary cell cultures, or *in vivo*, it is difficult to be certain that the effect of a putative regulatory substance is due to a direct action on myoblasts and whether all or only a subpopulation of myoblasts are responding. This problem does not arise when myoblast cell lines are used. However, when myoblast cell lines are used, it is not possible to be certain that the results obtained are applicable to normal myoblasts as all myoblasts cell lines have either abnormal proliferative or fusion characteristics.

Cossu and colleagues have recently developed an *in vitro* assay for myoblast proliferation using myoblasts under both clonal and mass cell culture conditions.[81] This approach has many of the benefits associated with the use of muscle cell lines and is more likely to give physiologically relevant information. Ultimately, all results obtained from *in vitro* investigations will need to be confirmed by *in vivo* experimentation.

Somatomedins (Insulin-like Growth Factors)

The somatomedins (SM) are a family of growth factors, which includes the peptides referred to as multiple stimulating activity (MSA), insulin-like growth factors (IGF), and the SM. The exact correspondence between these peptides has not been fully elucidated, but it is thought that IGF-I and SM-C are identical and that IGF-II may correspond to one of the MSAs. SM-A is a mixture of IGF-I and II.[82] IGF-I and -II are distinct proteins with different biological functions. The main function of IGF-I is as a mediator of growth hormone function, although IGF may have some nongrowth hormone-related functions.[83] The function of IGF-II is less well understood, but has been suggested to be a fetal growth factor due to the high levels of IGF-II in fetal plasma and tissues.[84] IGF-II cross-reacts with IGF-I receptors, but IGF-I appears to have only limited ability to bind to the IGF-II receptor.[85,86]

Most studies with SM have been done using L6 myoblasts.[87] SM stimulate the rate of both proliferation and fusion of L6 myoblasts, as well as of protein synthesis and amino acid and 2-deoxyglucose uptake.[87-90] In most but not all studies, IGF-I is a more potent affector of L6 myoblasts than IGF-II.[87,90-92] Exposure of L6 myoblasts to insulin or IGF stimulates the release of an IGF-binding protein which has been speculated to enhance the action of IGF.[93] The manner in which myoblasts regulate their responsiveness to various growth factors is only beginning to be studied, but is obviously of great importance to our understanding of the action of various hormones on muscle formation.

L6 myoblasts express both IGF-I and IGF-II receptors,[90,94] but the actions of IGF-I and IGF-II have been suggested to be mediated exclusively through the IGF-I receptor on the basis of the activity of specific IGF-I analogues[90] and the lack of effect of antibodies to IGF-II receptors.[92] Conversely, the stimulation of 2-deoxyglucose uptake by IGF-II has been suggested to be mediated via the IGF-II receptor on the basis of the relative affinities of IGF-II for its own receptor and that of IGF-I.[94] The L6 clone, or the culture conditions used, is atypical in the latter study as IGF-II has a much lower cross-reactivity with the IGF-I receptor than is normally reported and the cells respond to physiological concentrations of insulin. Muscle cells in culture, including L6 myoblasts, normally only respond to pharmacological levels of insulin (see the section on insulin, below). The nature of the IGF-receptors on L6 myoblasts is further complicated by a recent report that the insulin receptor on L6 myoblasts is also a high-affinity IGF-I receptor and is immunologically different to the insulin receptor found in mature skeletal muscle.[95] At present it is unclear whether the expression of this unusual receptor is due to the transformed nature of the L6 myoblasts or is typical of myoblasts in general.[95] The determination of whether IGF-II receptors mediate any of the responses of SM on muscles is important as *in vivo* developing skeletal muscle contains 55 times more IGF-II binding sites than IGF-I binding sites.[96] This is suggestive that IGF-II has an important role in muscle development, but it should also be noted that developing muscle also contains more IGF-I receptors than mature skeletal muscle.[96] The hypothesis that IGF-I receptors mediate cell proliferation and fusion is thus still tenable. IGF-II receptors may have as yet an unidentified function which is unrelated to cell proliferation or fusion.

SM have also been reported to influence normal chicken myoblasts[97] and human[98] and rat[99] myosatellite cells. The responses of these cells to SM have not been extensively investigated, but appear to similar to that observed with L6 myoblasts. The proliferative effects of SM on rat myosatellite cells appear to involve both the IGF-I and IGF-II receptors.[97]

There are three sources of IGF which may influence myogenesis *in vivo*, blood-borne IGF, locally produced IGF, and motoneuron-derived IGF. In the rat, plasma levels of IGF-II are high around birth,[84,100] when myotube formation is occurring.[19,24] The concentration of IGF-I during this period is lower than in the adult,[100,101] but it may still cause partial activation of IGF-I receptors.

A local source of IGF is suggested by the presence of high levels of IGF-II mRNA in rat muscles around birth.[102,103] IGF-I transcripts can also be detected in developing muscle, but at a lower level than IGF-II.[104] The source of the IGF within developing muscle is unclear. IGF-I and IGF-II transcripts have been suggested to be restricted to the connective tissue of developing human muscle[104] and, conversely, to myoblasts in developing rat muscle[105] on the basis of *in situ* hybridization (see also Reference 106). The techniques used in both of these studies are, however, probably incapable of giving such precise localization. An electron microscope study with a nonradioactively labeled probe will probably be required to resolve this issue.

A dual source of SM is a possibility given that both myoblasts[107] and fibroblasts[108] can secrete SM *in vitro*. The secretion of SM by fibroblasts is stimulated by placental lactogen[109] and the secretion by myoblasts inhibited by growth hormone.[107] The plasma concentration of growth hormone[109] is, however, too low during the period of myotube production to influence SM secretion. The effects of growth hormone on muscle development appear to be entirely due to an influence on muscle growth rather than formation.

The possibility that neuron-derived SM may influence muscle formation is more speculative. IGF-I occurs in the axons and Schwann cells of the sciatic nerve, and when the nerve is crushed, IGF-I accumulates proximal to the crush, indicating that it is normally transported down the sciatic nerve.[110,111] At present it is unknown whether immature motoneurons also produce SM and whether SM in the sciatic nerve are released within muscle.

Fibroblast Growth Factor

Acidic and basic fibroblast growth factors (FGF) are growth factors which influence the development of a wide range of cells. Depending on the cell type, FGF can have proliferative, antiproliferative, trophic, or other actions.[112,113] FGF is both a potent mitogen and inhibitor of fusion of mouse MM14 myoblasts,[114-117] with the basic form being 30-fold more potent than the acidic form.[116] MM14 myoblasts will not fuse while FGF is present in the culture medium. Fusion only occurs once the FGF in the medium has been consumed and can be prevented by continued addition of FGF. In the absence of FGF, MM14 myoblasts permanently withdraw from the mitotic cycle and fuse. The loss of the capacity to divide during differentiation has been postulated to result from a loss of cell surface FGF receptors.[117] The mechanism by which FGF inhibits fusion is unknown. The possiblity that FGF inhibits fusion as a simple consequence of stimulating mitosis is an attractive idea, but one which seems to be incorrect.[87,118] Studies of other regulatory substances indicate quite clearly that the regulation of myoblast proliferation and myoblast differentiation is not linked: FGF stimulates mitosis and inhibits differentiation; IGF stimulates proliferation and fusion; transforming growth factor-β (TFG) inhibits fusion, but does not stimulate mitosis.

The response of various myoblast cell lines to FGF is variable. For instance, the differentiation of L6 myoblasts, which are used for the study of SM appears not to be influenced by FGF.[87] This illustrates a problem with the use of cell lines, namely, it is difficult to know whether they are exhibiting properties which are characteristic of myosatellite cells, a subpopulation of myoblasts, or of transformed cells. As indicated below, the response of chicken myoblasts to FGF has some characteristics not seen with any of the cell lines. It is unknown whether the properties of mammalian myoblasts and myosatellite cells are similar to chicken myoblasts, L6 myoblasts, or MM14 myoblasts.

Bovine,[119] chicken,[120,121] and mouse[122] myoblasts and rat satellite cells[123-125] have been reported to be responsive to FGF. The FGF used in most of these studies was impure and the relative potencies of the acidic and basic forms of FGF on normal myoblasts are unknown. Seed and Hauschka have recently studied the responsiveness of chicken myoblasts to FGF and have observed that chicken myoblasts vary in their response to FGF.[54] The differentiation of all myoblasts was slowed, but not inhibited as occurs with MM14 cells. Furthermore, a subpopulation of clones grew in the absence of FGF, but required FGF to differentiate. Thus, these myoblasts require FGF for differentiation, although an agent other than FGF may trigger their differentiation. Myoblasts of this type were observed in both the early and late subclasses of myoblasts.[54]

It is difficult to judge whether FGF may be an important regulator of myogenesis *in vivo* due to the lack of available information about its occurrence in developing muscles. FGF can be detected in developing limbs, albeit at a low level, but it is not known whether this FGF is localized in the developing muscles, bones, or skin.[126] Mature skeletal muscles, spinal cord, and peripheral nerves contain an FGF-like substance,[121,127] which raises the possibility that there is either a local source of FGF within muscle or a motoneuron-derived source. Investigations of the level of FGF in developing nerve and muscle are required.

Transforming Growth Factor-β

Transforming growth factor-β (TGF) belongs to a family of related growth factors which influence the proliferation and differentiation of many cell types.[128,129] TGF has recently been shown to inhibit the fusion of various myoblast cell lines[130-133] and rat myosatellite cells,[125,134] although which members of the TGF family are active has yet to be investigated.

TGF and FGF both inhibit fusion of myoblasts, but have distinctly different effects on myoblast proliferation. As discussed above, FGF stimulates the proliferation of some, but not all, myoblasts. TGF, on the other hand, has been reported to inhibit the proliferation of rat skeletal myoblasts[125] and to have no effect on the proliferation of L6 myoblasts.[130] TGF, like IGF, is a potent stimulator of amino acid uptake in L6 myoblasts. The effects of TGF and IGF on amino acid uptake are additive beyond the maximum attainable with IGF alone, indicating that they may be exerting their effects through different mechanisms.[133]

Prostaglandins

The prostaglandins are a family of chemically related compounds which are derived from essential fatty acids and are responsible for mediating and synchronizing the action of many localized groups of cells.[135] The initial evidence that prostaglandins may influence myogenesis came from a series of investigations of chicken myogenesis by Zalin.[136] Myoblasts *in vitro* synthesize prostaglandin E_1 (PGE_1)[138,139] and are also responsive it.[136,137,140] Addition of PGE_1 to cultures of chicken myoblasts[140,141] or PGE_2 to cultures of human myoblasts[138] causes the myoblasts to precociously fuse. Conversely, inhibitors of prostaglandin synthesis, such as aspirin or indomethacin, delay the fusion of chicken myoblasts,[136,141] possibly by inhibiting prostaglandin-induced changes in the membrane of the myoblast which are essential for myoblast-myoblast adhesion and membrane fusion.[142,143] On the basis of these observations, it can be speculated that the function of PGE may be to trigger myoblast fusion once the local myoblast density is sufficient to form large multinucleated myotubes. When the density of myoblasts is high, the local production of PGE will be high and myoblast fusion may be favored, whereas when the density of myoblasts is low, the production of PGE will be low and myoblasts would be expected to remain in the mitotic cycle rather than to fuse. The plausibility of this hypothesis is supported by observations that the onset of myoblast fusion *in vitro* is influenced by myoblast cell density.[144]

There are three studies which have produced results different from those discussed in the previous paragraph. Allen et al.[124] have reported that PGE_1 does not affect myotube formation and Schutzle et al.[145] and Steiner et al.[146] obtained negative results with inhibitors of prostaglandin formation. The reasons for these discrepancies in the literature are unresolved, but may be a reflection of the diversity of myoblast types. Allen et al. studied rat myosatellite cells and Steiner et al. employed a mouse myogenic cell line, MM14, the properties of which may or may not be similar to mouse myoblasts. Schutzle et al.[145] used a similar source of myoblasts to Zalin and others (11- or 12-d old chicken embryo muscles), but used twice the concentration of chicken embryo extract. The significance of this difference is unknown since the way in which PGE interacts with various growth factors has yet to be investigated.

The possibility that PGE triggers myotube formation has received some support from *in vivo* investigations. The levels of PGE_1 and PGE_2 in developing chicken muscles are an order of magnitude higher just prior to the onset of secondary myogenesis than after myotube formation is complete.[147] Furthermore, addition of PGE_1 or PGE_2 to developing chicken embryos reduces the number of myotubes formed in their muscles, a result which is consistent with PGE inducing myoblasts to prematurely leave the mitotic cycle.[147,148]

The action of prostaglandins is, however, probably not limited to switching myoblasts from a mitotic to a fusion mode. Reducing the levels of PGE with inhibitors of prostaglandin synthesis would be expected to delay the exit of myoblasts from the mitotic cycle and thus produce more myoblasts. However, prolonged inhibition of prostaglandin synthesis in vitro reduces human myoblast density[138] and in vivo reduces the number of myotubes formed in chicken embryo muscles.[148] One explanation of these observations is that a prostaglandin is acting as a myoblast mitogen. Preliminary observations suggest that mitogenic potential PGF_α and PGI_2 need to be investigated. Chicken embryo muscles contain high levels of PGF_α[147] and $PGF_{2\alpha}$ increases myoblast production in cultures of human myoblasts.[138] Administration of high levels of PGI_2, but not $PGF_{1\alpha}$ or $PGF_{2\alpha}$, to chicken embryos increases the size of their muscles, whereas administration of an inhibitor of PGI_2 production reduces the number of muscle fibers in the chickens' muscles.[147]

Insulin

Insulin has been considered a prime candidate for regulating the growth of human fetuses. Infants exposed to high levels of insulin during embryogenesis, as a consequence of maternal diabetes, Beckwith-Wiedemann syndrome, or Nesidioblastosis, are overgrown and obese, whereas those infants with congenital absence of pancreas or insulin resistance are growth retarded.[149-152] It is unclear, however, whether the altered muscle size in these syndromes is due to an effect on muscle formation or on muscle growth or whether the effects on growth are secondary to changes in fetal nutrition or due to a direct action of insulin on developing muscle. Insulin is present in human fetal circulation from the 12th week of gestation.[153]

Insulin produces several effects on cultured myoblasts and myotubes. In serum free cultures, insulin increases the rate of proliferation of chicken and bovine myoblasts.[97,154,155] The extent of this increase is much less, however, than that produced by serum plus chicken embryo extract[154] and only half of that produced by FGF. As occurs with the SM, insulin also stimulates the rate of fusion of myoblasts.[89,156,157] It also increases the stability and differentiation of myotubes in serum free cultures, although it cannot completely substitute for serum.[155,158] Part, but not all, of these effects of insulin may be a consequence of its metabolic effects, which include stimulation of amino acid uptake into myotubes and myoblasts,[159] stimulation of glucose uptake into myotubes, but not myoblasts,[160] and inhibition of protein breakdown.[161]

All of the above effects of insulin only occur when supraphysiological levels are used. In the chicken embryo, plasma insulin levels increase from 1 to 4×10^{-11}; M during the period of myotube production.[162,163] The level of insulin which is required to produce an effect in vitro is generally around $10^{-7} M$. The significance of this discrepancy has yet to be fully resolved. High-affinity (Kd 2×10^9 and 10^{10})[164] insulin receptors have been detected in developing chicken muscle,[165] cultured myoblasts, and myotubes,[164] but not in myosatellite cells.[166] These receptors would be saturated by levels of insulin which have no observable effect on muscle cultures. This discrepancy is also observed when other insulin-sensitive tissues are cultured, and it has been suggested that the biochemical processes between insulin binding and a cellular effect are less effectively linked in vitro than in vivo. Alternatively, insulin may be producing its in vitro effects by cross-reacting with SM receptors.[159,164]

Thyroid Hormones

Thyroxine stimulates chicken myoblasts,[155] but not rat myosatellite cells,[124] to proliferate in serum free cultures. However, the levels of Triiodothyronine in chicken embryos[167] are low until hatching, which is several days after myotube formation has ceased.[29] Furthermore, the effect of thyroxine on myoblast proliferation is small and its interaction with insulin to maintain the stability of myotubes[155] may be of greater importance. Thyroid hormones presumably do not have a large effect on human muscle formation or growth since infants born with hypothyroidism are normal in size.[168]

Glucocorticosteroids

Dexamethasone, a synthetic glucocorticosteroid,[169] stimulates the proliferation of rat myosatellite cells *in vitro*[124] (see also Guerriero and Florini[170]). Paradoxically, for a substance which can promote cell division, dexamethasone also suppresses protein production in both developing[171] and regenerating[172] muscle. *In vitro*, the effect of dexamethasone can be reversed by simultaneous administration of insulin.[161]

Glucocorticosteroids are probably not involved in myotube formation in normal fetuses as fetal plasma glucocorticosteroids levels[173] are too low to exert a significant influence. Glucocorticosteroids may, however, be of relevance in some pathological situations, as elevated maternal steroid levels are known to affect fetal development.[174]

Testosterone

Testosterone also stimulates myoblast proliferation *in vitro*,[175] but the effect is minor compared to that of FGF, insulin, or the SM and has not been detected in all studies.[119] The effect of testosterone may, however, be sufficient to account for the small sex differences[176] observed in the number of nuclei in muscles of newborn humans.

Catecholamines

Catecholamines induce chicken myoblasts to fuse precociously *in vitro*.[177,178] The effects of catecholamines and PGE_1 on *in vitro* myogenesis are similar, but not identical. PGE_1 and catecholamines both cause a transient rise in cAMP, which may be one of the early events of fusion. The increase in cAMP produced by isoproterenol (a β-receptor agonist) is smaller than that produced by PGE_1. Further, the response to isoproterenol increases as fusion occurs, whereas that of PGE_1 diminishes.[177,178]

The pharmacology of the catecholamine response is complex. The capacity of noradrenaline (NE) or isoproterenol to increase cAMP is blocked by the β-receptor antagonist propranolol. Phentolamine, an α-receptor agonist, is inhibitory prior to fusion, but potentiates the effects of isoproterenol after fusion.[177,178] NE is an α-agonist with a weak β-action, whereas adrenaline (E) has mixed α- and β-effects.[179] Thus, the effect of E relative to NE decreases as myoblast fusion occurs *in vitro*.[178] Interestingly, the ratio of E to NE in the plasma of rats is high late in gestation,[180] when myotube formation is occurring, and begins to diminish around parturition,[180] when myotube formation is complete in most muscles. Although both E and NE have the same effect on cAMP, they may also produce other divergent effects. E increases the level of glucose-1,6-biphosphate, a metabolic regulator, in myotubes, but not in myoblasts.[181] However, when the β-receptor is blocked, E has the opposite effect, reducing the level of this regulator in both myoblasts and myotubes.[181] The effect of NE is unknown, but would be expected to be the same as E plus a β-receptor antagonist, i.e., opposite to the effect of E alone.

The existence of β-receptors on myoblasts has been a matter of controversy. β-Receptors are found on chicken myoblasts[178] and rat myogenic cell lines,[182-184] but not on pigeon myoblasts.[185] Human myosatellite cells have also been reported to be insensitive to catecholamines.[186] Both pigeon[185] and human[186] myotubes are, however, responsive to noradrenergic stimulation. The absence of β-receptors on myoblasts of some species does not necessarily imply that catecholamines are unimportant, as during secondary myotube formation influences may be transmitted via primary myotubes.

Platelet-Derived Growth Factor

Platelet-derived growth factor (PDGF) acts as an autocrine growth factor for many cells of mesenchymal origin and has been postulated to be involved in myogenesis.[187] Myoblasts *in vitro* synthesize and release an A, but not a B, chain PDGF-like molecule.[187] The effect of PDGF on myoblast cell division is unknown, but it may be negative, as a myoblast cell line, MM14DY, is unresponsive to PDGF.[115] PDGF does, however, cause myoblasts to migrate toward it.[188] If myoblasts secrete PDGF *in vivo* and if PDGF is chemotaxic to myoblasts *in vivo*, then a simple consequence of this would be that myoblasts would be drawn to each other. Thus, PDGF may be involved in the condensation of immature myoblasts into muscles and the coalescing of terminally differentiated myoblasts just prior to their fusion.

Leukotrienes

Mouse MM14 myoblasts respond to inhibitors of leukotriene production in a manner analogous to the way chicken myoblasts respond to inhibitors of prostaglandin production.[146] Leukotrienes are, like prostaglandins, derived from essential fatty acids, and the possible involvement of these compounds in mammalian myogenesis needs to be examined.

Proopiomelanocortin-Derived Peptides

Two proopiomelanocortin (POMC)-derived peptides, adrenocorticotropin (ACTH) and melanocyte-stimulating hormone (MSH), cause a two- to threefold increase in the proliferation of mouse myoblasts and myosatellite cells *in vitro*.[81] At present there is no known source of ACTH or MSH in developing muscle. Circulatory levels of ACTH are likely to be elevated after muscle injury, suggesting that ACTH might have a role in stimulating muscle regeneration.[81]

Growth Hormone

Growth hormone does not appear to be involved in muscle formation in either human fetuses or experimental animals.[87,150] The function of growth hormone is related to the control of growth rather than to regulation of tissue formation.

Interactions Between Hormones

As discussed above, a large number of substances are capable of influencing the proliferation and/or fusion of myoblasts *in vitro*. The properties of myoblasts *in vivo* are likely to be a reflection of interactions between some, or all, of these putative regulatory substances as well as neural influences. Most cells express receptors for a variety of growth factors, and in many instances the presence of one growth factor alters the response of a cell to another growth factor.[112] The way in which the various hormones might interact to control the proliferation of myoblasts has only begun to be investigated. Most studies to date have concentrated on the action of a substance in isolation of other regulatory influences. A notable exception to this has been the studies of rat myosatellite cells by Allen and colleagues.[124,125] They have shown that FGF, IGF, and TGF in combination promote mitosis and strongly inhibit fusion, whereas FGF and IGF together promote proliferation and fusion.[125] On the basis of these observations, they have speculated that FGF, IGF, and TGF in combination may be responsible for the period of rapid cell proliferation that occurs immediately after muscle injury.[125] Similarly, the hiatus in myotube formation between primary and secondary myotube formation may be produced by a similar combination of hormones. The available information about the action of various regulatory factors *in vivo* is, however, too scant to suggest a substantive hypothesis about how the multiphasic nature of *in vivo* myogenesis may be generated.

Basal Lamina

During secondary myotube formation, myoblasts proliferate and fuse on the surface of primary myotubes. This association may, in part, be due to collagens, laminin, and other proteins on the surface of the primary myotubes. Collagens and laminin are important constituents of the endomysium and basal lamina surrounding a muscle fiber[189-191] and small patches of collagen can be detected on the surface of myotubes at an early stage of their development.[192] *In vitro*, myoblast motility, proliferation, and fusion are dependent upon the surface of the culture medium being similarly coated with collagen[193,194] or laminin.[195,196] The type of collagen does not appear to be important, but the effect is not nonspecific as most noncollagenous proteins do not promote muscle differentiation.[194]

MUSCLE SIZE

Fiber Production

In laboratory and farm animals, the size of a muscle is related to the number of fibers within it, rather than the size of each fiber.[48,197,198] In human and wild populations of animals, muscle fiber size will differ between individuals due to nutritional factors[199,200] and work load.[201] However, within a given environment, larger people will have more muscle fibers than smaller people. As myotube production is complete by midgestation, the ultimate size of a person's muscles will be greatly influenced by events during early embryogenesis.

In pigs, differences in fiber numbers between individuals are due to differences in the production of secondary rather than primary myotubes.[48,202] The number of primary myotubes differs only slightly between individuals. The size of primary myotubes does vary, however, and this has been suggested to be an important determinant of the number of secondary myotubes which can be formed.[48]

Fiber Loss

Cell death is a feature of most developing tissues and in some structures, such as the nervous system, can be extensive.[203-205] The possibility that cell death occurs within developing muscle has received scant attention. Small numbers of degenerating myotubes have been observed during the early stages of myogenesis: in 10- to 16-week-old human fetuses[206-208] and 7-d-old chicken embryos.[10] The degenerating myotubes are characterized by extensive degenerative changes in their nuclei and sarcoplasms, with loss of cell membrane, the formation of vesicles, swollen mitochondria, and changes in the peripheral myofibrils, which become an amorphous electron-dense mass.[206] Additionally, there are muscles where extensive[209] or complete degeneration occurs during development. This extensive cell death appears to remove muscles which have no function in the adult and may be a completely different phenomenon to the occurrence of small numbers of degenerating myotubes in normal developing muscles.

Although the number of degenerating cells observed within a developing muscle is small, this is not necessarily an indication that cell degeneration is unimportant in myogenesis.[206] Degenerating cells are rapidly removed from developing tissues and for this reason they are only rarely seen, even in tissues undergoing massive degeneration. In the lateral motor column of *Xenopus laevis,* for instance, 70% of all motoneurons formed degenerate, yet during the peak of degeneration, only 3% of motoneurons show degenerative changes.[210] If extensive cell death is a part of myogenesis, then its occurrence must be limited to around the time of muscle formation, as quantitative studies have indicated that an extensive fiber loss does not occur after this period.[211]

REFERENCES

1. **Ontell, M. A.,** The growth and metabolism of developing muscle, in *Biochemical Development of the Fetus and Neonate,* Jones, C. T., Ed., Elsevier, Amsterdam, 1982, 213.
2. **Chevallier, A., Kieny, M., Mauger, A., and Sengel, P.,** Developmental fate of the somitic mesoderm in the chick embryo, in *Vertebrate Limb and Somite Morphogenesis,* Ede, D. A., Hinchcliffe, J. R., and Balls, M., Eds., Cambridge University Press, New York, 1977, 421.

3. Christ, B. and Cihak, R., *Development and Regeneration of Skeletal Muscles*, S. Karger, Basel, 1986.
4. Wolpert, L. and Hornbruch, A., Positional signalling along the anteroposterior axis of the chick wing. The effect of multiple polarising region grafts, *J. Embryol. Exp. Morphol.*, 63, 145, 1981.
5. Lance-Jones, C., The morphogenesis of the thigh of the mouse with reference to tetrapod homologies, *J. Morphol.*, 162, 275, 1979.
6. McLachlan, J. and Wolpert, L., The spatial pattern of muscle development in chick limb, in *Development and Specialization of Skeletal Muscle*, Goldspink, D. F., Ed., Cambridge University Press, Cambridge, 1980, 1.
7. Christ, B., Jacob, H. J., and Jacob, M., Differentiating abilities of avian somatopleural mesoderm, *Experientia*, 35, 1376, 1979.
8. Kaehn, K., Jacob, J., Christ, B., Hinrichsen, K., and Poelmann, R. E., The onset of myotome formation in the chick, *Anat. Embryol.*, 177, 191, 1988.
9. Straus, W. L. and Rawles, M. E., An experimental study of the origin of the trunk musculature and ribs in the chick, *Am. J. Anat.*, 92, 471, 1953.
10. Christ, B., Jacob, M., and Jacob, H. J., On the origin and development of the ventrolateral abdominal muscles in the avian embryo. An experimental and ultrastructural study, *Anat. Embryol.*, 166, 87, 1983.
11. Chevallier, A., Role of the somitic mesoderm in the development of the thorax in bird embryos. II. Origin of thoracic and appendicular musculature, *J. Embryol. Exp. Morphol.*, 49, 73, 1979.
12. Sevel, D., A reappraisal of the origin of human extraocular muscles, *Opthalmology*, 88, 1330, 1981.
13. Beresford, B., Le Lievre, C., and Rathbone, M. P., Chimera studies of the origin and formation of the pectoral muscle in the avian embryo, *J. Exp. Zool.*, 205, 321, 1978.
14. Kieny, M. and Chevallier, A., Existe-t-il une relation spatiale entre le niveau d'origine des cellules somitiques myogenes et leur localisation terminale dans l'aile?, *Arch. Anat. Microsc. Morphol. Exp.*, 69, 35, 1980.
15. Beresford, B., Brachial muscles in the chick embryo: the fate of individual somites, *J. Embryol. Exp. Morphol.*, 77, 99, 1983.
16. Lance-Jones, C., The somitic level of origin of embryonic chick hindlimb muscles, *Dev. Biol.*, 126, 394, 1988.
17. Hunt, E. A., The differentiation of chick limb buds in chorio-allantoic grafts, with special reference to the muscles, *J. Exp. Zool.*, 62, 57, 1932.
18. Hamburger, V., The development and innervation of transplanted limb primordia of chick embryo, *J. Exp. Zool.*, 80, 347, 1939.
19. Kelly, A. M. and Zacks, S. I., The histogenesis of rat intercostal muscle, *J. Cell Biol.*, 42, 135, 1969.
20. Ontell, M., Neonatal muscle: an electron microscopic study, *Anat. Rec.*, 189, 669, 1977.
21. Draeger, A., Weeds, A. G., and Fitzsimons, R. B., Primary, secondary and tertiary myotubes in developing skeletal muscle: a new approach to the analysis of human myogenesis, *J. Neurol. Sci.*, 81, 19, 1987.
22. Ontell, M. A. and Kozeka, K., Neonatal muscle growth: a quantitative study, *Am. J. Anat.*, 152, 539, 1984.
23. Ontell, M. A., Hughes, D., and Bourke, D., Morphometric analysis of the developing mouse soleus muscle, *Am. J. Anat.*, 181, 279, 1988.
24. Ontell, M. A. and Dunn, R. F., Neonatal muscle growth: a quantitative study, *Am. J. Anat.*, 152, 539, 1978.
25. Ross, J. J., Duxson, M. J., and Harris, A. J., Formation of primary and secondary myotubes in rat lumbrical muscles, *Development*, 100, 383, 1987.
26. McLennan, I. S., Quantitative relationships between motoneuron and muscle development in *Xenopus laevis*: implications for motoneuron cell death and motor unit formation, *J. Comp. Neurol.*, 27, 19, 1988.
27. McLennan, I. S., Neural dependence and independence of myotube production in chicken hindlimb muscles, *Dev. Biol.*, 98, 287, 1983.
28. Harris, A. J., Embryonic growth and innervation of rat skeletal muscles. I. Neural regulation of muscle fibre numbers, *Philos. Trans. R. Soc. (London) Ser. B*, 293, 257, 1981.
29. McLennan, I. S., The development of the pattern of innervation in chicken hindlimb muscles: evidence for specification of nerve-muscle connections, *Dev. Biol.*, 97, 229, 1983.
30. Stockdale, F. E. and Miller, J. B., The cellular basis of myosin heavy chain isoform expression during development of avian skeletal muscles, *Dev. Biol.*, 123, 1, 1987.
31. Hoh, J. F. Y., Hughes, S., Hale, P. T., and Fitzsimons, R. B., Immunocytochemical and electrophotectic analyses of changes in myosin gene expression in cat limb fast and slow muscles during postnatal development, *J. Muscle Res. Cell Motil.*, 90, 30, 1988.
32. Fidzianska, A., Human ontogenesis. I. Ultrastructural characteristics of developing human muscle, *J. Neuropathol. Exp. Neurol.*, 39, 476, 1980.
33. Church, J. C. T., Satellite cells and myogenesis; a study in fruit-bat web, *J. Anat.*, 105, 419, 1969.

34. **Kikuchi, T. and Ashmore, C.**, Developmental aspects of the innervation of skeletal muscle fibres in the chick embryo, *Cell Tissue Res.*, 171, 233, 1976.
35. **Russell, R. G. and Oteruelo, F. T.**, An ultrastructural study of skeletal muscle in the bovine fetus, *Anat. Embryol.*, 162, 403, 1981.
36. **Platzer, A. C.**, The ultrastructure of normal myogenesis in the limb of the mouse, *Anat. Rec.*, 190, 639, 1978.
37. **Ontell, M. and Kozeka, K.**, The organogenesis of murine striated muscle: a cytoarchitectural study, *Am. J. Anat.*, 171, 133, 1984.
38. **Ontell, M., Bourke, D., and Hughes, D.**, Cytoarchitecture of the fetal murine soleus muscle, *Am. J. Anat.*, 181, 267, 1988.
39. **Campion, D. R., Fowler, S. P., Hausman, G. J., and Reagan, J. O.**, Ultrastructural analysis of skeletal muscle development in the fetal pig, *Acta Anat.*, 110, 277, 1981.
40. **Gamble, H. J., Fenton, J., and Allsop, G.**, Electron microscopic observations on human fetal striated muscle, *J. Anat.*, 126, 567, 1978.
41. **Tomanek, R. J. and Colling-Saltin, A. S.**, Cytological differentiation of human skeletal muscle, *Am. J. Anat.*, 149, 227, 1977.
42. **Dennis, M. J., Ziskind-Conhaim, L., and Harris, A. J.**, Development of neuromuscular junctions in rat embryos, *Dev. Biol.*, 81, 266, 1981.
43. **Bennett, M. V. L., Spray, D. C., and Harris, A. L.**, Gap junctions and development, *Trends Neurosci.*, 4, 159, 1981.
44. **Stickland, N. C.**, Muscle development in the human fetus as exemplified by m. sartorius: a quantitative study, *J. Anat.*, 132, 557, 1981.
45. **Campion, D. R.**, The muscle satellite cell: a review, *Int. Rev. Cytol.*, 87, 225, 1984.
46. **Rowe, R. W. D. and Goldspink, G.**, Muscle fibre growth in five different muscles in both sexes of mice. I. Normal mice, *J. Anat.*, 104, 519, 1969.
47. **Stickland, N. C.**, A quantitative study of muscle development in the bovine foetus *(Bos indicus)*, *Anat. Histol. Embryol.*, 7, 193, 1978.
48. **Wigmore, P. M. C. and Stickland, N. C.**, Muscle development in large and small pig fetuses, *J. Anat.*, 137, 235, 1983.
49. **Bonner, P. H. and Hauschka, S. D.**, Clonal analysis of vertebrate myogenesis. I. Early developmental events in the chick limb, *Dev. Biol.*, 37, 317, 1974.
50. **White, N. K., Bonner, P. H., Nelson, D. R., and Hauschka, S. D.**, Clonal analysis of vertebrate myogenesis. IV. Medium-dependent classification of colony-forming cells, *Dev. Biol.*, 44, 346, 1975.
51. **Hauschka, S. D.**, Clonal analysis of vertebrate myogenesis. II. Environmental influences upon human muscle differentiation, *Dev. Biol.*, 37, 345, 1974.
52. **Rutz, R., Haney, C., and Hauschka, S.**, Spatial analysis of limb bud myogenesis: a proximal distal gradient of muscle colony-forming cells in chick embryo leg buds, *Dev. Biol.*, 90, 399, 1982.
53. **Rutz, R. and Hauschka, S.**, Spatial analysis of limb bud myogenesis: elaboration of the proximodistal gradient of myoblasts requires the continuing presence of apical ectodermal ridge, *Dev. Biol.*, 96, 366, 1983.
54. **Seed, J. and Hauschka, S. D.**, Clonal analysis of vertebrate myogenesis. VIII. Fibroblast growth factor (FGF)-dependent and FGF-independent muscle colony types during chick wing development, *Dev. Biol.*, 128, 40, 1988.
55. **Miller, J. B. and Stockdale, F. E.**, Developmental origins of skeletal muscle fibers: clonal analysis of myogenic cell lineages based on expression of fast and slow myosin heavy chains, *Proc. Natl. Acad. Sci. U.S.A.*, 83, 3860, 1986.
56. **Schafer, D. A., Miller, J. B., and Stockdale, F. E.**, Cell diversification within the myogenic lineage: in vitro generation of two types of myoblasts from a single myogenic progenitor cell, *Cell*, 48, 659, 1987.
57. **Schweitzer, J. S., Dichter, M. A., and Kaufman, S. J.**, Fibroblasts modulate expression of thy-1 on the surface of skeletal myoblasts, *Exp. Cell Res.*, 172, 1, 1987.
58. **Womble, M. D. and Bonner, P. H.**, Developmental fate of a distinct class of chick myoblasts after transplantation of cloned cells into quail embryos, *J. Embryol. Exp. Morphol.*, 58, 119, 1980.
59. **Seed, J. and Hauschka, S. D.**, Temporal separation of the migration of distinct myogenic precursor populations into the developing chick wing bud, *Dev. Biol.*, 106, 389, 1984.
60. **Harris, A. J.**, personal communication.
61. **Cossu, G., Ranaldi, G., Senni, M. I., Molinaro, M., and Vivarelli, E.**, Early mammalian myoblasts are resistant to phorbol ester-induced block of differentiation, *Development*, 102, 65, 1988.
62. **Cossu, G., Cicinelli, P., Fieri, C., Coletta, M., and Molinaro, M.**, Emergence of TPA-resistant "satellite" cells during muscle histogenesis of human limb, *Exp. Cell Res.*, 160, 403, 1985.
63. **Cossu, G., Eusebi, F., Grassi, F., and Wanke, E.**, Acetylcholine receptor channels are present in undifferentiated satellite cells but not in embryonic myoblasts in culture, *Dev. Biol.*, 123, 43, 1987.

64. **Chevallier, A., Pauto, M. P., Harris, A. J., and Kieny, M.**, On the nonequivalence of skeletal muscle satellite cells and embryonic myoblasts, *Arch. Anat. Microsc. Morphol. Exp.*, 75, 161, 1987.
65. **Hauschka, S. D., Rutz, R., Linkhart, T. A., Clegg, C. H., Merrill, G. F., Haney, C. M., and Lim, R. W.**, Skeletal muscle development. I. Developmental changes in muscle colony-forming cell type and location during vertebrate limb development, in *Disorders of the Motor Unit*, Schotland, D. L., Ed., John Wiley & Sons, New York, 1982, 903.
66. **Ahmed, Y. Y.**, The effect of a muscle relaxant on the growth and differentiation of skeletal muscle in the chick embryo, *Anat. Rec.*, 155, 133, 1966.
67. **Drachman, D. B.**, Is acetylcholine the trophic neuromuscular transmitter, *Arch. Neurol.*, 17, 206, 1967.
68. **Phillips, W. D. and Bennett, M. R.**, Differentiation of fiber types in wing muscles during embryonic development: effect of neural tube removal, *Dev. Biol.*, 106, 457, 1984.
69. **Ross, J. J., Duxson, M. J., and Harris, A. J.**, Neural determination of muscle fibre numbers in embryonic rat lumbrical muscles, *Development*, 100, 395, 1987.
70. **Bonner, P. H.**, Nerve-dependent changes in clonable myoblast populations, *Dev. Biol.*, 66, 207, 1978.
71. **Bonner, P. H.**, Differentiation of chick embryo myoblasts is transiently sensitive to functional denervation, *Dev. Biol.*, 76, 79, 1980.
72. **Bonner, P. H. and Adams, T. R.**, Neural induction of chick myoblast differentiation in culture, *Dev. Biol.*, 90, 175, 1982.
73. **McArdle, J. J.**, Molecular aspects of the trophic influence of nerve on muscle, *Prog. Neurobiol.*, 21, 135, 1983.
74. **Popiela, H.**, Trophic effects of adult peripheral nerve extract on muscle cell growth and differentiation *in vitro*, *Exp. Neurol.*, 62, 405, 1978.
75. **Markelonis, G. J., Bradshaw, R. A., Oh, T. A., Johnson, J. L., and Bates, O. J.**, Sciatin is a transferrin-like polypeptide, *J. Neurochem.*, 39, 315, 1982.
76. **Markelonis, G. J., Oh, T. H., Park, L. P., Cha, C. Y., Sofia, S. A., Kim, J. W., and Azari, P.**, Synthesis of the transferrin receptor by cultures of embryonic chicken spinal neurons, *J. Cell Biol.*, 100, 8, 1985.
77. **Knaack, D. and Poleski, T.**, Ascorbic acid mediates acetylcholine receptor increase induced by brain extract on myogenic cells, *Proc. Natl. Acad. Sci U.S.A.*, 82, 575, 1985.
78. **Ager, A., Gordon, J. L., Moncada, S., Pearson, J. D., Salmon, J. A., and Trevethick, M. A.**, Effect of isolation and culture on prostaglandin synthesis by porcine aortic endothelial and smooth muscle cells, *J. Cell. Physiol.*, 110, 9, 1982.
79. **Hill, C. E., McLennan, I. S., and Hendry, I. A.**, Development of sympathetic neurones *in vivo*: an investigation of the possible role of glucocorticosteroids in regulating transmitter type, *Aust. J. Exp. Biol. Med. Sci.*, 63, 439, 1985.
80. **Adamson, E. D.**, Oncogenes in development, *Development*, 99, 449, 1987.
81. **Cossu, G., Cusella-De Angelis, M. G., Senni, M. I., De Angelis, L., Vivarelli, E., Vella, S., Bouche, M., Boitani, C., and Molinaro, M.**, Adrenocorticotropin is a specific mitogen for mammalian myogenic cells, *Dev. Biol.*, 131, 331, 1989.
82. **Rothstein, H.**, Regulation of the cell cycle by somatomedins, *Int. Rev. Cytol.*, 78, 127, 1982.
83. **Van Wyk, J. J.**, The somatomedins: biological actions and physiological control mechanisms, in *Hormonal Proteins and Peptides*, Li, C. H., Ed., Academic Press, New York, 1984, 81.
84. **Moses, A. C., Nissley, S. P., Short, P. A., Rechler, M. M., White, R. M., Knight, A. B., and Higa, O. Z.**, Increased levels of multiple stimulating activity, an insulin-like growth factor, in fetal rat serum, *Proc. Natl. Acad. Sci. U.S.A.*, 77, 1649, 1980.
85. **Roth, R. A.**, Structure of the receptor for insulin-like growth factor II: the puzzle amplified, *Nature (London)*, 239, 1269, 1988.
86. **Rosenfeld, R. G., Conover, C. A., Hodges, D., Lee, P. D. K., Misra, P., Hintz, R. L., and Li, C. H.**, Heterogeneity of insulin-like growth factor-I affinity for the insulin-like growth factor-II receptor: comparison of natural, synthetic and recombinant DNA-derived insulin-like growth factor-I, *Biochim. Biophys. Res. Commun.*, 143, 199, 1987.
87. **Florini, J. R.**, Hormonal control of muscle growth, *Muscle Nerve*, 10, 577, 1987.
88. **Merrill, G. F., Florini, J. R., and Dulak, N. C.**, Effects of multiple stimulating activity (MSA) on AIB transport into myoblast and myotube cultures, *J. Cell. Physiol.*, 93, 173, 1977.
89. **Ewton, D. Z. and Florini, J. R.**, Effects of the somatomedins and insulin on myoblast differentiation *in vitro*, *Dev. Biol.*, 86, 31, 1981.
90. **Ewton, D. Z., Falen, S. L., and Florini, J. R.**, The type II insulin-like growth factor (IGF) receptor has low affinity for IGF-I analogs: pleiotypic actions of IGFs on myoblasts are apparently mediated by the type I receptor, *Endocrinology*, 120, 115, 1987.
91. **Ballard, F. J., Francis, G. L., Ross, M., Bagley, C. J., May, B., and Wallace, J. C.**, Natural and synthetic forms of insulin-like growth factor-I (IGF-I) and the potent derivative, destripeptide IGF-I: biological activities and receptor binding, *Biophys. Biochem. Res. Commun.*, 149, 398, 1987.

92. **Kiess, W., Haskell, J. F., Lee, L., Greenstein, L. A., Miller, B. E., Aarons, A. L., Rechler, M. M., and Nissley, S. P.,** An antibody that blocks insulin-like growth factor (IGF) binding to the type II IGF receptor is neither an agonist nor an inhibitor of IGF-stimulated biologic responses in L6 myoblasts, *J. Biol. Chem.*, 262, 12,745, 1987.
93. **McCusker, R. H. and Clemmons, D. R.,** Insulin-like growth factor binding protein secretion by muscle cells: effect of cellular differentiation and proliferation, *J. Cell. Physiol.*, 137, 505, 1988.
94. **Beguinot, F., Kahn, C. R., Moses, A. C., and Smith, R. J.,** Distinct biologically active receptors for insulin, insulin-like growth factor-I and insulin-like growth factor-II in cultured skeletal muscle cells, *J. Biol. Chem.*, 260, 15,892, 1985.
95. **Burant, C. F., Treutelaar, M. K., Allen, K. D., Sens, D. A., and Buse, M. G.,** Comparison of insulin and insulin-like growth factor I receptors from rat skeletal muscle and L6 myocytes, *Biochim. Biophys. Res. Commun.*, 147, 100, 1987.
96. **Alexandrides, T., Moses, A. C., and Smith, R. J.,** Developmental expression of receptors for insulin, insulin-like growth factor I (IGF-I), and IGF-II in rat skeletal muscle, *Endocrinology*, 124, 1064, 1989.
97. **Schmid, C., Steiner, T., and Froesch, E. R.,** Preferential enhancement of myoblast differentiation by insulin-like growth factors (IGF I and IGF II) in primary cultures of chicken embryonic cells, *FEBS Lett.*, 161, 117, 1983.
98. **Shimizu, M., Webster, C., Morgan, D. O., Blau, H. M., and Roth, R. A.,** Insulin and insulin-like growth factor receptors and responses in cultured human cells, *Am. J. Physiol.*, 251, E611, 1986.
99. **Dodson, M. V., Allen, R. E., and Hossner, K. L.,** Ovine somatomedin, multiple-stimulating activity, and insulin promote skeletal muscle satellite cell proliferation *in vitro*, *Endocrinology*, 117, 2357, 1985.
100. **Daughaday, W. H., Parker, K. H., Borowsky, S., Trivedi, B., and Kapadia, M.,** Measure of somatomedin-related peptides in fetal, neonatal and maternal rat serum by insulin-like growth factor (IGF) I radioimmunoassay, IGF II radioreceptor assay (RRA), and multiplication-stimulating activity RRA after acid-ethanol extraction, *Endocrinology*, 110, 575, 1982.
101. **Handelsman, D. J., Spaliviero, J. A., Scott, C. D., and Baxter, R. C.,** Hormonal regulation of the peripubertal surge of insulin-like growth factor-I in the rat, *Endocrinology*, 120, 491, 1987.
102. **Brown, A. L., Graham, D. E., Nissley, S. P., Hill, D. J., Strain, A. J., and Rechler, M. M.,** Developmental regulation of insulin-like growth factor II mRNA in different rat tissues, *J. Biol. Chem.*, 261, 13,144, 1986.
103. **Romanus, J. A., Yang, Y. W.-H., Adams, S. O., Sofair, A. N., Tseng, L. Y.-H., Nissley, S. P., and Rechler, M. M.,** Synthesis of insulin-like growth factor II (IGF-II) in fetal rat tissues: translation of IGF-II ribonucleic acid and processing of pre-pro-IGF-II, *Endocrinology*, 122, 709, 1988.
104. **Han, V. K. M., D'Ercole, A. J., and Lund, P. K.,** Cellular localization of somatomedin (insulin-like growth factor) messenger RNA in the human fetus, *Science*, 236, 193, 1987.
105. **Stylianopoulou, F., Efstratiadis, A., Herbert, J., and Pintar, J.,** Pattern of the insulin-like growth factor II gene expression during rat embryogenesis, *Development*, 103, 497, 1988.
106. **Beck, F., Samani, N. J., Penschaw, J. D., Thorley, B., Tregear, G. W., and Coghlan, J. P.,** Histochemical localization of IGF-I and -II mRNA in the developing rat embryo, *Development*, 101, 175, 1987.
107. **Hill, D. J., Grace, C. J., Fowler, L., Holder, A. T., and Milner, R. D. G.,** Cultured fetal rat myoblasts release peptide growth factors which are immunologically and biologically similar to somatomedin, *J. Cell. Physiol.*, 119, 349, 1984.
108. **Adams, S. O., Nissley, S. P., Greenstein, L. A., Yang, Y. W. H., and Rechler, M. M.,** Synthesis of multiple-stimulating activity (rat insulin-like growth factor II) by rat embryo fibroblasts, *Endocrinology*, 112, 979, 1983.
109. **Kelly, P. A., Tsushima, T., Shiu, R. P. C., and Friesen, H. G.,** Lactogenic and growth hormone-like activities in pregnancy determined by radioreceptor assay, *Endocrinology*, 99, 765, 1976.
110. **Andersson, I., Billig, H., Frylund, L., Hansson, H.-A., Isaksson, O., Isgaard, J., Nilsson, A., Rozell, B., Skottner, A., and Stemme, S.,** Localization of IGF-I in adult rats. Immunohistochemical studies, *Acta Physiol. Scand.*, 126, 311, 1986.
111. **Hansson, H.-A., Rozell, B., and Skottner, A.,** Rapid axoplasmic transport of insulin-like growth factor I in the sciatic nerve of adult rats, *Cell Tissue Res.*, 247, 241, 1987.
112. **Sporn, M. B. and Roberts, A. B.,** Peptide growth factors are multifunctional, *Nature (London)*, 332, 217, 1988.
113. **Gospodarowicz, D., Neufeld, G., and Schweigerer, L.,** Fibroblast growth factor: structure and biological properties, *J. Cell. Physiol.*, Suppl. 5, 15, 1987.
114. **Linkhart, T. A., Clegg, C. H., and Hauschka, S. D.,** Control of mouse myoblast commitment to terminal differentiation by mitogens, *Prog. Clin. Biol. Res.*, 66B, 263, 1981.
115. **Linkhart, T. A., Clegg, C. H., and Hauschka, S. D.,** Myogenic differentiation in permanent clonal mouse myoblast cell lines: regulation by macromolecular growth factors in the culture medium, *Dev. Biol.*, 86, 19, 1981.

116. **Clegg, C. H., Linkhart, T. A., Olwin, R. B., and Hauschka, S. D.**, Growth factor control of skeletal muscle differentiation: commitment to terminal differentiation occurs in G1 and is repressed by fibroblast growth factor, *J. Cell Biol.*, 105, 949, 1987.
117. **Olwin, B. B. and Hauschka, S. D.**, Cell surface fibroblast growth factor and epidermal growth factor receptors are permanently lost during skeletal muscle terminal differentiation in culture, *J. Cell Biol.*, 107, 761, 1988.
118. **Spizz, G., Roman, D., Strauss, A., and Olson, E. A.**, Serum and fibroblast growth factor inhibit myogenic differentiation through a mechanism dependent on protein synthesis and independent of cell proliferation, *J. Biol. Chem.*, 261, 9483, 1986.
119. **Gospodarowicz, D., Weseman, J., Moran, J. S., and Lindstrom, J.**, Effect of fibroblast growth factor on the division and fusion of bovine myoblasts, *J. Cell Biol.*, 70, 395, 1976.
120. **Kardami, E., Spector, D., and Strohman, R. C.**, Myogenic growth factor present in skeletal muscle is purified by heparin-affinity chromatography, *Proc. Natl. Acad. Sci. U.S.A.*, 82, 8044, 1985.
121. **Kardami, E., Spector, D., and Strohman, R. C.**, Selected muscle and nerve extracts contain an activity which stimulates myoblast proliferation and which is distinct from transferrin, *Dev. Biol.*, 112, 353, 1985.
122. **Linkhart, T. A., Clegg, C. H., and Hauschka, S. D.**, Control of mouse myoblast commitment to terminal differentiation by mitogens, *J. Supramol. Struct.*, 14, 483, 1980.
123. **Bischoff, R.**, Proliferation of muscle satellite cells on intact myofibers in culture, *Dev. Biol.*, 115, 129, 1986.
124. **Allen, R. E., Dodson, M. V., Luiten, L. S., and Boxhorn, L. K.**, A serum-free medium that supports the growth of cultured skeletal muscle satellite cells, *In Vitro Cell Dev. Biol.*, 21, 636, 1985.
125. **Allen, R. E. and Boxhorn, L. K.**, Regulation of skeletal muscle satellite cell proliferation and differentiation by transforming growth factor-beta, insulin-like growth factor I, and fibroblast growth factor, *J. Cell. Physiol.*, 138, 311, 1989.
126. **Seed, J., Olwin, B. B., and Hauschka, S. D.**, Fibroblast growth factor levels in the whole embryo and limb bud during chick development, *Dev. Biol.*, 128, 50, 1988.
127. **Logan, A. and Logan, S. D.**, Distribution of fibroblast growth factor in the central and peripheral nervous systems of various mammals, *Neurosci. Lett.*, 69, 162, 1986.
128. **Sporn, M. B., Roberts, A. B., Wakefield, L. M., and de Crombrugghe, B.**, Some recent advances in the chemistry and biology of transforming growth factor-beta, *J. Cell Biol.*, 105, 1039, 1987.
129. **Rizzino, A.**, Transforming growth factor-b: multiple effects on cell differentiation and extracellular matrices, *Dev. Biol.*, 130, 411, 1988.
130. **Florini, J. R., Roberts, A. B., Ewton, D. Z., Falen, S. L., Flanders, H. C., and Sporn, M. B.**, Transforming growth factor-β. A very potent inhibitor of myoblast differentiation, identical to the differentiation inhibitor secreted by buffalo rat liver cells, *J. Biol. Chem.*, 261, 16,509, 1986.
131. **Massague, J., Cheifetz, S., Endo, T., and Nadal-Ginard, B.**, Type β transforming growth factor is an inhibitor of myogenic differentiation, *Proc. Natl. Acad. Sci. U.S.A.*, 83, 8206, 1986.
132. **Olson, E. N., Sternberg, E., Hu, J. S., Spizz, G., and Wilcox, C.**, Regulation of myogenic differentiation by type b transforming growth factor, *J. Cell Biol.*, 103, 1799, 1986.
133. **Florini, J. R. and Ewton, D. Z.**, Actions of transforming growth factor-b on muscle cells, *J. Cell. Physiol.*, 135, 301, 1988.
134. **Allen, R. E. and Boxhorn, L. K.**, Inhibition of skeletal muscle satellite cell differentiation by transforming growth factor-beta, *J. Cell. Physiol.*, 133, 567, 1987.
135. **Samuelsson, B., Goldyne, M., Granstrom, E., Hamberg, M., Hammarstrom, S., and Malmsten, C.**, Prostaglandins and thromboxanes, *Annu. Rev. Biochem.*, 47, 997, 1978.
136. **Zalin, R. J.**, The role of hormones and prostanoids in the *in vitro* proliferation and differentiation of human myoblasts, *Exp. Cell Res.*, 172, 265, 1987.
137. **Hausman, R. E., Dobi, E. T., Woodford, E. J., Petrides, S., Ernst, M., and Nichols, E. B.**, Prostaglandin binding activity and myoblast fusion in aggregates of avian myoblasts, *Dev. Biol.*, 113, 40, 1986.
138. **Zalin, R. J.**, Prostaglandins and myoblast fusion, *Dev. Biol.*, 59, 241, 1977.
139. **Zalin, R. J.**, The cell cycle, myoblast differentiation and prostaglandin as a developmental signal, *Dev. Biol.*, 71, 274, 1979.
140. **Hausman, R. E. and Velleman, S. G.**, Prostaglandin E_1 receptors on chick embryo myoblasts, *Biochem. Biophys. Res. Commun.*, 103, 213, 1981.
141. **David, J. D. and Higginbotham, C.-A.**, Fusion of chick embryo skeletal myoblasts: interactions of prostaglandin E1, adenosine 3':5'monophosphate, and calcium influx, *Dev. Biol.*, 82, 308, 1981.
142. **Hausman, R. E. and Berggrun, D. A.**, Prostaglandin binding does not require direct cell-cell contact during chick myogenesis, *Exp. Cell Res.*, 168, 457, 1987.
143. **Santini, M. T., Indovina, P. L., and Hausman, R. E.**, Prostaglandin dependence of membrane order changes during myogenesis *in vitro*, *Biophys. Biochem. Acta*, 938, 489, 1988.
144. **Konigsberg, I. R.**, The culture environment and its control of myogenesis, in *Regulation of Cell Proliferation and Differentiation*, Nichols, W. W. and Murphy, D. G., Eds., Elsevier, Amsterdam, 1977, 779.

145. Schutzle, U. B., Wakelam, M. J. O., and Pette, D., Prostaglandins and cyclic-AMP stimulate creatine kinase synthesis but not fusion in cultured embryonic muscle cells, *Biochim. Biophys. Acta*, 805, 204, 1984.
146. Steiner, S., Manley, G., and Adams, T., Effect of inhibitors of the lipoxygenase pathway on mouse myoblast fusion, *Exp. Cell Res.*, 155, 289, 1984.
147. McLennan, I. S., unpublished data, 1989.
148. McLennan, I. S., Hormonal regulation of myoblast proliferation and myotube production *in vivo*: influence of prostaglandins, *J. Exp. Zool.*, 241, 237, 1987.
149. Avery, M. A. and Taeusch, H. W., *Schaffer's Diseases of the Newborn*, 5th ed., W. B. Saunders, Philadelphia, 1984.
150. Underwood, L. E. and D'Ercole, A. J., Insulin and insulin-like growth factors/somatomedins in fetal and neonatal development, *Clin. Endocrinol. Metab.*, 13(1), 69, 1984.
151. Roe, T. F., Kershner, A. K., Weitzman, J. J., and Madrigal, L. E., Beckwith's syndrome with extreme organ hyperplasia, *Pediatrics*, 56, 648, 1973.
152. D'Ercole, A. J., Underwood, L. E., Groelke, J., and Plet, A., Leprechaunism: studies of the relationship among hyperinsulinism, insulin resistance, and growth retardation, *J. Clin. Endocrinol. Metab.*, 48, 495, 1979.
153. Kaplan, S. L., Grumbach, M. M., and Shepard, T. H., The ontogenesis of human-fetal hormones, I. Growth hormone and insulin, *J. Clin. Invest.*, 51, 3080, 1972.
154. De La, Haba, G., Cooper, G. W., and Elting, V., Hormonal requirements for myogenesis of striated muscle *in vitro*: insulin and somatotropin, *Proc. Natl. Acad. Sci. U. S. A.*, 56, 1719, 1966.
155. Kumegawa, M., Ikeda, E., Hosoda, S., and Takuma, T., *In vitro* effects of thyroxine and insulin on myoblasts from chick embryo skeletal muscle, *Dev. Biol.*, 79, 493, 1980.
156. Pinset, C. and Whalen, R. G., Induction of myogenic differentiation in serum-free medium does not require DNA synthesis, *Dev. Biol.*, 108, 284, 1985.
157. Turo, K. A. and Florini, J. R., Hormonal stimulation of myoblast differentiation in the absence of DNA synthesis, *Am. J. Physiol.*, 243, C278, 1982.
158. De La, Haba, G., Cooper, G. W., and Elting, V., Myogenesis of striated muscle *in vitro*: hormone and serum requirements for the development of glycogen synthetase in myotubes, *J. Cell. Physiol.*, 72, 21, 1968.
159. Farfel, Z., Karlish, S., and Prives, J., A transient increase in amino acid transport modulated by insulin in differentiating muscle cells, *J. Cell. Physiol.*, 98, 279, 1979.
160. Schudt, C., Gaertner, U., and Pette, D., Insulin action on glucose transport and calcium fluxes in developing muscle cells *in vitro*, *Eur. J. Biochem.*, 68, 103, 1976.
161. Ballard, F. L. and Francis, G. L., Effects of anabolic agents on protein breakdown in L6 myoblasts, *Biochem. J.*, 210, 243, 1983.
162. Benzo, C. A. and Green, T. D., Functional differentiation of the chick endocrine pancreas: insulin storage and secretion, *Anat. Rec.*, 180, 491, 1974.
163. de Pablo, F., Roth, J., Hernandez, E., and Pruss, R. M., Insulin is present in chicken eggs and early chicken embryos, *Endocrinology*, 111, 1909, 1984.
164. Sandra, A., and Przybylski, R. J., Ontogeny of insulin binding during skeletal myogenesis in vitro, *Dev. Biol.*, 68, 546, 1979.
165. Bassas, L., de Pablo, F., Lesniak, M. A., and Roth, J., The insulin receptors of chicken embryos show tissue-specific structural differences which parallel those of the insulin-like growth factor I receptors, *Endocrinology*, 121, 1468, 1987.
166. Shimizu, M., Webster, C., Morgan, D. O., Blau, H. M., and Roth, A. A., Insulin and insulinlike growth factor receptors and responses in culture human muscle cells, *Am. J. Physiol.*, 251, E611, 1986.
167. Kuhn, E. R., Decuypere, E., and Rudas, P., Hormonal and environmental interactions on thyroid function in the chicken embryo and post-hatching chick, *J. Exp. Zool.*, 232, 653, 1984.
168. Anderson, H. J., Studies of hypothyroidism in children, *Acta Paediatr. Scand.*, 125, 11, 1961.
169. Munck, A. and Leung, K., Glucocorticoid receptors and mechanism of action, *Mod. Pharmacol. Toxicol.*, 8, 311, 1977.
170. Guerriero, V. and Florini, J. R., Dexamethasone effects on myoblast proliferation and differentiation, *Endocrinology*, 106, 1198, 1980.
171. Kelly, F. J., McGrath, J. A., Goldspink, D. F., and Cullen, M. J., A morphological/biochemical study on the actions of corticosteroids on rat skeletal muscle, *Muscle Nerve*, 9, 1, 1986.
172. Steiss, J. E., Effect of high-and low-dose dexamethasone on regeneration of minced skeletal muscle autografts in rats, *Exp. Neurol.*, 93, 300, 1986.
173. Martin, C. E., Cake, M. H., Hartmann, P. E., and Cook, I. F., Relationship between foetal progesterone and parturition in the rat, *Acta Endocrinol.*, 84, 167, 1977.
174. Johnakait, C. M., Bohn, M. C., and Black, I. B., Maternal glucocorticoid hormones influence neurotransmitter phenotypic expression in embryos, *Science*, 210, 551, 1980.

175. **Powers, M. L. and Florini, J. R.**, A direct effect of testosterone on muscle cells in tissue culture, *Endocrinology*, 97, 1043, 1975.
176. **Cheek, P. B. and Hill, D. E.**, Muscle and liver cell growth: role of hormones and nutritional factors, *Fed. Proc.*, 29, 1503, 1970.
177. **Curtis, D. H. and Zalin, R. J.**, Regulation of muscle differentiation: stimulation of myoblast fusion *in vitro* by catecholamines, *Science*, 214, 1355, 1981.
178. **Curtis, D. H. and Zalin, R. J.**, The differentiation of avian skeletal muscle in culture: changes in responsiveness of adenyl cyclase to prostaglandin E_1 and adrenergic agonists, *J. Cell. Physiol.*, 123, 219, 1985.
179. **Levitski, A.**, Catecholamine receptors, *Rev. Physiol. Biochem. Pharmacol.*, 82, 1, 1978.
180. **Legrand, C. and Maltier, J. P.**, Fetal and maternal plasma adrenaline and noradrenaline concentrations in the late pregnant rat: effects of intrauterine isotonic saline or uterine handling, *Acta Endocrinol.*, 96, 541, 1981.
181. **Wakelam, M. J. and Pette, D.**, The control of glucose, 1,6-bi-phosphate by developmental state and hormone stimulation in cultured muscle tissue, *Biochem. J.*, 204, 765, 1982.
182. **Atlas, D., Hanski, E., and Levitzki, A.**, Eighty thousand β-adrenoreceptors in a single cell, *Nature (London)*, 268, 144, 1977.
183. **Schonberg, M., Bilezikian, J. P., Apfelbaum, M., and Benn, R. C.**, Beta-adrenergic receptors and myogenesis, *J. Cyclic Nucleotide Res.*, 4, 55, 1978.
184. **Pochet, R. and Schmitt, H.**, Re-evaluation of the number of specific β adrenergic receptors on muscle cells, *Nature (London)*, 277, 58, 1979.
185. **Parent, J. B., Tallman, J. F., Henneberry, R. C., and Fishman, P. H.**, Appearance of β-adrenergic receptors and catecholamine-responsive adenylate cyclase activity during fusion of avian embryonic muscle cells, *J. Biol. Chem.*, 255, 7782, 1980.
186. **Mawatari, S., Miranda, A., and Rowland, L. P.**, Adenyl cyclase abnormality in Duchenne muscular dystrophy: muscle cells in culture, *Neurology*, 26, 1021, 1978.
187. **Sejersen, T., Betsholtz, C., Sjolund, M., Heldin, C.-H., Westermark, B., and Thyberg, J.**, Rat skeletal myoblasts and arterial smooth muscle cells express the gene for the A chain but not the B chain (c-sis) of platelet-derived growth factor (PDGF) and produce a PDGF-like protein, *Proc. Natl. Acad. Sci. U.S.A.*, 83, 6844, 1986.
188. **Venkatasubramanian, K. and Solursh, M.**, Chemotactic behavior of myoblasts, *Dev. Biol.*, 104, 428, 1984.
189. **Bailey, A. J., Shellswell, G. B., and Duance, V. C.**, Identification and change of collagen types in differentiating myoblasts and developing chick muscle, *Nature (London)*, 278, 67, 1979.
190. **Kluhl, U., Timpl, R., and von der Mark, K.**, Synthesis of type IV collagen and laminin in cultures of skeletal muscle cells and their assembly on the surface of myotubes, *Dev. Biol.*, 93, 344, 1982.
191. **Sanes, J. R., Schachner, M., and Covault, J.**, Expression of several adhesive macromolecules (N-cam, L1, J1, NILE, Uvomorulin, Laminin, fibronectin and a heparin sulfate proteoglycan) in embryonic, adult, and denervated muscle, *J. Cell Biol.*, 102, 420, 1986.
192. **Chiu, A. U. and Sanes, J. R.**, Development of basal lamina in synaptic and extrasynaptic portions of embryonic rat muscle, *Dev. Biol.*, 103, 456, 1984.
193. **Hauscha, S. D. and Konigsberg, I. R.**, The influence of collagen on the development of muscle clones, *Proc. Natl. Acad. Sci. U.S.A.*, 55, 119, 1966.
194. **Ketley, J. N., Orkin, R. W., and Martin, G. R.**, Collagen in developing chick muscle *in vivo* and *in vitro*, *Exp. Cell Res.*, 99, 261, 1976.
195. **Foster, R. F., Thompson, J. M., and Kaufman, S. J.**, A laminin substrate promotes myogenesis in rat skeletal muscle cultures: analysis of replication and development using antidesmin and anti-BrdUrd monoclonal antibodies, *Dev. Biol.*, 122, 11, 1987.
196. **Ocalan, M., Goodman, S. L., Kuhl, U., Hauschka, S. D., and von der Mark, K.**, Laminin alters cell shape and stimulates motility and proliferation of murine skeletal myoblasts, *Dev. Biol.*, 125, 158, 1988.
197. **Luff, A. R. and Goldspink, G.**, Large and small muscles, *Life Sci.*, 6, 1821, 1967.
198. **Penney, R. K., Prentis, P. F., Marshall, P. A., and Goldspink, G.**, Differentiation of muscle and the determination of ultimate tissue size, *Cell Tissue Res.*, 228, 375, 1983.
199. **Bedi, K. S., Birzgalis, A., Mahon, M., Smart, S., and Wareham, A.**, Early undernutrition in rats. I. Quantitative histology of skeletal muscles from underfed and refed adult animals, *Br. Nutr.*, 47, 417, 1982.
200. **Wilson, S. J., Ross, J. J., and Harris, A. J.**, A critical period for formation of secondary myotubes defined by prenatal undernourishment in rats, *Development*, 102, 815, 1988.
201. **Holly, R. G., Barnett, J. G., Ashmore, C. R., Taylor, R. G., and Mole, P. A.**, Stretch-induced growth in chicken wing muscles: a new model of stretch hypertrophy, *Am. J. Physiol.*, 238, C62, 1980.
202. **Aberle, E. D.**, Myofiber differentiation in skeletal muscles of newborn runt and normal weight pigs, *J. Anim. Sci.*, 59, 1651, 1984.

203. **Glucksman, A.**, Cell death in normal vertebrate ontogeny, *Biol. Rev. Cambridge Philos. Soc.*, 26, 59, 1951.
204. **Oppenheim, R. W.**, Neuronal cell death and some related regressive phenomena during neurogenesis, in *Studies in Developmental Neurobiology. Essays in Honor of Viktor Hamburger,* Cowan, W. M., Ed., Oxford University Press, Oxford, 1981, 74.
205. **Prestige, M. C.**, The control of cell number in the lumbar ventral horns during the development of *Xenopus laevis* tadpoles, *J. Embryol. Exp. Morphol.*, 18, 359, 1967.
206. **Webb, J. N.**, The development of human skeletal muscle with particular reference to muscle cell death, *J. Pathol.*, 106, 221, 1972.
207. **Webb, J. N.**, Cell death in developing skeletal muscle: histochemistry and ultrastructure, *J. Pathol.*, 123, 175, 1977.
208. **Korenji-Both, A. and Marosan, G.**, Patterns of neuromuscular disease: as related to stages of normal embryogenesis in voluntary muscle, *Am. J. Pathol.*, 95, 359, 1979.
209. **Hayes, V. E. and Hikida, R. S.**, Naturally occurring degeneration in chick muscle development: ultrastructure of the M. complexus, *J. Anat.*, 122, 67, 1976.
210. **Prestige, M. C.**, The control of cell number in the lumbar ventral horns during the development of *Xenopus laevis* tadpoles, *J. Embryol. Exp. Morphol.*, 18, 359, 1967.
211. **Stickland, N. C.**, The arrangement of muscle fibers and tendons in two muscles used for growth studies, *J. Anat.*, 136, 175, 1983.

DIFFERENTIATION OF MUSCLE FIBERS: MYOFIBRILS AND CONTRACTILITY

Ian S. McLennan

PHYSIOLOGICAL DEVELOPMENT

Introduction

Nerve-directed muscle contraction occurs at a very early stage of muscle development. In the chicken, for instance, bursts of hindlimb activity in which antagonistic muscles often alternate with synergistic muscles can be detected by 8 d *in ovo*.[1,2] The muscles of the hindlimb at this age contain fewer than 10% of their fibers, all of which belong to the primary generation.[3-5] Mature patterns of muscle activation and voluntary motor performance occur at a later stage of development.[6,7]

The physiological characteristics of the earliest spontaneous muscle activity have received little attention. Most studies have concentrated on neonates, which, with the exception of the rat, have their full complement of muscle fibers. Myogenesis in the rat is complete during the first few days post partum.[5] Neonatal muscle contraction has many quantitative differences to that of the adult, and these are discussed below.

Tension Generation

The maximum isometric and tetanic twitch tension is low at birth and increases considerably during the first few months of life.[8,9] The increase in the maximum tetanic tension is proportionally greater than that of isometric tension, leading to a decrease in the twitch-to-tetanus ratio.[9] The development of muscle strength appears to be a consequence of the maturational increase in fiber size and myofibrillar content which occurs during this period[10] (see below). Changes in the efficacy of the synaptic junction and the coupling of excitation to contraction may also be important, particularly in determining the twitch-to-tetanus ratio. For instance, the tension generated in newborn rat soleus is not increased when the frequency of stimulation is raised above 40 Hz, whereas in the adult, maximum tension is not generated until the frequency of the stimulation reaches 200 Hz.[9]

Speed of Contraction

The muscles of neonatal mammals contract and relax more slowly than those of the adult. In newborn cats[11-14] and rats,[9,15] the twitch contraction time (time to reach peak tension) and the rate of relaxation are similar in presumptive fast and slow muscles. During the first month of life, the speed of these processes increases in fast muscles, whereas in slow muscles, there is a transient increase, followed by a small decrease in the speed of contracture.[11-15] The increase in the twitch contraction time is in part due to a decrease in the latency between excitation and the onset of contraction[16,17] and in part to an increase in the velocity of sarcomere shortening.[9]

The speed of contracture is regulated by the muscle's innervation. If a presumptive fast muscle is innervated by a nerve which normally innervates a slow muscle, then it differentiates with a slower than normal speed of contracture. Conversely, the speed of contracture of a slow muscle is increased when innervated by fast motoneurons.[18,19] The differing effect of fast and slow motoneurons on muscle differentiation is related to their different frequency of activation. Electrically changing the pattern of activation of a muscle is, for instance, sufficient to cause a complete transformation of its biochemical and electrophysiological characteristics.[20-22] This clear demonstration of neural regulation of the speed of contracture has lead to the discounting of myogenic influences. However, the situation is probably more complex than this. As discussed later, each fiber has an inherent disposition for a particular fiber type and during the establishment of innervation of the muscle, there appears to be a matching of motoneurons to immature fibers with compatible inherent characteristics. The slow speed of contracture of presumptive fast and slow muscles may in part be due to the presence of embryonic and neonatal myosins, although other factors such as the immaturity of the myofibrils and tubular system may also be important.

Fatigue and Potentiation

In the adult, the capacity to maintain tension during tetanic stimulation varies from muscle to muscle. This is also true of immature muscles, although nonextreme stimuli need to be used to demonstrate this phenomenon. For instance, if kitten muscles are stimulated at 40 Hz for 330 ms each second, then the resistance to fatigue of the soleus and gastrocnemius appears to be mature in 1-week-old kittens, whereas a mature response does not occur in the anterior tibial muscle until 7 weeks.[12,13] If, however, a more strenuous tetanus of 40 Hz for 30 s is used, the soleus, medial gastrocnemius, and anterior tibial muscles of 1-week-old kittens all fatigue to a similar extent.[12,13]

Muscle tension during repeated stimulation is influenced by two counteracting forces, fatigue and potentiation.[23] The apparent maturity of the soleus during an intermittent tetanus has been attributed to the occurrence of greater facilitation in the younger kittens, which masks their greater fatigability.[13] Developmental changes in the effect of a prior tetanus on the twitch tension have been observed in the cat[13,24,25] and other mammals[26] and vary from muscle to muscle. For instance, a prior tetanus potentiates the twitch tension to a greater extent in the plantaris of young rather than mature hamsters. In the hamster soleus, however, twitch tension is depressed by a prior tetanus at all ages.[26]

The greater loss of tension during tetanus in developing muscles could be related to the immaturity of the motoneurons, the synaptic junction, or the muscle fibers. The maturation of innervation of the muscle and synaptic sites has been reviewed elsewhere.[27,28] An important factor in the capacity of a muscle to maintain tension is its ability to pump Ca^{2+}. In mature muscle, fatigue is thought to occur as a consequence of the buildup in Ca^{2+} in the transverse tubules.[29] As outlined below, the tubular system in mammals matures postnatally. A further factor may be the lack of maturity of the metabolism of the muscle. During a tetanus, the energy demand on the cell will be high due to the utilization of ATP to maintain the intracellular ion balance. The energetics of immature muscle cells are reviewed in the next chapter.

Membrane Characteristics

Membrane Potential

The membrane potential of fast and slow muscles increases as the muscle matures,[30-33] possibly as a consequence of a thyroxine-mediated increase in Na,K ATPase activity. *In vitro*, thyroxine increases the transmembrane resting potential and Na,K ATPase activity. If the increase in the ATPase activity is blocked with ouabain, then the change in membrane potential is also blocked.[34,35] Further, the *in vivo* Na,K ATPase activity of mature and immature muscles is known to be influenced by changes in the level of thyroid hormones.[36,37]

Action Potential

The speed of the action potential changes as the muscle matures. The action potential in neonatal muscles is slow and increases during the first few weeks post partum.[33] For instance, in the rat extensor digitorium longus, the rate of potential change during the action potential increases from 147 to 232 V/s between the 5th and 30th days post partum. During this period, the duration of the action potential decreases from 1.80 to 1.43 ms.[33] These changes in the characteristics of the potential are thought to be largely due to changes in the kinetics of the ion channels associated with the acetylcholine receptor.[38]

Acetylcholine Receptor Channels

Synapse formation occurs at an early stage of myotubal differentiation. Acetylcholine receptors are present prior to innervation and rapidly accumulate at the site of neuromuscular contact.[27,39] The accumulation of receptors at immature synaptic sites occurs prenatally[40] and is thus not responsible for the maturational changes in synaptic currents. Changes in the properties of the receptors and their associated ion channels appear to be more important in this respect. In mature muscles, the properties of synaptic and nonsynaptic receptors differ, with the nonsynaptic receptors having a greater mean channel open time (t).[41] The gating time of synaptic receptors in prenatal and early neonatal rat muscles is similar to that of mature nonsynaptic receptors (t = 4.5 ms; 21°C) and decreases to that of mature synaptic receptors (t = 1.4 ms) during the first few weeks post partum.[42,43] The change in t appears to be due to an alteration in the type of channel present, rather than to a change in the properties of preexisting channels. Single-channel analysis of cultured *Xenopus* muscle has indicated that there are two channels at all stages of development, but as differentiation occurs, the proportion of high-q (single-channel conductance) channels increases over the low-q channels. The low-q channels have a mean channel open time which is approximately three times longer than the high-q channels.[38] The occurrence of high-q and low-q channels has also been detected in immature mammalian[44] and amphibian muscles.[45]

MORPHOGENESIS

Myofibrils

Myofibrillar assembly occurs very shortly after the formation of a myotube and is a sequential process. Thin filaments initially predominate and are arranged in loose aggregates, oriented in the longitudinal axis of the myotube. Thick filaments are rapidly incorporated with the thin filaments to give a hexagonal array. Additional thick and thin filaments are added to the lattice and Z disks begin to form between neighboring sarcomeres. M bands are the last morphological feature to develop. Myofibrillar assembly generally occurs at the periphery of myotubes, with myofibrils in the center of the cytoplasm only rarely being observed. In myofibers, myofibrillar proliferation occurs through the growth and longitudinal splitting of existing myofibrils rather than *de novo* assembly (human;[46] laboratory animals[47-50]).

The alignment of developing myofibrils is along the long axis of the myotube at all stages of development.[48] The Z disks and M bands of neighboring myofibrils do not, however, become aligned with each other until the myofiber stage.[50] It is generally accepted that the cytoskeleton of the cell is responsible for laterally integrating myofilaments and aligning them along the axis of the muscle. The assembly and protein composition of the cytoskeleton and Z disks of the cell has been extensively studied and this work has been reviewed elsewhere.[51,52]

Our knowledge of the factors which regulate myofibrillar assembly and alignment is extremely limited. As discussed below, the filaments during development are composed of developmental isozymes which are sequentially eliminated and replaced by their mature forms. At present it is unknown whether the occurrence of these developmental isozymes is important for the initial assembly of myofibrils or how these immature forms are eliminated as the myofibrils mature.

Innervation of the muscle is important for the normal development of myofibrils. When denervated or paralyzed, the rate and extent of myofibrillar assembly are reduced, and if maintained, myofibrillar disorganization occurs.[53,54] This effect is more pronounced after denervation than paralysis, which suggests that both depolarization and a neurotrophic substance are involved in this phenomenon.[54] Autocrine factors may also be important, as the myofibrils in the muscles of chicken embryos treated with inhibitors of prostaglandin synthetase are incompletely formed and are not aligned along the long axis of the fiber.[55]

Internal Membrane Systems

The sarcoplasmic reticulum (SR) and transverse tubular system play an important role in the excitability and activation of muscle fibers. Depolarization is thought to spread along the T-tubules and to stimulate, through the triadic junctions, the release of Ca^{2+} from the SR. The SR also has an extensive Ca^{2+} transport capacity which is responsible for maintaining the intracellular Ca^{2+} concentration below that of the extracellular.[56,57] The maturation of the triads has been postulated to be responsible for the developmental decrease in the latency between excitation and contraction[16,17] and the increase in membrane resting potential during development has been related to the Ca^{2+} uptake capacity of the SR.[17]

Sarcoplasmic Reticulum

The sarcoplasmic reticulum (SR) evolves from rough endoplasmic reticulum[58,59] as the result of the insertion of Ca^{2+} ATPase and other macromolecules into its membrane.[57,60] *In vivo*, SR begins to form prior to the myofiber stage, which is after the formation of the initial myofibrils.[16,46,48,49,59] Its subsequent elaboration and biochemical differentiation parallel the developmental increase in myofibrillar proteins,[57] and in the chicken pectoralis, this is not complete until 1 month after hatching.[61] The increase in its capacity to transport Ca^{2+}, which occurs during its development,[61,62] has been postulated to be regulated by both the innervation of the muscle and intracellular Ca^{2+} concentration,[57] and during the later stages of development, the increase has been postulated to be regulated by circulatory catecholamines.[63]

During the early stages of its development, SR forms junctions with the sarcolemma. Once T-tubular development occurs, these junctions are lost and terminal cisternal development occurs.[16,17] Initially the SR spreads over several sarcomeres, but as triadic contacts become more extensively established, the SR network differentiates into two portions, in register with the A and I bands.[17] The morphological development of SR *in vitro* differs slightly from that described above, possibly as a consequence of the absence of neural influences *in vitro* (see Ontell[64]).

T-Tubules

T-Tubules develop after SR proliferation and appear to be formed from invaginating sarcolemma.[46,58,59] Triads begin to become frequent at the myofiber stage, although initially they are poorly aligned and only a few are localized at A-I junctions.[16,46,50,59] The maturation of traids in rat muscles is temporally correlated with the decrease in their excitation-contraction latency,[16] which is consistent with the contention that the slow coupling between excitation and contraction in developing muscles is due to the immaturity of the triadic junctions.[16,17]

FIBER TYPE

ATPase Histochemistry
Expression During Development

In the adult, muscle fibers can be subdivided into two basic types on the basis of their myosin ATPase staining characteristics. After preincubation of the muscle in an alkaline buffer, type I fibers stain lightly and type II fibers stain intensely, whereas the reverse pattern of staining occurs after acid preincubation.[65] The ATPase fiber types have been studied extensively in mature muscles, as they are a reliable indicator of the speed of contracture of the fiber. Type I fibers contract more slowly than type II fibers.[19,66]

All immature muscle fibers stain with a similar intensity after alkaline preincubation, which in the adult is characteristic of type II fibers.[3,66-68] This has been interpreted to imply that fibers initially develop fast characteristics, with some then transforming into slow fibers postnatally. However, more recent studies on chicken embryo muscles have shown that distinct embryonic fiber types, types I_{EMB} and II_{EMB}, can be detected on the basis of their acid-stable staining characteristics.[3,69] The acidic pH at which the ATPase within types I_{EMB} and II_{EMB} fibers becomes inactivated is different and also differs from all the known mature fiber types. This is an indication that the various embryological fiber types contain ATPases which differ from each other and from the isozymes found in mature muscle fibers.[3] Distinct embryological isozymes of myosin are known to occur (see next section) and probably account for the differences between the various embryonic and mature fiber types. The existence of embryonic isozymes in chicken and mammalian muscles implies that the use of adult classifications, such as types I and II, to describe immature muscles is of dubious value and may lead to erroneous conclusions being drawn, the most notable of these being that muscles are undifferentiated with respect to fiber type until late in their development.

In the chicken embryo, embryological fiber types can be detected shortly after the first myotubes are formed.[3,69,70] This stage of development corresponds to the seventh week of gestation in humans (see Chapter 1). In most muscles, type I_{EMB} myotubes mature into type I fibers and type II_{EMB} myotubes mature into type II fibers. The maturation of the fiber types involves sequential changes in the acid lability of the ATPases[3] which correspond well to similar changes occurring in the characteristics of the myosin molecules. The embryological fiber types of primary and secondary myotubes have a slightly different sequence of maturation. The first maturational change in the staining characteristics of primary myotubes occurs around the time secondary myotubes are beginning to form. The pH lability and immunological properties of the newly formed secondary myotubes are the same as that of the primary myotubes present at the time of their formation, i.e., the most immature staining characteristics seen in the first formed primary myotubes are not observed in secondary myotubes.[3]

Acquisition of Fiber Type

The fiber type aquired by a fiber is very predictable. Within a given muscle, the mature ATPase characteristics of a fiber are related to the time of formation of the fiber,[71] particularly whether it is of primary or secondary origin,[71-75] and its position within the muscle.[71] In the chicken sartorius, for instance, type I fibers are derived from primary myotubes in the ventral lateral half of the muscle and from the last-formed secondary myotubes on the ventral lateral border.[71] Primary and secondary myotubes frequently have different fiber types, with secondary myotubes never developing type I characteristics unless all the primary myotubes surrounding them are also type I.[71]

In the adult, the fiber type and speed of contraction of a fiber are related to the properties of its innervating motoneuron.[19,66] In contrast, neural regulation of fiber type does not appear to occur during the initial stages of muscle formation. Muscles become innervated shortly after their formation, and if denervated the ATPase fiber type of their myotubes is normal,[69,70] indicating that the initial expression of fiber type is probably intrinsic to the myotube. Motoneurons do, however, have a trophic influence on myosin accumulation. In particular, the intensity of the acid labile stain of these fibers is decreased by denervation[69] which can erroneously give the appearance of neural regulation of type I_{EMB} expression.[3,69]

Clonal analysis of developing human and chicken muscles has revealed that they contain a number of distinct myoblast varieties which can be separated on the basis of their trophic requirements or the type of myosin which they synthesize.[5] The intrinsic characteristics of a fiber have been suggested to be due to the type of myoblasts from which the fiber was formed and evidence for this has been the subject of recent commentary.[76,77]

In considering the recent evidence that fibers have intrinsic properties, it should not be forgotten that as the fiber matures, its fiber type becomes influenceable by the pattern of its activation. However, in most instances, immature fast or slow myotubes differentiate into mature fast or slow fibers, respectively. This has been interpreted to imply that immature motoneurons selectively innervate myotubes whose intrinsic fiber type is the same as that which the mature motoneuron will endeavor to impose upon it.[71]

The last stages of differentiation of immature type I but not type II fibers are dependent upon functional innervation of the muscle. For instance, denervation of newborn rat soleus[68] or curarization of embryonic chicken anterior latissimus dorsi[78] causes the muscle to develop as a mosaic of type I and type II fibers rather than being uniformly type I. The reason why only some of the fibers are affected is unknown, but may reflect the embryological type of the various fibers. In the above studies, the muscles will have reduced numbers of secondary myotubes, as a consequence of paralyzing the developing muscle. Some secondary myotubes would, however, have been formed due to the timing of the denervation of the rat soleus and the dose of curare used in the chicken study.[4,79-81] The greater size and maturity of type I fibers in denervated rat soleus[67,68] suggest that they may be of primary myotubal origin and that those with type II characteristics may be of secondary myotubal origin.

In some muscles, a proportion of the fibers change their fiber type from type I to type II or vice versa. Such changes occur at all stages of development, ranging from the period when only embryological fiber types are being expressed to early adulthood.[69,71,72,82] Although such transformations of fiber types clearly occur, many previous and recent reports describing transformation of type II to type I fibers in the neonate are probably erroneous. In many instances, the fibers assumed to be type II, on the basis of the alkaline stability of their ATPases, are probably type I_{EMB} and the apparent transformation is merely due to the maturation of the fiber.

Contractile Proteins
Expression During Development

A common feature of the proteins associated with the contractile apparatus is the occurrence of multiple isomeric forms. In general, each isozyme is associated with a particular fiber type, fast vs. slow, or a particular muscle type, skeletal vs. cardiac. The isozymes which occur in developing fibers are frequently different to those which occur in the adult. For instance, isozymes which in the adult are only found in fast or cardiac muscle can be detected in presumptive slow skeletal fibers. Additionally, several isozymes have recently been detected which are mainly, but not exclusively, expressed in developing muscle. The characteristics of the contractile protein isozymes found in developing muscle fibers are summarized in Table 1 and have also been recently reviewed.[76]

The nature of the myosin heavy chain in developing muscle fibers has received considerable attention since, in the adult, the characteristics of this molecule are important determinants of the speed of contraction of the fiber[117] and ATPase staining.[118,119] To date, two developmental forms of the myosin heavy chain, HC_{EMB} and HC_{NEO}, have been identified in various laboratory species (see Table 1). There are multiple isozymes of HC_{EMB}, which appear to be differentially expressed in presumptive fast or slow fibers.[88,89] The developmental myosins are immunologically similar to the mature forms and early reports[100,120] of the occurrence of HC_{FAST}, HC_{SLOW}, and $HC_{CARDIAC}$ in developing muscle fibers are now thought to be due to the nonspecificity of the antibodies used in these studies. The changes in the myosin isozyme expression as fast and slow fibers mature have been reviewed elsewhere[76] and only papers on human muscle will be commented on in this review.

In humans only one developmental isozyme of the myosin heavy chain has been identified,[98] although others may exist. HC_{EMB} in humans, unlike in laboratory animals, is not lost until after birth.[98] Continued expression of HC_{EMB} does not normally occur beyond the first month except in various diseases involving muscular weakness, such as muscular dystrophy and infantile muscular atrophy.[121-123]

Table 1
CHARACTERISTICS OF CONTRACTILE PROTEINS IN DEVELOPING SKELETAL MUSCLES

Protein	Chickens	Mammals	Humans
Myosin heavy chain			
Developmental isozymes	83—90	91—97	98
Acquisition of mature isozymes	85—89	93—97	98
Myosin light chains			
Developmental isozyme	99	100, 101	98, 101
Slow isozymes in presumptive fast fibers	103—105		
Fast isozymes in presumptive slow fibers	105, 106	107	
Acquisition of mature isozymes	85, 103—106		98, 108
Tropomyosin			
Decrease in ratio of β to α isozyme	109—112	109	
Decrease in phosphorylation	112		
Tropin			
Fast isozyme in presumptive slow fibers		113	
Loss of leg isozyme from breast	110		
Actin			
Cardiac isozyme in skeletal muscle	114	115	
C-Protein			
Fast and slow leg isozymes in breast muscle	116		

REGULATORY INFLUENCES

The regulation of the type of contractile protein isozymes expressed by developing fibers is complex. The initial expression appears to be largely intrinsic to the immature muscle fiber, with external factors such as innervation and thyroxine influencing muscle fiber type during the later stages of development.

Denervation has differential effects on presumptive fast and slow fibers. In presumptive fast fibers, the transition of the myosin heavy chain from the embryonic to the neonatal and finally to the mature fast form will occur in the absence of innervation, although at a reduced rate.[124] Their pattern of myosin light chains is also normal.[78] In contrast, functional denervation of a presumptive slow muscle causes half of the fibers to develop type II characteristics. In concordance with the ATPase classification of these fibers, they cross-react with antibodies directed against myosin LC_{3F}, but not with antislow heavy chain.[78] Such observations are consistent with previous work which has shown that denervated slow muscles acquire some fast characteristics.[68,125]

The possible involvement of hormonal factors in regulating the maturation of the contractile system has received scant attention, even though other events during the initial stages of myogenesis are thought to be influenced by a wide range of circulatory and locally acting factors (see Chapter 1).

Thyroid hormones appear to be responsible for promoting the final stages of muscle fiber differentiation. The loss of myosin HC_{NEO} from rat muscles is temporally correlated with increased plasma level of thyroxine[95] and the replacement of HC_{NEO} with mature fast isozymes is inhibited in hypothyroidal rats and accelerated in hyperthyroidal rats.[96,126] The aquisition of adult myosins in slow muscles is influenced in only a proportion of muscle fibers.[126]

The involvement of thyroid hormones in the loss of the embryonic forms of myosin is less clear. The major loss of embryonic myosin occurs when plasma levels of thyroxine are very low.[95] However, embryonic myosin persists in hypothyroidal rats, suggesting that the loss of embryonic myosin is partially influenced by thyroxine.[96,126] Changes in ATPase histochemistry, which are a reflection of myosin heavy chain expression, have been reported to be correlated with changes in plasma levels of thyroxine,[127] although the available data do not support this contention. The loss of the embryonic type IIc phenotype has a similar time course to the loss of embryonic myosins, with most fibers being transformed in mice between birth and 8 d,[127] which corresponds to a period of development when the plasma levels of free T_3 and T_4 are very low.[127]

The effect of thyroid hormones on muscle differentiation is probably not restricted to myosin expression as hypo- or hyperthyroidism produces alterations in the electrical, mechanical, biochemical, and morphological characteristics of mature human[128-130] and animal skeletal muscles.[131-134] Interestingly, some of the effects of thyroid hormones on mature muscles have been suggested to be secondary to changes in the muscle responsiveness to catecholamines[135] and the possibility that this is occurring during development needs to be investigated.

REFERENCES

1. **Bekoff, A.**, Ontogeny of leg motor output in the chick embryo. A neural analysis, *Brain Res.*, 106, 271, 1976.
2. **Landmesser, L. and O'Donovan, M. J.**, Activation patterns of embryonic chick hindlimb muscles recorded *in ovo* and in an isolated spinal cord preparation, *J. Physiol. (London)*, 347, 189, 1984.
3. **McLennan, I. S.**, The development of the pattern of innervation in chicken hindlimb muscles: evidence for the specification of nerve-muscle connections, *Dev. Biol.*, 97, 229, 1983.
4. **McLennan, I. S.**, Neural dependence and independence of myotube production in chicken hindlimb muscles, *Dev. Biol.*, 98, 287, 1983.
5. **McLennan, I. S.**, Early development and fusion of muscle cells (Chapter 1 in this volume).
6. **Thelen, E. and Bradley, N. S.**, Motor development: posture and locomotion, in *Handbook of Human Growth and Developmental Biology*, Vol. I, Part B, Meisami, E. and Timiras, P. S., Eds., CRC Press, Boca Raton, FL, 1988, 221.
7. **Clark, J. E.**, Development of voluntary motor skill, in *Handbook of Human Growth and Developmental Biology*, Vol. I, Part B, Meisami, E. and Timiras, P. S., Eds., CRC Press, Boca Raton, FL, 1988, 237.
8. **Lennerstrand, G. and Hanson, J.**, The postnatal development of the inferior oblique muscle of the cat. I. Isometric twitch and tetanic properties, *Acta Physiol. Scand.*, 103, 132, 1978.
9. **Close, R.**, Dynamic properties of fast and slow skeletal muscles of the rat during development, *J. Physiol. (London)*, 173, 74, 1964.
10. **Goldspink, G.**, Growth of muscle, in *Development and Specialization of Skeletal Muscle*, Goldspink, G., Ed., Cambridge University Press, Cambridge, 1980, 19.
11. **Buller, A. J., Eccles, J. C., and Eccles, R. M.**, Differentiation of fast and slow muscles in the cat hindlimb, *J. Physiol. (London)*, 150, 399, 1960.
12. **Hammarberg, C. and Kellerth, J.-O.**, The postnatal development of some twitch and fatigue properties of single motor units in the ankle muscles of the kitten, *Acta Physiol. Scand.*, 95, 243, 1975.
13. **Hammarberg, C. and Kellerth, J.-O.**, The postnatal development of some twitch and fatigue properties of the ankle flexor and extensor muscles of the cat, *Arch. Physiol. Scand.*, 95, 166, 1975.
14. **Westerman, R. A., Lewis, D. M., Bagust, J., Edjtehadi, G. D., and Pallot, D.**, Communication between nerves and muscles: postnatal development in kitten hindlimb fast and slow twitch muscle, in *Memory and Transfer of Information*, Zippel, H. P., Ed., Plenum Press, New York, 1973, 255.
15. **Drachman, D. B. and Johnston, D. M.**, Development of a mammalian fast muscle: dynamic and biochemical properties correlated, *J. Physiol. (London)*, 234, 29, 1973.
16. **Edge, M. B.**, Development of apposed sarcoplasmic reticulum at the T system and sarcolemma and the change in orientation of triads in rat skeletal muscle, *Dev. Biol.*, 23, 634, 1970.
17. **Schiaffino, S. and Margareth, A.**, Coordinated development of the sarcoplasmic reticulum and T system during postnatal differentiation of rat skeletal muscle, *J. Cell Biol.*, 41, 855, 1969.
18. **Buller, A. J., Eccles, J. C., and Eccles, R. M.**, Interactions between motoneurones and muscles in respect to the characteristic speeds of their responses, *J. Physiol. (London)*, 150, 417, 1960.
19. **Buchthal, F. and Schmalbruch, H.**, Motor unit of mammalian muscle, *Physiol. Rev.*, 60, 90, 1980.

20. **Salmons, S. and Sreter, F. A.**, Significance of impulse activity in the transformation of skeletal muscle type, *Nature (London)*, 263, 30, 1976.
21. **Goldring, J. M., Kuno, M., Nunez, R., and Weakly, J. N.**, Do identical activity patterns in fast and slow motor axons exert the same influence on the twitch time of cat skeletal muscle?, *J. Physiol. (London)*, 321, 211, 1981.
22. **Sreter, F. A., Pinter, K., Jolesz, F., and Mabuchi, K.**, Fast to slow transfomration of fast muscles in response to long-term phasic stimulation, *Exp. Neurol.*, 75, 95, 1982.
23. **Euter, U. S. V. and Swank, R. L.**, Tension changes during tetanus in mammalian and avian muscle, *Acta Physiol. Scand.*, 1, 203, 1940.
24. **Nystrom, B.**, Effect of direct tetanization on twitch tension in developing cat leg muscles, *Acta Physiol. Scand.*, 74, 319, 1968.
25. **Nystrom, B.**, Mechanical and electrical responses to single shocks in developing cat leg muscles following tetanization, *Acta Physiol. Scand.*, 74, 207, 1968.
26. **Kowalchuk, N., McComas, A. J., and Corely, K.**, Physiologic and histologic features of muscle development in the hamster, *Exp. Neurol.*, 85, 41, 1984.
27. **Harvey, A. L.**, Actions of drugs on developing skeletal muscle, *Pharm. Ther.*, 11, 1 1980.
28. **Oppenheim, R. W.**, Neuronal cell death and some related regressive phenomena during neurogenesis: a selective historical review and progress report, in *Studies in Developmental Neurobiology*, Cowan, W. M., Ed., Oxford University Press, Oxford, 1980, 74.
29. **Bianchi, C. P. and Narayan, S.**, Muscle fatigue and the role of transverse tubules, *Science*, 215, 295, 1982.
30. **Fudel-Ospova, S. I. and Martynenko, O. A.**, Early ontogenic development of membrane potential of muscle fibers in rat, *Fed. Proc. Trans. Suppl.*, 23, 128, 1964.
31. **Hazlewood, C. F. and Nichols, B. L., Jr.**, *In vitro* investigation of resting muscle membrane potential in preweaning and weaning rats, *Nature (London)*, 213, 935, 1967.
32. **Harris, J. B. and Luff, A. R.**, The resting membrane potentials of fast and slow skeletal muscles in the developing mouse, *Comp. Biochem. Physiol.*, 33, 923, 1970.
33. **McArdle, J. J., Michelson, K., and D'Alonzo, A. J.**, Action potentials in fast- and slow-twitch mammalian muscles during reinnervation and development, *J. Gen. Physiol.*, 75, 655, 1980.
34. **Bannett, R. R., Sampson, S. R., and Shainberg, A.**, Influence of thyroid hormone on some electrophysiological properties of developing rat skeletal muscle cells in culture, *Brain Res.*, 294, 75, 1984.
35. **Shainberg, A., Brik, H., Bar-Shavit, R., and Sampson, S. R.**, Inhibition of acetylcholine receptor synthesis by thyroid hormones, *J. Endocrinol.*, 101, 141, 1984.
36. **Asano, Y., Liberman, U. S., and Edelman, I. S.**, Relationships between Na+ dependent respiration and (Na++K+) ATPase activity in rat skeletal muscle, *J. Clin. Invest.*, 57, 368, 1974.
37. **Moore, G. E., Harvey, S., Klandorf, H., and Goldspink, G.**, Muscle development in thyroidectomised chickens *(Gallus domesticus)*, *Gen. Comp. Endocrinol.*, 55, 195, 1984.
38. **Brehm, P., Kidokoro, Y., and Moody-Corbett, F.**, Acetylcholine receptor channel properties during development of *Xenopus* muscle cells in culture, *J. Physiol. (London)*, 357, 203, 1984.
39. **Kuromi, H., Brais, B., and Kidokoro, Y.**, Formation of acetylcholine receptor clusters at neuromuscular junction in *Xenopus* cultures, *Dev. Biol.*, 109, 165, 1985.
40. **Bevan, S. and Steinbach, J. H.**, The distribution of α-bungarotoxin binding sites on mammalian skeletal muscle developing *in vivo*, *J. Physiol. (London)*, 267, 195, 1977.
41. **Sakmann, B.**, Acetylcholine-induced ionic channels in rat skeletal muscle, *Fed. Proc.*, 37, 2654, 1978.
42. **Fischbach, G. D. and Schuetze, S. M.**, A post-natal decrease in acetylcholine channel open time at rat end-plates, *J. Physiol. (London)*, 303, 125, 1980.
43. **Sakmann, B. and Brenner, H. R.**, Change in synaptic channel gating during neuromuscular development, *Nature (London)*, 276, 401, 1978.
44. **Hamill, O. P. and Sakmann, B.**, Multiple conductance states of single acetylcholine receptor channels in embryonic muscle cells, *Nature (London)*, 294, 462, 1981.
45. **Brehm, P., Kullberg, R., and Moody-Corbett, F.**, Properties of non-junctional acetylcholine receptor channels on innervated muscle of *Xenopus laevis*, *J. Physiol. (London)*, 350, 631, 1984.
46. **Tomanek, R. J. and Colling-Saltin, A.-S.**, Cytological differentiation of human fetal skeletal muscle, *Am. J. Anat.*, 149, 227, 1977.
47. **Allen, E. R. and Pepe, F. A.**, Ultrastructure of developing muscle cells in the chick embryo, *Am. J. Anat.*, 116, 115, 1965.
48. **Fischman, D. A.**, An electron microscope study of myofibril formation in embryonic chick skeletal muscle, *J. Cell Biol.*, 32, 557, 1967.
49. **Kilarski, W. and Jakubowska, M.**, An electron microscope study of myofibril formation in embryonic rabbit skeletal muscle, *Z. Mikrosk. Anat. Forsch.*, 93, 1159, 1979.
50. **Platzer, A. C.**, The ultrastructure of normal myogenesis in the limb of the mouse, *Anat. Rec.*, 190, 639, 1978.

51. **Lazarides, E.**, Intermediate filaments as mechanical integrators of cellular space, *Nature (London)*, 283, 249, 1980.
52. **Lazarides, E., Granger, B. L., Gard, D. L., O'Connor, C. M., Breckler, J., Price, M., and Danto, S. I.**, Desmin- and vimentin-containing filaments and their role in the assembly of the Z disk in muscle cells, *Cold Spring Harbor Symp. Quant. Biol.*, XLVI, 351, 1981.
53. **Freeman, S. S., Engel, A. G., and Drachman, D. B.**, Experimental acetylcholine blockade of the neuromuscular junction. Effect on end plate and muscle fiber ultrastructure, *Ann. N.Y. Acad. Sci.*, 274, 46, 1976.
54. **Sohal, G. S. and Holt, R. K.**, Role of innervation in the embryonic development of skeletal muscle, *Cell Tissue Res.*, 210, 383, 1980.
55. **McLennan, I. S.**, Inhibition of prostaglandin synthesis produces a muscular dystrophy-like myopathy, *Exp. Neurol.*, 89, 616, 1985.
56. **Costantin, L. L.**, Contractile activation in skeletal muscle, *Prog. Biophys. Mol. Biol.*, 29, 199, 1975.
57. **Martonosi, A., Roufa, D., Ha, D. B., and Boland, R.**, The biosynthesis of sarcoplasmic reticulum, *Fed. Proc.*, 39, 2415, 1980.
58. **Ezerman, E. B. and Ishikawa, H.**, Differentiation of the sarcoplasmic reticulum and T system in developing chick skeletal muscle *in vitro*, *J. Cell Biol.*, 35, 405, 1967.
59. **Kelly, A. M.**, Sarcoplasmic reticulum and T tubules in differentiating rat skeletal muscle, *J. Cell Biol.*, 49, 335, 1971.
60. **Martonosi, A., Roufa, D., Boland, R., Reyes, E., and Tillack, T. W.**, Development of sarcoplasmic reticulum in cultured chicken muscle, *J. Biol. Chem.*, 252, 318, 1977.
61. **Tillack, T. W., Boland, R., and Martonosi, A.**, The ultrastructure of developing sarcoplasmic reticulum, *J. Biol. Chem.*, 249, 624, 1974.
62. **Boland, R., Martonosi, A., and Tillack, T. W.**, Developmental changes in the composition and function of sarcoplasmic reticulum, *J. Biol. Chem.*, 249, 612, 1974.
63. **Mazina, T. I. and Pevzner, R. A.**, Influence of catecholamines on adenosine triphosphatase activity of the sarcoplasmic reticulum of skeletal muscles during the ontogenesis of chickens, *J. Evol. Biochem. Physiol.*, 15, 17, 1979.
64. **Ontell, M. A.**, The growth and metabolism of developing muscle, in *Biochemical Development of the Fetus and Neonate*, Jones, C. T., Ed., Elsevier, Amsterdam, 1982, 213.
65. **Guth, L. and Samaha, F. J.**, Procedures for the histochemical demonstration of actinomyosin ATPase, *Exp. Neurol.*, 28, 365, 1970.
66. **Jolesz, F. and Sreter, F. A.**, Development, innervation and activity-pattern induced changes in skeletal muscle, *Annu. Rev. Physiol.*, 43, 531, 1981.
67. **Engel, W. K. and Karpati, G.**, Impaired skeletal muscle maturation following neonatal neurectomy, *Dev. Biol.*, 17, 713, 1968.
68. **Rubinstein, N. A. and Kelly, A. M.**, Myogenic and neurogenic contributions to the development of fast and slow twitch muscles in rat, *Dev. Biol.*, 62, 473, 1978.
69. **Phillips, W. D. and Bennett, M. R.**, Differentiation of fiber types in wing muscle during embryonic development: effect of neural tube removal, *Dev. Biol.*, 106, 457, 1985.
70. **Butler, J., Cosmos, E., and Brierley, J.**, Differentiation of muscle fiber types in aneurogenic brachial muscles of the chick embryo, *J. Exp. Zool.*, 224, 65, 1982.
71. **McLennan, I. S.**, The development of the pattern of innervation in chicken hindlimb muscles: evidence for specification of nerve-muscle connections, *Dev. Biol.*, 97, 229, 1983.
72. **Ashmore, C. R., Robinson, D. W., Rattray, P., and Doerr, L.**, Biphasic development of muscle fibers in the fetal lamb. *Exp. Neurol.*, 37, 241, 1972.
73. **Ashmore, C. R., Addis, P. B., Doerr, L., and Stokes, H.**, Development of muscle fibers in the complexus muscle of normal and dystrophic chicks, *J. Histochem. Cytochem.*, 21, 266, 1973.
74. **Kelly, A. M. and Rubinstein, N. A.**, Why are fetal muscles slow?, *Nature (London)*, 288, 266, 1980.
75. **Rubinstein, N. A. and Kelly, A. M.**, Development of muscle fiber specialization in the rat hindlimb, *J. Cell Biol.*, 90, 128, 1981.
76. **Stockdale, F. E. and Miller, J. B.**, The cellular basis of myosin heavy chain isoform expression during development of avian skeletal muscles, *Dev. Biol.*, 123, 1, 1987.
77. **Hoh, J. F. Y., Hughes, S., Hale, P. T., and Fitzsimons, R. B.**, Immunocytochemical and electrophoretic analyses of changes in myosin gene expression in cat limb fast and slow muscles during postnatal development, *J. Muscle Res. Cell Motility*, 9, 30, 1988.
78. **Gauthier, G. F., Ono, G. R., and Hobbs, A. W.**, Curare-induced transformation of myosin pattern in developing skeletal muscle fibers, *Dev. Biol.*, 105, 144, 1984.
79. **Harris, A. J.**, Embryonic growth and innervation of rat skeletal muscles. I. Neural regulation of muscle fiber numbers, *Philos. Trans. R. Soc. (London) Ser. B*, 293, 257, 1981.
80. **Pittman, R. and Oppenheim, R. W.**, Cell death of motoneurons in the chick embryo spinal cord. IV. Evidence that a functional neuromuscular interaction is involved in regulation of naturally occurring cell death and the stabilisation of synapses, *J. Comp. Neurol.*, 187, 425, 1979.

81. **McLennan, I. S.**, Relationship between muscle size and motoneurone survival in temporarily paralysed embryos, *J. Exp. Zool.*, 230, 239, 1984.
82. **Renaud, D., Gardanaut, M.-F., Rouaud, T., and Le Douarin, G. H.**, Influence of chronic spinal cord stimulation upon differentiation of B muscle fibres in a fast muscle (posterior latissimus dorsi) of the chick embryo, *Exp. Neurol.*, 80, 157, 1983.
83. **Masaki, T. and Yoshizaki, C.**, Differentiation of myosin in chick embryos, *J. Biochem.*, 767, 123, 1974.
84. **Rushbrook, J. I. and Stracher, A.**, Comparison of adult, embryonic and dystrophic myosin heavy chains from chicken muscle by sodium dodecyl sulfate/polyacrylamide gel electrophoresis and peptide mapping, *Proc. Natl. Acad. Sci. U.S.A.*, 76, 4331, 1979.
85. **Winkelman, D. A., Lowey, S., and Press, J. L.**, Monoclonal antibodies localize changes on myosin heavy chain isozymes during avian myogenesis, *Cell*, 34, 295, 1983.
86. **Bader, D., Masaki, T., and Fishman, D. A.**, Immunochemical analysis of myosin heavy chain during avian myogenesis *in vivo* and *in vitro*, *J. Cell Biol.*, 95, 763, 1982.
87. **Benfield, P. A., Lowey, S., LeBlanc, D. D., and Waller, G. S.**, Myosin isozymes in avian skeletal muscles. II. Fractionation of myosin isozymes from adult and embryonic chicken pectoralis muscle by immunoaffinity chromatography, *J. Muscle Res. Cell Motility*, 4, 717, 1983.
88. **Lowey, S., Benfield, P. A., LeBlanc, D. D., and Waller, G. S.**, Myosin isozymes in avian skeletal muscles. I. Sequential expression of myosin isozymes in developing chicken pectoralis muscles, *J. Muscle Res. Cell Motility*, 4, 695, 1983.
89. **Umeda, P. K., Kavinsky, C. J., Sinha, A. M., Hsu, H.-J., Jakovcic, S., and Rabinowitz, M.**, Cloned mRNA sequences for two types of embryonic myosin heavy chains from chick skeletal muscle, *J. Biol. Chem.*, 258, 5206, 1983.
90. **Bandman, E., Matsuda, R., and Strohman, R. C.**, Developmental appearance of myosin heavy and light chain isoforms *in vivo* and *in vitro* in chicken skeletal muscle, *Devl. Biol.*, 93, 508, 1982.
91. **Huszar, G.**, Developmental changes of the primary structure and histidine methylation in rabbit skeletal muscle myosin, *Nature (London) New Biol.*, 240, 260, 1972.
92. **Whalen, R. G., Sell, S. M., Butler-Browne, G. S., Schwartz, K., Bouveret, P., and Pinset-Harstrom, I.**, Three myosin heavy-chain isozymes appear sequentially in rat muscle development, *Nature (London)*, 292, 805, 1981.
93. **Butler-Browne, G. S. and Whalen, R. G.**, Myosin isozyme transitions occurring during the postnatal development of the rat soleus muscle, *Dev. Biol.*, 102, 324, 1984.
94. **Periasamy, M., Wieczorek, D. F., and Nadal-Ginard, B.**, Characterization of a developmentally regulated perinatal myosin heavy-chain gene expressed in skeletal muscle, *J. Biol. Chem.*, 259, 13,573, 1984.
95. **Gambke, B., Lyons, G. E., Haselgrove, J., Kelly, A. M., and Rubinstein, N. A.**, Thyroidal and neural control of myosin transitions during development of rat fast and slow muscles, *FEBS Lett.*, 156, 335, 1983.
96. **Gambke, B. and Rubinstein, N. A.**, A monoclonal antibody to the embryonic myosin heavy chain of rat skeletal muscle, *J. Biol. Chem.*, 259, 12,092, 1984.
97. **Lyons, G. E., Haselgrove, J., Kelly, A. M., and Rubinstein, N. A.**, Myosin transitions in developing fast and slow muscles of the rat hindlimb, *Differentiation*, 25, 168, 1983.
98. **Biral, D., Damiani, E., Margreth, A., and Scarpini, E.**, Myosin subunit composition in human developing muscle, *Biochem. J.*, 224, 923, 1984.
99. **Takano-Ohumuro, H., Obinata, T., Kawashima, M., Masaki, T., and Tanaka, T.**, Embryonic chicken skeletal, cardiac and smooth muscles express a common embryo-specific myosin light chain, *J. Cell Biol.*, 100, 2025, 1985.
100. **Gauthier, G. F., Lowey, S., Benfield, P. A., and Hobbs, A. W.**, Distribution and properties of myosin isozymes in developing avian and mammalian skeletal muscle fibers, *J. Cell Biol.*, 92, 471, 1982.
101. **Whalen, R. G., Butler-Browne, G. S., and Gros, F.**, Identification of a novel form of myosin light chain present in embryonic muscle tissue and cultured muscle cell, *J. Mol. Biol.*, 126, 415, 1978.
102. **Strohman, R. C., Micou-Eastwood, J., Glass, C. A., and Matsuda, R.**, Human fetal muscle and cultured myotubes derived from it contain a fetal-specific myosin light chain, *Science*, 221, 955, 1983.
103. **Hoh, Y. F. Y.**, Developmental changes in chicken skeletal myosin isoenzymes, *FEBS Lett.*, 98, 267, 1979.
104. **Stockdale, F. E., Raman, N., and Baden, H.**, Myosin light chains and the developmental origin of fast muscle, *Proc. Natl. Acad. Sci. U.S.A.*, 78, 931, 1981.
105. **Crow, M. T., Olson, P. S., and Stockdale, F. E.**, Myosin light-chain expression during avian muscle development, *J. Cell Biol.*, 96, 736, 1983.
106. **Rubinstein, N. A., Pepe, F. A., and Holtzer, H.**, Myosin types during the development of embryonic chicken fast and slow muscles, *Proc. Natl. Acad. Sci. U.S.A.*, 74, 4524, 1977.
107. **Sreter, F. A., Balint, M., and Gergely, J.**, Structural and functional changes of myosin during development, *Dev. Biol.*, 46, 317, 1974.
108. **Volpe, P., Biral, D., Damiani, E., and Margreth, A.**, Characterization of human muscle myosins with respect to the light chains, *Biochem. J.*, 195, 251, 1981.

109. **Roy, R. K., Sreter, F. A., and Sarkar, S.,** Changes in tropomyosin subunits and myosin light chains during development of chicken and rabbit striated muscles, *Dev. Biol.,* 69, 15, 1979.
110. **Matsuda, R., Obinata, T., and Shimada, Y.,** Types of troponin components during development of chicken skeletal muscle, *Dev. Biol.,* 82, 11, 1981.
111. **Matsuda, R., Bandman, E., and Strohman, R. C.,** Regional differences in the expression of myosin light chains and tropomyosin subunits during development of chicken breast muscle, *Dev. Biol.,* 95, 484, 1983.
112. **Montarras, D., Fiszman, M. Y., and Gros, G.,** Changes in tropomyosin during development of chick embryonic skeletal muscle *in vivo* and during differentiation of chick muscle cell *in vitro, J. Biol. Chem.,* 257, 545, 1982.
113. **Dhoot, G. K. and Perry, S. U.,** Distribution of polymorphic forms of troponin components and tropomyosin in skeletal muscle, *Nature (London),* 278, 714, 1979.
114. **Paterson, B. M. and Eldridge, J. D.,** α-Cardiac actin is the major sarcomeric isoform expressed n embryonic avian skeletal muscle, *Science,* 224, 1436, 1984.
115. **Minty, A. J., Alonso, S., Caravalti, M., and Buckingham, M. E.,** A fetal skeletal muscle actin mRNA in the mouse and its identity with cardiac actin mRNA, *Cell,* 30, 185, 1982.
116. **Obinata, T., Kitani, S., Masaki, T., and Fischman, D. A.,** Coexistence of fast-type and slow-type c-proteins in neonatal chicken breast muscle, *Dev. Biol.,* 105, 253, 1984.
117. **Barany, M.,** ATPase activity of myosin correlates with speed of muscle shortening, *J. Gen. Physiol. Suppl.,* 50, 197, 1967.
118. **Wagner, P. D. and Weeds, J.,** Studies on the role of myosin alkali light chain, *Mol. Biol.,* 109, 455, 1977.
119. **Sivaramakrishnan, M. and Burke, M.,** The heavy chain of vertebrate skeletal myosin subfraction I shows full enzymatic activity, *J. Biol. Chem.,* 257, 1102, 1982.
120. **Cantini, M., Sartore, S., and Schiaffino, S.,** Myosin types in cultured muscle cells, *J. Cell Biol.,* 85, 903, 1980.
121. **Fritzsimons, R. B. and Hoh, J. F. Y.,** Embryonic and foetal myosins in human skeletal muscle. The presence of foetal myosins in Duchenne muscular dystrophy and infantile spinal muscular atrophy, *J. Neurol. Sci.,* 52, 367, 1981.
122. **Fitzsimons, R. B. and Hoh, J. F. Y.,** Fetal myosins in skeletal muscle from a patient with myalgia and fatigue, *Lancet,* I, 480, 1982.
123. **Bandman, E.,** Continued expression of neonatal myosin heavy chain in adult dystrophic skeletal muscle, *Science,* 227, 780, 1985.
124. **Butler-Browne, G. S., Bugaisky, L. B., Cuenoud, S., Schwartz, K., and Whalen, R. G.,** Denervation of newborn rat muscles does not block the appearance of adult fast myosin heavy chain, *Nature (London),* 299, 830, 1982.
125. **Ishiura, S., Nonaka, I., Sugita, H., and Mikawa, T.,** Effect of denervation of neonatal sciatic nerve on the differentiation of myosin in a single fiber, *Exp. Neurol.,* 73, 487, 1981.
126. **Butler-Browne, G. S., Herlicoviez, D., and Whalen, R. G.,** Effects of hypothyroidism on myosin isozyme transitions in developing rat muscle, *FEBS Lett.,* 166, 71, 1984.
127. **D'Albis, A., Lenfant-Guyot, M., Janmot, C., Chanoine, C., Weinman, J., and Gallien, C. L.,** Regulation by thyroid hormones of terminal differentiation in the skeletal dorsal muscle. I. Neonatal mouse, *Dev. Biol.,* 123, 25, 1987.
128. **Hudgson, P. and Hall, R.,** Endocrine myopathies, in *Skeletal Muscle Pathology,* Mastaglia, F. L. and Walton, J. N., Eds., Churchill Livingstone, London, 1982, 393.
129. **Salviata, G., Zeviani, M., Betto, R., Nacamulli, D., and Busnardo, B.,** Effects of thyroid hormones on the biochemical specialisation of human muscle fibres, *Muscle Nerve,* 8, 363, 1985.
130. **Ono, S., Inouge, K., and Mannen, T.,** Myopathology of hypothyroid myopathy. Some new observations, *J. Neurol. Sci.,* 77, 237, 1987.
131. **Dulhunty, A. F., Gage, P. W., and Lamb, G. D.,** Differential effects of thyroid hormone on T-tubules and terminal cisternae in rat muscles: an electrophysiological and morphometric analysis, *J. Muscle Res. Cell Motility,* 7, 225, 1986.
132. **Everts, M. E. and Clausen, T.,** Effects of thyroid hormones on calcium content and ^{45}Ca exchange in rat skeletal muscle, *Am. J. Physiol.,* 251, E256, 1986.
133. **Angeras, U. and Hasselgren, P.-O.,** Protein degradation in skeletal muscle during experimental hyperthyroidism in rats and the effect of β-blocking agents, *Endocrinology,* 120, 1477, 1987.
134. **Heeley, D. H., Dhoot, G. K., and Perry, S. O.,** Factors determining the subunit composition of tropomyosin in mammalian skeletal muscle, *Biochem. J.,* 226, 461, 1985.
135. **Miller, J. L., Ismail, F., Waligora, J. K., and Gevers, W.,** Modulating influence of D,L-propanolol on tri-iodothyronine induced skeletal muscle protein degradation, *Endocrinology,* 117, 869, 1985.

DIFFERENTIATION OF MUSCLE FIBERS — BIOCHEMISTRY AND ENERGETICS

D. D. Hatcher and A. R. Luff

INTRODUCTION

The process of skeletal muscle myogenesis gives rise to small, but identifiable muscle fibers containing myofibrils, peripheral nuclei, and a few mitochondria. There is, however, a relatively poorly developed sarcoplasmic reticulum (SR) and transverse tubular system (T-system). Fibers are produced in at least two distinct phases. During late fetal development in the mouse, larger primary myotubes are surrounded by several smaller secondary myotubes (Figure 1). The secondaries subsequently dissociate from the cluster and develop as fibers.[1] The limb muscle of a newborn rat is capable of generating force, but relatively less than that of an adult animal.[2] Functional motor innervation is present and polyneuronal innervation is common, although this regresses over the first 2 weeks of postnatal life in the rat.[3]

The newly formed fiber continues to undergo considerable quantitative and qualitative changes. It increases in both length and cross-sectional area in appropriate proportion to the increasing size of the organism. An increase in fiber length is achieved by the addition of sarcomeres at the ends of the fibers.[4] Cross-sectional area expands mainly by an increase in myofibril number due to splitting of existing myofibrils.[5] Satellite cells adjacent to the fiber divide and fuse with the fiber to increase the nuclear content of the growing fiber.[6] The number of mitochondria proliferate, particularly in the slow-oxidative (type I) fibers, and the organization and amount of SR increases.[7] A sequential transition of myosin isozymes occurs, with specific embryonic and neonatal forms appearing during fetal and postnatal development.[8] Consistent changes occur in the regulatory proteins, SR, and in the enzyme systems relating to ATP production. These biochemical and histochemical changes form the basis of the alterations seen in the physiological and contractile properties of the muscle in early postnatal development, including an increase in the speed of contraction of fast-twitch muscles.[9,10]

Postnatal development of a muscle is initially determined by its genotype. However, as the innervation of the fibers becomes fully established, the innervating motoneuron exerts a significant influence on the specific characteristics of the fiber. During the life of an animal, the functional demands placed on a muscle will alter. Sufficient plasticity continues to exist within the motor system to allow for further appropriate changes in the characteristics of muscle fibers. This can lead to fiber hypertrophy, atrophy, or an alteration in fiber type.

CONTRACTILE PROTEINS

Myosin

The myosin molecule consists of two heavy chains (myosin heavy chain, MHC, 200,000 Da) and two pairs of light chains. There are several isozymes of skeletal muscle myosin. Adult slow-twitch muscle contains a slow-MHC and two specific light chains, $LC1_s$ and $LC2_s$, with molecular weights of 27,000 and 19,000 Da, respectively.[11] Adult fast-twitch muscle probably contains two fast MHC[12] and three light chains, the so-called alkali light chains, $LC1_f$ and $LC3_f$ with molecular weights of 25,000 and 16,000 Da, respectively,[13] and the so-called DTNB (5,5'-dithiobis-(2-nitrobenzoic acid) light chain, $LC2_f$, having a molecular weight of 18,000 Da.

Developmental Changes in Myosin

Biochemical and immunological studies over the past few years have shown that fetal

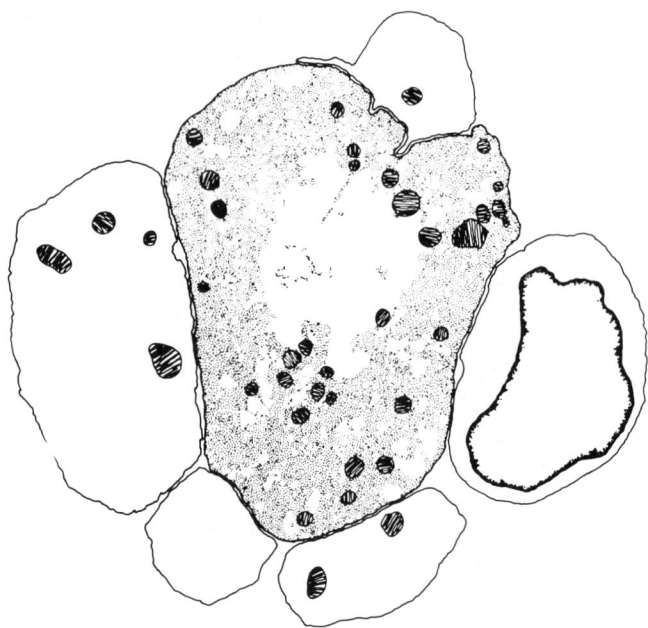

FIGURE 1. A large primary-generation myotube, containing myofibrils and glycogen, is surrounded by new generations of secondary cells. Drawn from an electron micrograph of Kelly and Zacks[43] of a cross section of rat intercostal muscle at 18 d of gestation.

and newborn skeletal muscles contain myosin subunits which are distinct from those found in adult muscles. In addition to the light chains found in adult muscles, Whalen et al.[14] have identified an embryonic light chain designated LC_{emb} (Figure 2). This is similar to $LC1_f$, but has a slightly different molecular weight and isoelectric point. Its existence has since been confirmed by others.[15,16] During rat muscle development, it is replaced fairly rapidly with $LC1_f$ during the first 2 weeks after birth.

The developmental pattern of change in the myosin light chains is probably better understood than that of the myosin heavy chains or native myosin. In the late fetal stages in the rat, all muscles contain $LC1_f$, LC_{emb}, and $LC2_f$. As development proceeds in fast-twitch muscle, LC_{emb} recedes[15] and $LC3_f$ appears at the expense of LC_{emb} and $LC1_f$.[15,17-19] In slow-twitch muscle 2 d after birth, there is a mixture of fast and slow light chains; thereafter the embryonic and fast light chain disappear rapidly.[17,20]

The possibility of a developmental form of the myosin heavy chains was discussed by Trayer et al.,[21] who identified 3-methyl histidine in adult muscle, but not in fetal muscle. In addition, Huszar[22] demonstrated an amino acid sequence in a myosin peptide obtained from neonatal rabbits which is distinct from that found in adult muscle. Subsequently, a MHC was identified in neonatal rabbits which is related to, but different from, either adult fast or slow MHC.[23,24] Whalen et al.[25] also noted that the transition of a fetal MHC to the adult form is slower than the $LC1_{emb}$ to $LC1_f$ transition. This might allow for the formation of a number of different myosin isozymes if all heavy and light chains could associate independently.[25] More recently, Whalen and colleagues[26] presented strong evidence for specific embryonic (MHC_{emb}) and neonatal (MHC_{neo}) myosin heavy chains appearing sequentially during early development. MHC_{emb} exists in fetal rat muscles[15,26] and is progressively replaced in the 2 weeks after birth by MHC_{neo}. However, MHC_{neo} bears no resemblance to adult slow MHC.[27] MHC_{neo} is itself replaced by the appropriate adult MHC and disappears about 3 weeks after birth in the rat. A very similar pattern of development has also been

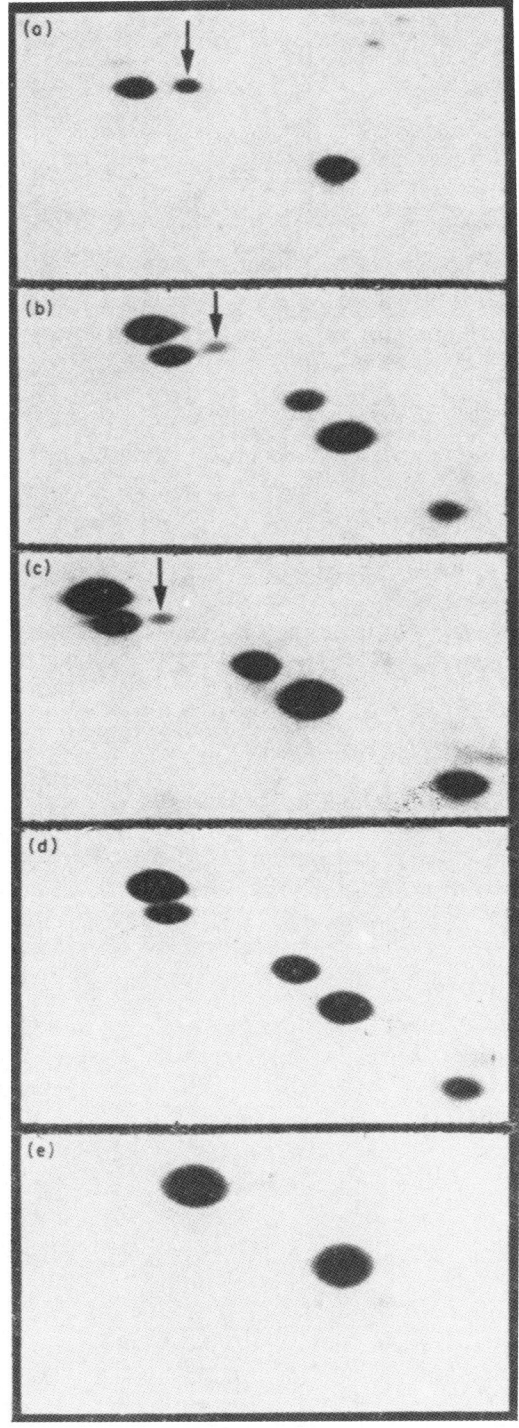

FIGURE 2. Analysis of embryonic, newborn, and adult rat myofibrils: (a) 20-d embryonic thigh muscle; (b) 3-d soleus; (c) 7-d soleus; (d) 12-d soleus; (e) adult soleus. Crude myofibrils were analyzed by two-dimensional gel electrophoresis. The vertical arrows indicate the LC_{emb} protein. The two spots in (e) correspond to the $LC1_s$ and $LC2_s$ proteins. (From Whalen, R. G., Butler-Browne, G. S., and Gros, F., *J. Mol. Biol.*, 126, 415, 1978. With permission.)

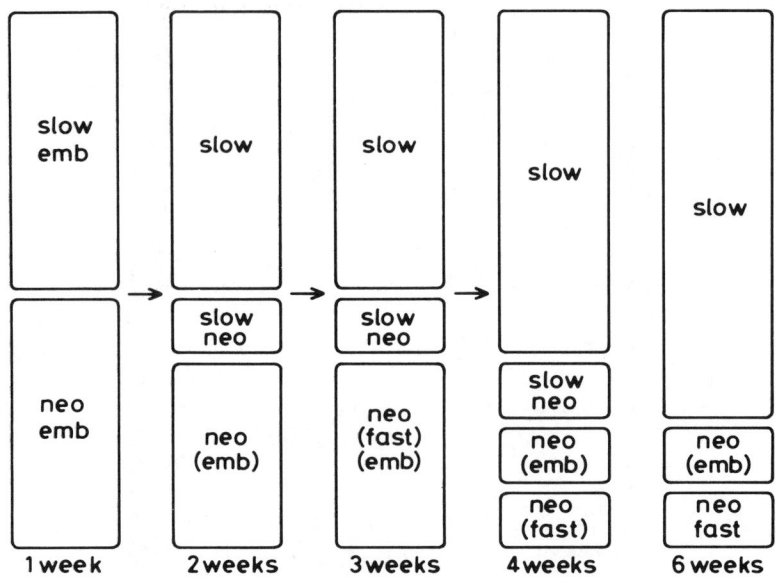

FIGURE 3. Diagrammatic representation of the developmental changes of myosin isozymes in rat soleus muscle. The size of the blocks is an approximate indication of the relative number of fibers. This is based on immunocytochemical analysis using antibodies raised against embryonic (emb), neonatal (neo), fast, and slow myosin. The postnatal age of the animals is indicated. (Drawn from results of Butler-Browne and Whalen.[8])

described for chick muscle.[28-30] Although the existence of the developmental myosin isozymes is generally accepted, there is, at the present time, some debate about the myosin isozyme sequence in muscles which are destined to become fast-twitch and slow-twitch.

Butler-Browne and Whalen[8] provide a detailed description of the developmental changes in myosin in rat soleus. Although this is a typical slow-twitch muscle, they consider that the developmental pattern is, in general, applicable to all muscles. Figure 3 shows a diagrammatic summary of their results: 1 week after birth, there are two groups of fibers distinguishable by their differing diameters, each comprising about half the total number. The group with the larger diameters contains embryonic and slow myosin, but probably has only slow myosin from 2 weeks onward. The small-diameter group contains embryonic and neonatal myosin at 1 week and some fibers subsequently accumulate fast myosin. The fact that two or more isozymes are found in any one fiber suggests that transitions are occurring. It also seems that even at 1 month when only fast and slow myosins are detectable, the proportion of fast fibers continues to diminish. Gauthier et al.[15] concluded that in early development all muscles, irrespective of their fate, will react with both antislow and antifast myosin antibody, suggesting that both fast and slow myosins are present. Whalen et al.[31] criticized these results on the grounds that the antibodies used by Gauthier et al.[15] are not sufficiently specific. It is certainly not at all clear what degree of cross-reactivity exists between antibodies raised to the adult forms and the fetal and neonatal myosins.

Kelly and colleagues,[16,20,32] while not disagreeing with some of the conclusions of Whalen and associates,[8,26] have nonetheless argued consistently that the primary myotubes of myogenesis are destined to become slow-twitch fibers while the secondary myotubes become fast-twitch fibers except in slow-twitch muscles where they are progressively converted to slow-twitch fibers. Specifically, they showed that in the 15-d rat fetus *all* myotubes and fibers react with antibody to fast myosin, but not to slow. By 17 d, antibody to slow myosin starts to bind only to primary myotubes. These fibers progressively decrease their binding of antifast myosin and become slow-twitch fibers. By 20 d of gestation, the destiny of individual myotubes is precisely determined.[20] Whalen et al.[26] maintained that the antibody to slow myosin, by which the slow fibers are identified in the late fetal muscles, is in fact cross-reacting with MHC_{emb}. The incidence and destiny of fetal slow fibers were investigated recently by Whalen et al.[31] in fast-twitch mouse muscles. They found that fibers which are identified as slow in the newborn do not necessarily continue as slow fibers in the adult. The contingent of slow fibers present at 3 to 4 weeks diminishes with further development. At the present time, therefore, there is clearly some disagreement concerning the early development pattern of muscle fibers.

Rubinstein and Kelly[16] maintain that the appearance of slow-type myosin in the primary myotubes of muscles in 17-d fetal rats is influenced by the initial neural contact which is assumed to exist at that time. In fact, denervated muscle fibers are incapable of the transition to slow-twitch fibers, but the transition from neonatal to fast-twitch fibers is unaffected.[17,33] It was also shown that preinnervated, cultured muscle fibers synthesize only fast myosin regardless of their destiny.[34] These results, therefore, support the general contention of Lyons and colleagues[20] that there is an intrinsic, neurally independent program of development which directs muscle to the fast phenotype. However, according to current hypotheses, slow fiber characteristics should only appear when the aggregate level of activity reaches some relatively high threshold.[35] Differences in electromyographic (EMG) firing pattern normally seen in the adult between fast and slow-twitch muscles are not detected until at least the first postnatal week.[36]

How the cell decides which myosin isozyme it will produce and when, or more precisely which gene or combination of genes will be expressed and to what extent, is emerging as one of the current problems of muscle development. It is clear that the myosin heavy chains are coded by a large multigene family which is regulated both spatially and temporally and in which the expression of specific genes can be readily modified. The organization of the heavy chain genomic sequences is unusual in that coding regions of individual members of the multigene family share small regions of homology marked by divergent regions.[37] It is likely that research in this area of muscle development will proceed very rapidly in the next few years.

Myosin ATPase Activity of Newborn Muscles

It has been long held from early work[38,39] that the myosin ATPase activity of newborn, fast-twitch muscle is only about one third that of adult muscle. It is these findings that led Perry[39] to suggest the existence of a discrete fetal form of myosin, correctly as it turned out, but partly for the wrong reasons. The importance attached to values for myosin ATPase activity relates to the fact that it is generally considered to be the rate-limiting step in the cross-bridge cycle. This low level of activity in the newborn muscles was considered[40] to extend the correlation between speed of contraction and myosin ATPase activity that existed for adult muscles.[41]

The earlier estimates appear to have been incorrect as the myosin ATPase activity in neonatal muscle is very susceptible to inactivation by oxidation of sulfhydryl groups.[42] When the assays are done in the presence of a protective thiol reagent, the myosin ATPase activity of neonatal fast-twitch muscles is only slightly less than adult muscles.[19,23] The light chain pattern (see previous section) further confirms that neonatal muscles can be regarded biochemically as fast-twitch. At the present time, there is some uncertainty on the precise role, if any, that light chains play in determining the characteristics of myosin in mammalian skeletal muscle. For example, it was shown that the type and amount of light chain present in a fiber do not directly determine the maximum speed of shortening of skinned neonatal fibers.[44]

The paradox remains that, physiologically, neonatal fast-twitch muscles are slow (Table 1). The nexus that exists between speed of shortening and myosin ATPase activity in adult

Table 1
MAXIMUM SPEED OF SARCOMERE SHORTENING IN NEONATAL AND ADULT FAST-TWITCH (EDL AND FDL) AND SLOW-TWITCH (SOLEUS) MUSCLES MEASURED AT 35 OR 37°C

Animal and muscle	Maximum speed of sarcomere shortening ($\mu m\ s^{-1}$)		Adult body weight (kg)
	Neonate	Adult	
Mouse			
EDL	28.2	60.5	0.02—0.03
Soleus	30.6	31.7	
Rat			
EDL	17.3	42.7	0.2—0.25
Soleus	19.3	18.2	
Cat			
FDL	22.3	36.7	0.2—0.3
Soleus	31.1	14.9	

Note: Compiled from Close[116] (mouse), Close[9] (rat), and Hatcher and Luff[10] (cat).

muscles is clearly not applicable to neonatal muscles. It has been suggested that in stimulating some of these muscles via the nerve, only some of the fibers are activated.[16] However, the speed of shortening of a newborn rat extensor digitorum longus (EDL) obtained with direct stimulation[2] is little different to that obtained with nerve stimulation.[9] The relative slowness of neonatal muscles may result from inadequate activation related to the sparseness of the SR and transverse tubular system,[7] although this should not directly affect speed of shortening. It is interesting to note that in a recent study of the fast-twitch psoas muscle, there is very little difference in the unloaded maximum speed of shortening between fibers from newborn and adult rabbits, although the variability between individual fibers was high, extending over a threefold range.[44]

Thin Filament Proteins

Skeletal muscle contains two regulatory proteins, troponin and tropomyosin, attached to the actin filament. Troponin is a complex consisting of equimolar amounts of three proteins designated troponin C which binds calcium, troponin T which binds the complex to tropomyosin, and troponin I which together with tropomyosin inhibits the capacity of actin to bind with myosin. There are two slightly different forms of the troponin components known as "fast" and "slow" which occur predominantly in fast-twitch and slow-twitch muscles, respectively.[45,46] Two subunits of tropomyosin have been identified and designated as α and β; the relative ratios of these are different in fast- and slow-twitch muscles.[47] The molar ratios of these tropomyosins in adult rabbits have been found to be 3.8:1 in fast-twitch muscle and 1.1:1 in slow-twitch muscle.

In fetal rats and during the first few days of postnatal development, all muscle fibers appear to be capable of synthesizing both the "fast" and "slow" forms of troponin,[48] and 5 d after birth, fibers considered to be presumptive slow-twitch fibers (type I) on the basis of histochemical myosin ATPase activity show a clear diminution in the content of "fast"-type troponin. At 10 to 11 d, the differences between the fibers are well established.[48] It is interesting that this is very much faster than the changes in myosin isozymes (see previous section).

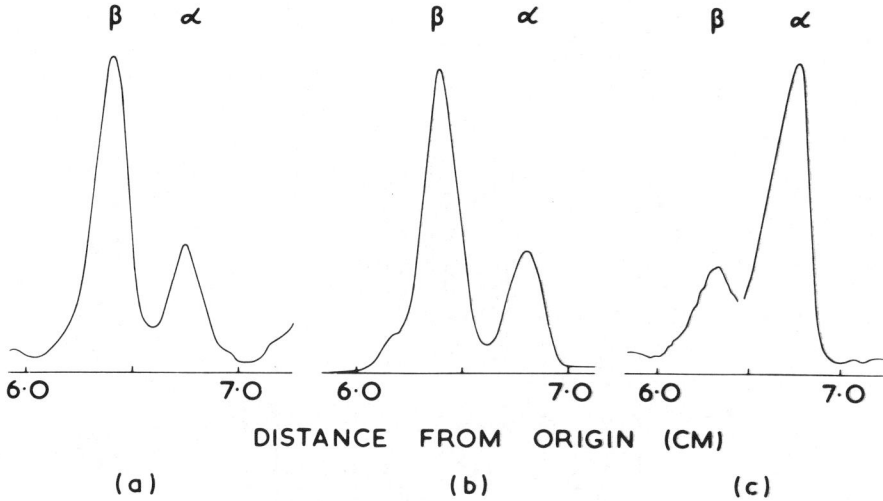

FIGURE 4. Densitometric scans of electrophoretograms of tropomyosin isolated from rabbit muscle at different stages of development. (a) From whole eviscerated carcass of 20-d fetus; (b) from longissimus dorsi of 24-d fetus; (c) from longissimus dorsi 30 d after birth. (From Amphlett, G. W., Syska, H., and Perry, S. V., *FEBS Lett.*, 63, 22, 1976. With permission.)

In late fetal development in rabbits, the tropomyosin subunit ratio is about 1:2 (α to β) which is very different to that found in any adult muscle[18,49] (Figure 4). Thereafter, the proportion of the α-subunit increases, not steadily but in two distinct phases,[49] such that by about 4 weeks after birth, the respective adult pattern is established. At the present time, there is no evidence for the existence of an embryonic form of either tropomyosin of troponin.

True to its conservative nature, there is probably no difference between actin from adult fast- and slow-twitch muscle and only recent evidence suggests that some minor changes may occur during development. During the early stages of muscle fiber formation, there is an accumulation of messenger RNA (mRNA) for a fetal actin which is slightly different from that for adult skeletal muscle actin, although it is very similar, if not identical, to cardiac actin. It is considered likely that this mRNA for fetal actin is expressed. The transition to the adult form does not seem to be coordinated with the myosin transitions.[50]

HISTOCHEMISTRY

The capacity of a muscle to resist fatigue, usually defined as a reduction in performance of the muscle, is used to assist in determining the metabolic basis of a muscle contracting for prolonged periods. The use of fatigue as a criterion for characterizing motor units has been systematically applied in several cat muscles.[51,52] Hammarberg and Kellerth[53] studied the development of motor units in the kitten using fatigue resistance as one of their criteria (Figure 5). In the youngest animal which was approximately 7 d old, the vast majority of the motor units identified are fatigue resistant, but by 42 d, the adult distribution of fatigue indices is seen. It is interesting to note that even in the youngest animals, some of the motor units investigated are sensitive to fatigue. Hatcher and Luff[10] found that the developing fast-twitch flexor digitorum longus muscle (FDL) is fatigue resistant at birth, but by 28 d, adult fatigue characteristics are well established.

At birth, limb muscles generally have similar isometric, force-velocity and fatigue properties. During subsequent development, the properties of individual muscles diverge, reflecting changes in the underlying metabolic and morphological processes. To a certain extent, these processes may be identified and quantified by examining the histochemical

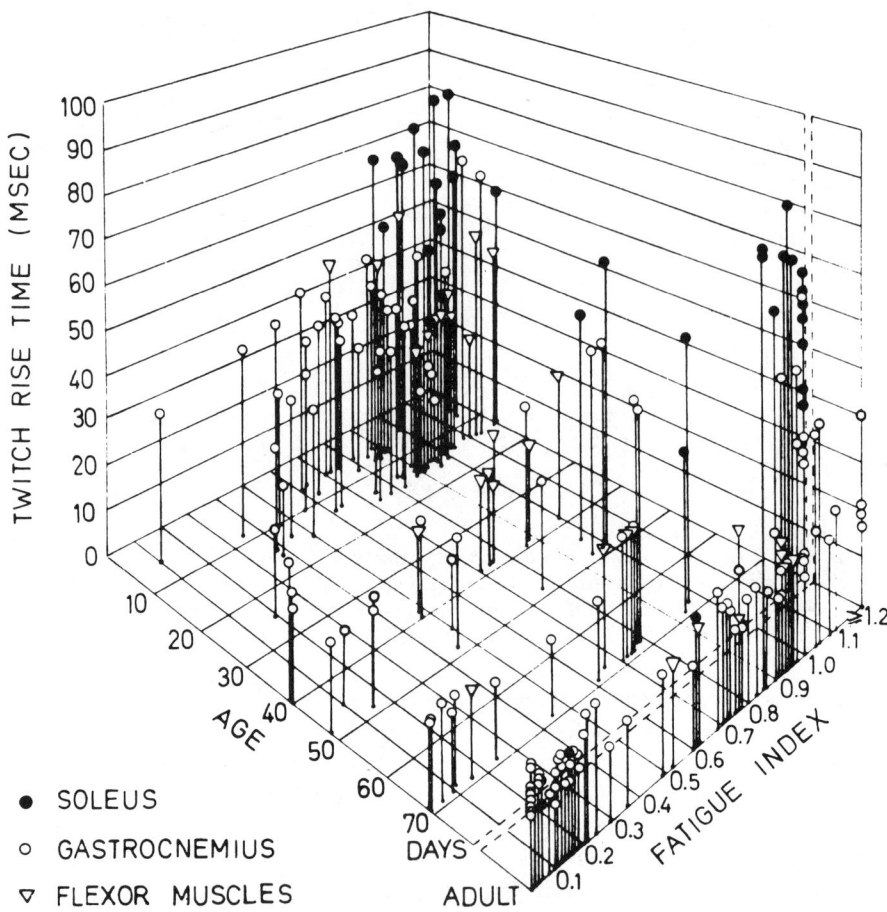

FIGURE 5. Three-dimensional plot showing the relationship between the contraction time (vertical y-axis), "fatigue index" (horizontal z-axis), and postnatal age (horizontal x-axis) for motor units of different muscles during postnatal development of the kitten. A high "fatigue index" indicates fatigue resistance. (From Hammarberg, C. and Kellerth, J.-O., *Acta Physiol. Scand.*, 95, 243, 1975. With permission.)

profile of the developing muscle, whereby changes in various enzymes and metabolic substrates can be visualized.

The vast majority of muscles in the adult animal are composed of a mixture of different "fiber types" which have different properties and characteristics.[54-56] These muscles fibers may be of different sizes[57] and histochemical composition.[56,58-60] The histochemistry of developing muscle has been extensively described by many workers.[55,61-66] During postnatal development, the trend is similar in all species; what differences exist result mainly from the relative state of maturity at birth. For example, there are few histochemical differences between fast- and slow-twitch muscles of the rat and mouse at birth, yet there is obvious differentiation in the muscles of the cat at birth.[55] In the newborn guinea pig and human,[67] muscle fiber differentiation is apparent, although adult patterns are not yet established. In the human quadriceps, there is an even distribution between type 1 (slow-twitch) and type 2 (fast-twitch) as well as 20% of fibers which cannot be classified into the major categories (type 2C).[67] The major alteration in fiber type distribution during development is that the type 2C fibers differentiate into type 1 fibers. After the first year, only subtle alterations in the fiber type distribution can be observed.[67,68] In the following sections, the developmental changes in various enzymes and substrates that have been observed histochemically are described in more detail.

OXIDATIVE SYSTEM ENZYMES

Both succinate dehydrogenase (SDH) and nicotamide adenine dinucleotide (NADH) are important indices of oxidative metabolism and as such they provide an important marker for oxidative capacity during development. SDH was shown previously to be bound to mitochondira,[69] as is NADH which may also be found in the microsomal fraction.[70]

On the basis of NADH staining, Ho et al.[71] concluded that there is no fiber differentiation in the plantaris and soleus muscles of the rat at birth. All fibers stain uniformly dark and are classed as type 2A. By day 11, differences between fibers are observed in plantaris and relative changes in staining intensity continue up to 140 d. At this stage, there are approximately equal numbers of type 2A and type 2B with 10% type 1 (Figure 6A). Soleus on the other hand is undifferentiated at day 11 on the basis of NADH staining, but different fiber types can be identified at day 6 on their myosin ATPase activity. Again, as with plantaris, the fiber type distribution continues until day 140 when 83% of soleus (SOL) muscle fibers are classed as type 1 and 17% are classed as type 2A (Figure 6B). Cooper et al.[72] concluded that in pig muscle at birth, all fibers stain for NADH, but with varying degrees of intensity with the largest fibers staining the darkest. At 1 week, all fibers can be distinguished. By 4 weeks, the gradual reduction in NADH staining in some fibers is virtually completed, giving intensity patterns similar to the adult.

Changes in the oxidative capacity of muscle fibers are observed during the postnatal period. While there may be slight differences when changes in the level of oxidative capacity for any one fiber type occur, the general trend is clear. At birth all muscle fibers are equally reliant upon oxidative metabolism; subsequently, some fibers reduce their dependence upon oxidative metabolism and increase their glycolytic metabolic activity.

LIPID

The intensity of lipid staining has been used in conjunction with staining for other enzymes during development.[55,66] Sudan black, which stains fat, phospholipids, lipoproteins, and nucleoproteins,[66] is normally used.[73] In both hindlimb and forelimb muscles[55] of the kitten, differences between fibers in staining intensity occur at less than 1 week. With increasing age, the degree of differentiation is greater, such that by 10 weeks the adult pattern has emerged. However, they also found that in some fibers from animals less than 1 week, lipid staining is intense and SDH staining is light. In soleus, both dark and light staining intensities are observed up to 6 weeks of age, but by 10 weeks uniform staining intensities are seen. Nystrom[55] obtained different results in hindlimb muscles and found no differences in lipid staining intensity in the newborn kitten. By 3 weeks of age, two distinct muscle fiber types can be distinguished in the gastrocnemius and soleus muscles. At 6 to 7 weeks, soleus still produces two distinct staining intensities while gastrocnemius produces three.

PHOSPHORYLASE

Glycogen phosphorylase staining intensities, demonstrated using iodine stains, are low in all fibers of the longissimus and diaphragm muscles of the pig at birth.[65,72] By 10 d, different staining intensities are detected indicating different levels of activity, and at day 13, adult staining characteristics prevail. In newborn kitten soleus and gastrocnemius,

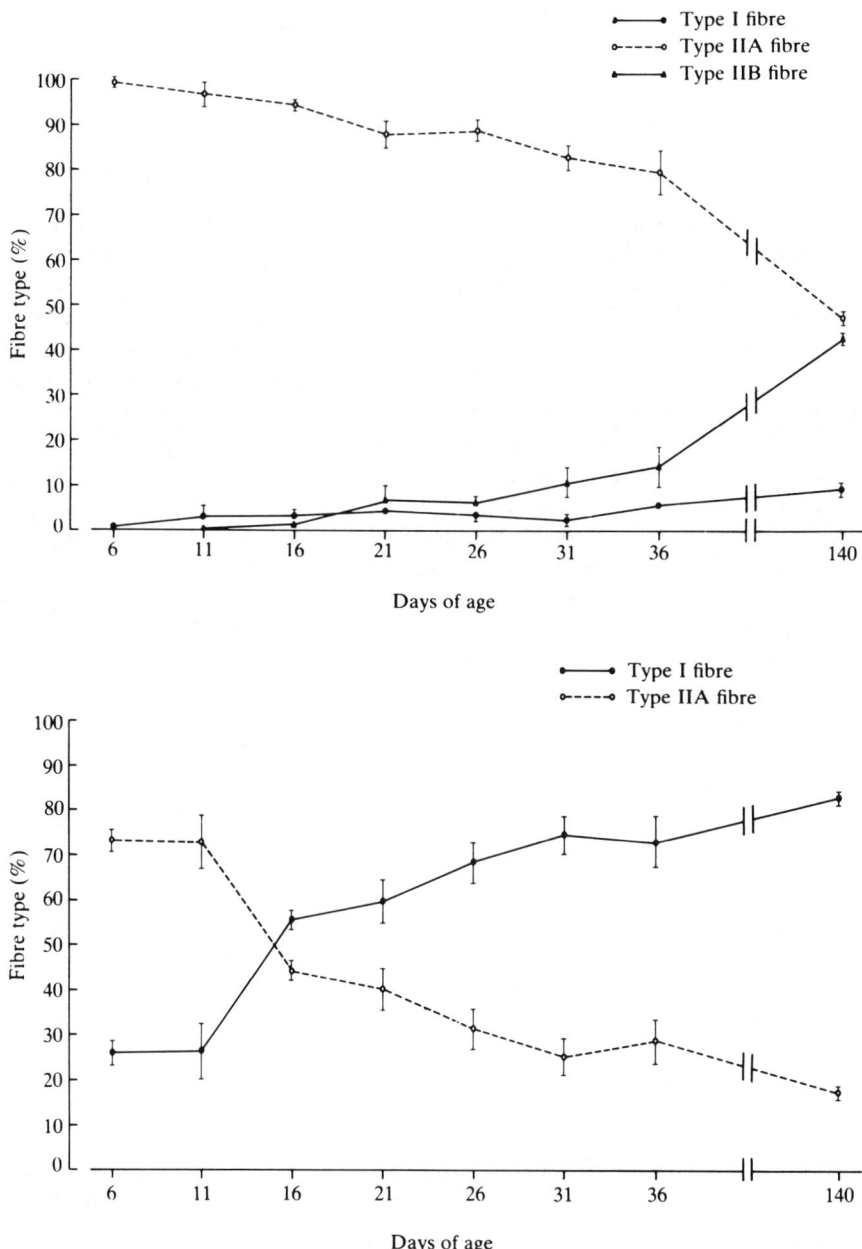

FIGURE 6. Changes in proportions of muscle fiber types during development in rat plantaris (A) and soleus (B). Categorization into the various types was based on NADH content and myosin ATPase activity of the fiber. Each point represents the mean and standard deviation (SD) for muscles from six rats. (From Ho, K. W., Heusner, W. W., Van Huss, J., and Van Huss, W. D., *J. Embryol. Exp. Morphol.*, 76, 37, 1983. With permission.)

Nystrom[55] found individual differences in staining intensities. In the forelimb of the animal, the differentiation appears to be more advanced than in the hindlimb. Adult staining patterns are observed in the gastrocnemius by 6 to 7 weeks of age. Hammarberg and Kellerth[53] attempted to characterize developing kitten muscle using phosphorylase as a marker, but were unable to obtain reliable staining patterns in the very young animal.

GLYCOGEN

Nystrom[55] noted very little difference in the glycogen staining intensities of fibers in the gastrocnemius and soleus muscles. By day 6, differentiation of the fibers can be observed and by 15 d in the gastrocnemius, three distinct levels of staining exist which relate to type I, IIA, and IIB fibers. Typical adult patterns are observed by 6 weeks.

SARCOPLASMIC RETICULUM (SR)

Calcium released from the terminal cisternae of the SR within the muscle fiber is probably the crucial regulatory step in the excitation-contraction coupling process.[74] During a single twitch, approximately 91 nmol g^{-1} of calcium is released from the terminal cisternae and rapidly reaccumulated during the relaxation phase.[75] The SR is a highly differentiated membrane system and is the major internal membrane system of the muscle fibers. Tubes of SR are arranged longitudinally within the fiber and about the radially oriented transverse tubular system (T-system) located approximately at the end of each A band. The SR is thought to be derived from the rough endoplasmic reticulum during development.[76] In the adult, the SR is continuous transversely across the fiber and longitudinally between adjacent T-tubules.

During development, as the myofibrils increase in number, parallel changes occur in the SR and T-tubule system[7,77,78] (Figure 7). In the immature fiber, the SR is sparsely distributed through the fiber both longitudinally and transversely. The specialized arrangement of the tubules in the region of the A and I bands seen in the adult is not apparent in the immature fibers. The immature triads, which in the adult consist of a central element formed from the T-system and two lateral elements formed from the terminal segments of the SR, are generally aligned longitudinally and are far fewer in number than in the adult.[79] In the 3-d-old rat, both longitudinally and obliquely orientated triads can be seen. The SR is generally disorganized, but between 10 to 15 d, it becomes more compact longitudinally at the level of the Z and M lines of the fibril. At other levels of the sarcomere it is longitudinally aligned. Corresponding alterations in the alignment of the triads occur during this postnatal period. They become predominantly transverse although longitudinal triads are still present.

There appears to be no difference in the sequence of development of the SR and T-system, both in degree and orientation between different muscle fiber types, although the rate at which it occurs is quantitatively different between muscle fiber types.[7] The volume increase of the SR with postnatal development is less in muscle fibers of the mouse soleus than in the EDL fibers. The soleus fibers achieve adult values by about 30 d postnatally, whereas in EDL approximately 60 d is required.

The SR is a key step in the excitation-contraction process.[80,81] Functionally, the role of the SR is to control the state of activation of the contractile apparatus, which it does by regulating the calcium concentration in the sarcoplasm surrounding the contractile filaments and myofibrils. The SR has two separate functions. Firstly, calcium is pumped by the SR into its interior and held, reducing the free calcium concentration in the sarcoplasmic space to approximately 10^{-7} M.[82,83] This results in a very low level of calcium binding to the troponin and hence inactivation of the contractile proteins. Secondly, following stimulation, the SR releases calcium which binds to the troponin, removing the inhibition on the contractile proteins. The SR pumps calcium continuously with calcium release occurring at the terminal cisternae.[75,84] In fact, approximately two thirds of total calcium in the muscle fiber is localized in the terminal cisternae in resting frog muscle fibers.

The SR must contain enough calcium to totally saturate the troponin following stimulation. To achieve this, many muscles have specialized calcium-binding proteins which allow

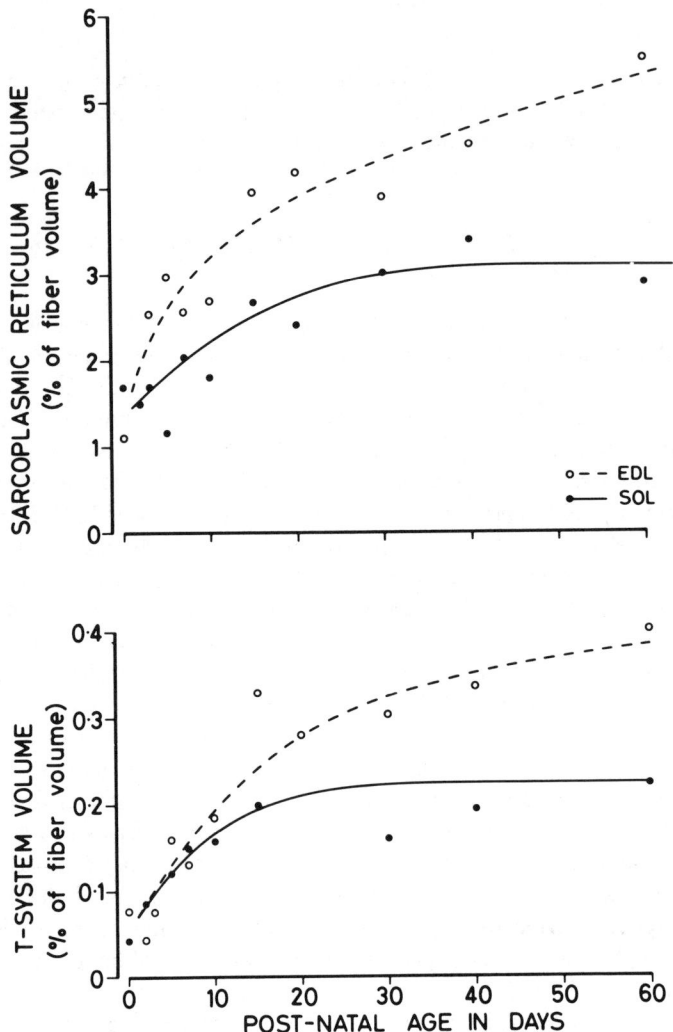

FIGURE 7. Developmental changes in the volume of SR (upper) and the transverse T-system (lower) in muscle fibers from the EDL (upper line) and soleus (lower line) of the mouse. Volumes are expressed as a percentage of fiber volume. (Redrawn from Luff and Atwood.[7])

the amount of calcium held to be increased. These proteins, such as parvalbumin[85] and calsequestrin, have binding constants similar to that of troponin.[86,87] Calsequestrin, a major extrinsic protein associated with the SR, is found in the terminal cisternae in intact fibers and in purified heavy SR fractions.[88-90] It binds calcium and functions as a calcium buffering agent which aids the pumping and retention of calcium by the SR.

The complexity and degree of development of the SR appear to be correlated with the contractile properties of the muscle.[91-93] Mature fast-twitch muscle fibers have a very extensive SR and T-tubule system allowing rapid and complete conduction of the depolarizing potential to the terminal cisternae of the SR thereby causing the release of calcium.[94] Fast-twitch muscles also have highly active calcium-activated ATPase (Ca^{2+}-ATPase) pumps, thereby enabling rapid calcium sequestration.[95] On the other hand, slow-twitch muscle fibers have a less extensive SR network[96] with slow calcium sequestration.

In adult rabbit white muscle, the calcium pump protein has a molecular weight of about

100,000 Da which accounts for 60 to 70% of the total SR protein,[97-99] while the calcium-binding proteins are between 32,000 and 55,000 Da. The amount of 100,000 Da protein increases throughout development. For example, from the 25-d-old rabbit embryo to 10-d postnatal animal, the amount of 100,000-Da calcium transport ATPase protein increases from 11 to 25% of the total SR protein.[100] In the 10-d chick embryo, 2% of the identified SR protein are 100,000 Da.[101] One can presume that during early embryonic life, skeletal muscle activity would be minimal, hence there is little demand placed on the muscles and therefore little demand on the calcium-activated ATPase in the SR. Consequently, it would appear that the rate and degree of development of this system would depend on the degree of maturity at birth.

The increased amount of 100,000-Da protein is reflected by the rate of calcium transport and calcium-sensitive ATPase activity[102] (Figure 8); the amount of 100,000-Da protein observed with SDS polyacrylamide gels; the concentration of phosphoprotein formed from ATP;[102] and further structural alteration in the SR membrane.[103] The observed increase in the 100,000-Da protein must surely be as a result of increased incorporation of newly formed protein into a preexisting membrane/phospholipid layer. This incorporation would continue until adult concentrations of the calcium ATPase are achieved, i.e., approximately 800 per square micrometer in 10-d-old chick embryos rising to 16,000 per square micrometer of surface area in adult muscle.[104,105]

In addition to the calcium transport ATPase, the SR contains the two extrinsic proteins calsequestrin and "high-affinity calcium-binding protein".[106-109] In the early embryo, the content of both these proteins is very nearly at adult levels even when the calcium-activated ATPase is very low. Consequently, their concentration does not alter even when the calcium ATPase is undergoing massive alterations in activity and concentration. This suggests that while both the calcium-activated ATPase and extrinsic proteins are apparently necessary for the regulated control of the contractile apparatus, both are independently regulated. Perhaps the extrinsic proteins supply some form of protection to the calcium ATPase of the contractile machinery itself during the early stages of development.

In cultured muscle cells,[110] the accumulation of calcium-activated ATPase appears to precede the activity of the actomyosin. The synthesis of calcium-activated ATPase also parallels the fusion of the myoblasts into multinucleated myotubes.[110-112] The two extrinsic calcium-binding proteins are synthesized well before fusion occurs and precede calcium ATPase synthesis.[109,113] Jorgensen et al.,[89] using antibody staining, detected calsequestrin localized at discrete sites in the perinuclear region of mononucleate cells prior to fusion. With further development, the calsequestrin is distributed throughout the myotube.

Sarzala et al.[100] found that in developing embryonic and newborn SR there are a few proteins in the 40,000- to 45,000- and 50,000 to 54,000-Da range. One of these is hydrophobic, being similar in character to the Ca^{2+} ATPase, while the other is similar to the water-soluble fraction found in adult SR. They concluded that this protein distribution resembles that of slow-twitch adult muscles. Similar results were observed with the Mg^{2+}-dependent ATPase.

ENERGETICS

Muscle is a chemomechanical converter. It utilizes energy from various stored sources and converts it into movement of force. A description and understanding of the energy costs involved in this conversion, and the eventual thermodynamic efficiency, can provide a detailed insight into the integrative molecular and cellular physiology of the cell itself. Kushmerick[114] recently reviewed the energetics of adult skeletal muscle, although the energetics of developing skeletal muscle is very poorly documented. As a means of obtaining the whole picture of muscle development, the techniques of energetic evaluation are a valuable tool. The only systematic investigation in this area is that of Wendt and Gibbs.[115]

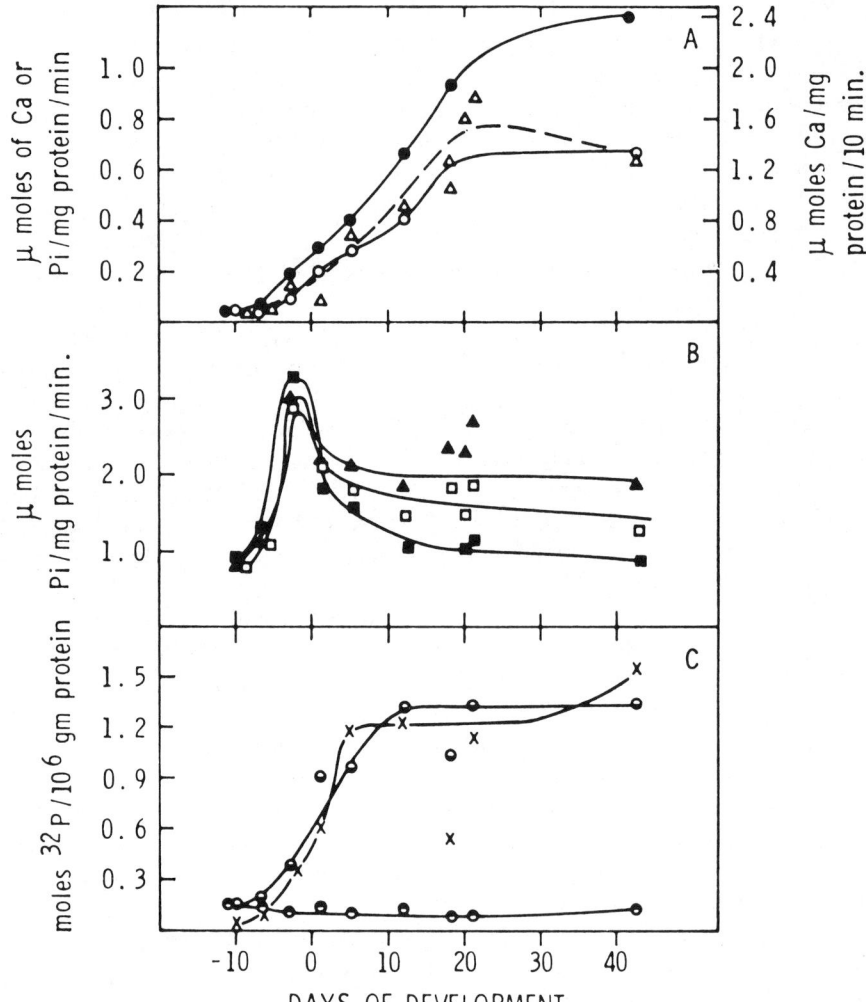

FIGURE 8. Calcium uptake (A), ATPase activity (B), and phosphoprotein concentration (C) of leg muscle microsomes during development. The rate of extra ATPase was calculated as the difference between measurements made in mediums A and B (Figure 8B) and is presented in Figure 8A for comparison with the rate of calcium transport. (A) ○—○, Calcium uptake after 1 min of incubation; ●—●, calcium uptake after 10 min of incubation; △- - -△, extra ATPase. (B) ▲—▲, ATPase activity in medium A (0.1 M KCl, 10 mM imidazole, 5 mM ATP, 5 mM MgCl$_2$, 0.5 mM EGTA, and 0.45 mM CaCl$_2$); □—□, ATPase activity in medium B (medium A minus CaCl$_2$); ■—■, ATPase activity in medium C (medium A, but 5 mM CaCl$_2$). (C) ◓—◓, phosphoprotein concentration in medium I; ◓ — ◓, phosphoprotein concentration in medium II; X — X, phosphoprotein concentration in medium III. (From Boland, R., Martonosi, A., and Tillack, T. W., *J. Biol. Chem.*, 249, 612, 1974. With permission.)

They examined the energetics of the rat fast-twitch EDL and slow-twitch soleus muscles at postnatal ages ranging from 7 to 41 d (Figure 9). At 7 d of age, the isometric heat developed by the EDL during a tetanic contraction is greater than that of the soleus. With further development, the soleus muscle heat production does not significantly alter, although the heat production of EDL rapidly increases up to 16 to 20 d when it levels off. At this age, the EDL value is similar to that of the adult. Interestingly enough, both muscles have similar efficiencies at birth; that of the soleus muscle increases from birth to 16 to 20 d, while EDL remains constant and is significantly less than soleus (Figure 10).

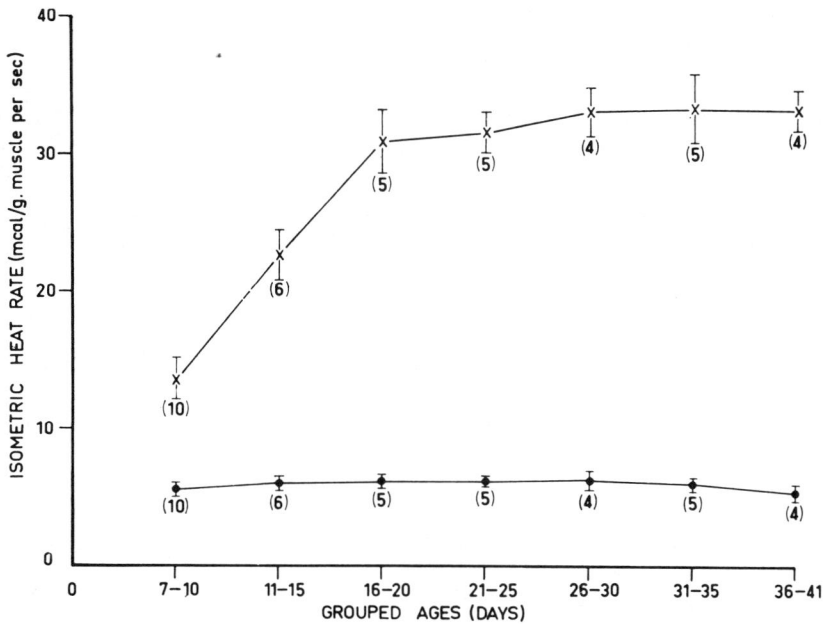

FIGURE 9. Average initial heat rates, during isometric tetani, of fast-twitch EDL (x) and slow-twitch soleus (●) plotted against ages of rats. Ages have been grouped and points are means for each group. Standard errors (SE) of means are represented by vertical bars and the number of experiments in each group is shown in parentheses. (From Wendt, I. R. and Gibbs, C. L., *Am. J. Physiol.*, 226, 642, 1974. With permission.)

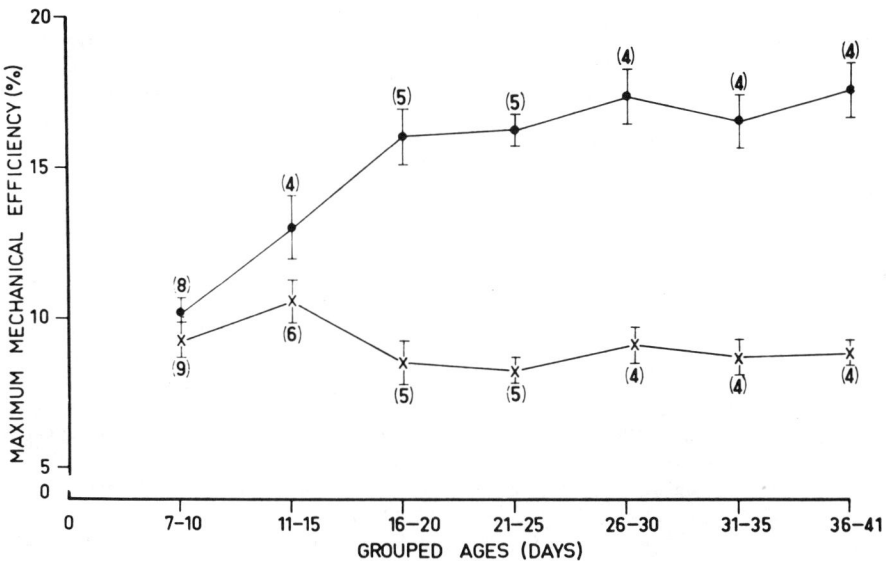

FIGURE 10. Maximum mechanical efficiency of fast-twitch EDL (x) and slow twitch soleus (●) obtained under experimental conditions plotted against age of rats. Ages are grouped and points are mean ± SE from the number of experiments shown in parentheses for each group. (From Wendt, I. R. and Gibbs, C. L., *Am. J. Physiol.*, 226, 642, 1974. With permission.)

The rapid increase in EDL heat production up to 16 to 20 d parallels increases in the SR as described in the previous section. It is interesting to note that the large changes seen in the SR following birth do not alter the mechanical efficiency of fast-twitch muscles, whereas slow-twitch muscles have relatively small changes in the SR and significant increases in efficiency. Unfortunately, there has been no investigation into the relationship between the tension-dependent and the tension-independent heat production at varying ages of development. This energetic study confirms the close relationship between the maturation of contractile[9,10] properties, the histochemical profile, and the biochemical and anatomical changes observed in the SR.

REFERENCES

1. **Ontell, M. and Kozeka, K.**, The organogenesis of murine striated muscle: a cytoarchitectural study, *Am. J. Anat.*, 171, 133, 1984.
2. **Close, R.**, The relation between intrinsic speed of shortening and duration of the active state of muscle, *J. Physiol.*, 180, 542, 1965.
3. **Bennett, M. R. and Pettigrew, A. G.**, The formation of synapses in striated muscle during development, *J. Physiol.*, 241, 515, 1974.
4. **Williams, P. and Goldspink, G.**, Longitudinal growth of striated muscle fibres, *J. Cell Sci.*, 9, 751, 1971.
5. **Goldspink, G.**, Alterations in myofibril size and structure during growth, exercise, and changes in environmental temperature, in *Handbook of Physiology, Section 10: Skeletal Muscle*, Peachey, L. D., Ed., American Physiological Society, Bethesda, MD, 1983, 539.
6. **Moss, F. P. and LeBlond, C. P.**, Nature of dividing nuclei in skeletal muscle of growing rats, *J. Cell Biol.*, 44, 459, 1970.
7. **Luff, A. R. and Atwood, H. L.**, Changes in the sarcoplasmic reticulum and transverse tubular system of fast and slow skeletal muscles of the mouse, *J. Cell Biol.*, 51, 369, 1972.
8. **Butler-Browne, G. S. and Whalen, R. G.**, Myosin isozyme transitions occurring during the post-natal development of the rat soleus muscle, *Dev. Biol.*, 102, 324, 1984.
9. **Close, R.**, Dynamic properties of fast and slow skeletal muscles of the rat during development, *J. Physiol.*, 173, 74, 1964.
10. **Hatcher, D. D. and Luff, A. R.**, Force-velocity properties of fast-twitch and slow-twitch muscles of the kitten, *J. Physiol.*, 367, 377, 1985.
11. **Weeds, A. G.**, Light chains from slow-twitch muscle myosin, *Eur. J. Biochem.*, 66, 157, 1976.
12. **Zweig, S. E.**, The muscle specificity and structure of two closely related fast-twitch white muscle myosin heavy chain isozymes, *J. Biol. Chem.*, 256, 11,847, 1981.
13. **Weeds, A. G. and Lowey, S.**, Substructure of the myosin molecule. II. The light chains of myosin, *J. Mol. Biol.*, 61, 701, 1971.
14. **Whalen, R. G., Butler-Browne, G. S., and Gros, F.**, Identification of a novel form of myosin light chain present in embryonic muscle tissue and cultured muscle cells, *J. Mol. Biol.*, 126, 415, 1978.
15. **Gauthier, G. F., Lowey, S., Benfield, P. A., and Hobbs, A. W.**, Distribution and properties of myosin isozymes in developing avian and mammalian skeletal muscle fibres, *J. Cell Biol.*, 92, 471, 1982.
16. **Rubinstein, N. A. and Kelly, A. M.**, Development of muscle fiber specialization in the rat hindlimb, *J. Cell Biol.*, 90, 128, 1981.
17. **Rubinstein, N. A. and Kelly, A. M.**, Myogenic and neurogenic contributions to the development of fast and slow-twitch muscles in rat, *Dev. Biol.*, 62, 473, 1978.
18. **Roy, R. K., Sreter, F. A., and Sarkar, S.**, Changes in tropomyosin subunits and myosin light chains during development of chicken and rabbit striated muscles, *Dev. Biol.*, 69, 15, 1979.
19. **Syrovy, I. and Gutmann, E.**, Differentiation of myosin in soleus and extensor digitorum longus muscle in different animal species during development, *Pfluegers Arch.*, 369, 85, 1977.
20. **Lyons, G. E., Haselgrove, J., Kelly, A. M., and Rubinstein, N. A.**, Myosin transitions in developing fast and slow muscles of the rat hindlimb, *Differentiation*, 25, 168, 1983.
21. **Trayer, I. P., Harris, C. I., and Perry, S. V.**, 3-Methyl-histidine and adult and foetal forms of skeletal muscle myosin, *Nature (London)*, 217, 452, 1968.
22. **Huszar, G.**, Developmental changes of the primary structure and histidine methylation in rabbit skeletal muscle myosin, *Nature (London) New Biol.*, 240, 260, 1972.

23. **Sreter, F. A., Balint, M., and Gergely, J.**, Structural and functional changes of myosin during development: comparison with adult fast, slow and cardiac myosin, *Dev. Biol.*, 46, 317, 1975.
24. **Hoh, J. F. Y. and Yeoh, G. P. S.**, Rabbit skeletal muscle isoenzymes from foetal fast-twitch and slow-twitch muscles, *Nature (London)*, 280, 321, 1979.
25. **Whalen, R. G., Schwartz, K., Bouveret, P., Sell, S. M., and Gros, F.**, Contractile protein isozymes in muscle development: identification of an embryonic form of myosin heavy chain, *Proc. Natl. Acad. Sci. U.S.A.*, 76, 5197, 1979.
26. **Whalen, R. G., Sell, S. M., Butler-Browne, G. S., Schwartz, K., Bouveret, P., and Pinset-Harstrom, I.**, Three myosin heavy-chain isozymes appear sequentially in rat muscle development, *Nature (London)*, 292, 805, 1981.
27. **Bugaisky, L. B., Butler-Browne, G. S., Sell, S. M., and Whalen, R. G.**, Structural differences in the subfragment 1 and rod portions of myosin isoenzymes from adult and developing rat skeletal muscles, *J. Biol. Chem.*, 259, 7212, 1984.
28. **Bandman, E., Matsuda, R., and Strohman, R. C.**, Developmental appearance of myosin heavy and light chain isoforms *in vivo* and *in vitro* in chicken skeletal muscle, *Dev. Biol.*, 93, 508, 1982.
29. **Bader, D., Masaki, T., and Fischman, D. A.**, Immunochemical analysis of myosin heavy chain during avian myogenesis *in vivo* and *in vitro*, *J. Cell Biol.*, 95, 763, 1982.
30. **Winkelmann, D. A., Lowey, S., and Press, J. L.**, Monoclonal antibodies localise changes on myosin heavy chain isozymes during avian myogenesis, *Cell*, 34, 295, 1983.
31. **Whalen, R. G., Johnstone, D., Bryers, P. S., Butler-Browne, G. S., Ecob, M. S., and Jaros, E.**, A developmentally regulated disappearance of slow myosin in fast-type muscles of the mouse, *FEBS Lett.*, 177, 51, 1984.
32. **Kelly, A. M. and Rubinstein, N. A.**, Why are fetal muscles slow?, *Nature (London)*, 288, 266, 1980.
33. **Butler-Browne, G. S., Bugaisky, L. B., Cuenoud, S., Schwartz, K., and Whalen, R. G.**, Denervation of newborn rat muscle does not block the appearance of adult fast myosin heavy chain, *Nature (London)*, 299, 830, 1982.
34. **Rubinstein, N. A. and Holtzer, H.**, Fast and slow muscles in tissue culture synthesize only fast myosin, *Nature (London)*, 280, 323, 1979.
35. **Salmons, S.**, The response of skeletal muscle to different patterns of use — some new developments and concepts, in *Plasticity of Muscle*, Petter, D. Ed., Walter DeGruyter, Berlin, 1980, 387.
36. **Navarrete, R. and Vrbova, G.**, Changes of activity patterns in slow and fast muscles during post-natal development, *Dev. Brain Res.*, 8, 11, 1983.
37. **Wydro, R. M., Nyugen, H. T., Gubits, R. M., and Nadal-Ginard, B.**, Characterization of sarcomeric myosin heavy chain genes, *J. Biol. Chem.*, 258, 670, 1983.
38. **de Villafranca, G. W.**, Adenosinetriphosphatase activity in developing rat muscle, *J. Exp. Zool.*, 127, 367, 1954.
39. **Perry, S. V.**, Biochemical adaptation during development and growth in skeletal muscle, in *Symposium on Physiology and Biochemistry of Muscle as a Food*, Briskey, E. J., Cassens, R. G., and Trautman, J. C., Eds., University of Wisconsin Press, Madison, 1966, 537.
40. **Barany, M., Tucci, A. F., Barany, K., Volpe, A., and Reckard, T.**, Myosin of newborn rabbits, *Arch. Biochem. Biophys.*, 111, 727, 1965.
41. **Barany, M.**, ATPase activity of myosin correlated with speed of muscle shortening, *J. Gen. Physiol.*, 50, 197, 1967.
42. **Dow, J. and Stracher, A.**, Changes in properties of myosin associated with muscle development, *Biochemistry*, 10, 1316, 1971.
43. **Kelly, A. M. and Zacks, S. I.**, The fine structure of motor end-plate morphogenesis, *J. Cell Biol.*, 42, 154, 1969.
44. **Julian, F. J., Moss, R. L., and Wagner, G. S.**, Mechanical properties and myosin light chain composition of skinned muscle fibres from adult and new-born rabbits, *J. Physiol.*, 311, 201, 1981.
45. **Dhoot, G. K., Gell, P. G. H., and Perry, S. V.**, The localisation of the different forms of troponin I in skeletal and cardiac muscle cells, *Exp. Cell Res.*, 117, 357, 1978.
46. **Dhoot, G. K., Frearson, N., and Perry, S. V.**, Polymorphic forms of troponin T and troponin C and their localisation in striated muscle types, *Exp. Cell Res.*, 122, 339, 1979.
47. **Cummins, P. and Perry, S. V.**, Chemical and immunochemical characteristics of tropomyosins from striated and smooth muscle, *Biochem. J.*, 141, 43, 1974.
48. **Dhoot, G. K. and Perry, S. V.**, The components of the troponin complex and development in skeletal muscle, *Exp. Cell Res.*, 127, 75, 1980.
49. **Amphlett, G. W., Syska, H., and Perry, S. V.**, The polymorphic forms of tropomyosin and troponin I in developing rabbit skeletal muscle, *FEBS Lett.*, 63, 22, 1976.
50. **Minty, A. J., Alonso, S., Caravatti, M., and Buckingham, M. E.**, A fetal skeletal muscle actin mRNA in the mouse and its identity with cardiac actin mRNA, *Cell*, 30, 185, 1982.

51. **Edstrom, L. and Kugelberg, E.**, Histochemical composition, distribution of fibres and fatiguability of single motor units. Anterior tibial muscle of the rat, *J. Neurol. Neurosurg. Psychiatry*, 31, 424, 1968.
52. **Burke, R. E., Levine, D. N., Tsairis, P., and Zajac, F. E.**, Physiological types and histochemical profiles in motor units of the cat gastrocnemius, *J. Physiol.*, 234, 723, 1973.
53. **Hammarberg, C. and Kellerth, J.-O.**, The postnatal development of some twitch and fatigue properties of single motor units in the ankle muscles of the kitten, *Acta Physiol. Scand.*, 95, 243, 1975.
54. **Nystrom, B.**, Mechanical and electrical responses to single shocks in developing cat leg muscles following tetanization, *Acta Physiol. Scand.*, 74, 207, 1968a.
55. **Nystrom, B.**, Histochemistry of developing cat muscles, *Acta Neurol. Scand.*, 44, 405, 1968b.
56. **Dubowitz, V. and Pearse, A. G. E.**, A comparative histochemical study of oxidative enzyme and phosphorylase activity in skeletal muscle, *Histochemie*, 2, 105, 1960.
57. **Goldspink, G.**, Biochemical and physiological changes associated with the postnatal development of biceps brachii, *Comp. Biochem. Physiol.*, 7, 157, 1962.
58. **Padykula, H. A.**, The localization of succinic dehydrogenase in tissue sectors of the rat, *Am. J. Anat.*, 91, 107, 1952.
59. **Ogata, T.**, A histochemical study of the red and white muscle fibres. I. Activity of the succinoxydase system in muscle fibers, *Acta Med. Okayama*, 12, 216, 1958.
60. **Brooke, M. H. and Kaiser, K. K.**, The use and abuse of muscle histochemistry, *Ann. N.Y. Acad. Sci.*, 228, 121, 1974.
61. **Dubowitz, V.**, Enzyme histochemistry of developing human muscle, *Nature (London)*, 211, 884, 1966.
62. **Beatty, C. H., Basinger, G. M., and Boelk, R. M.**, Differentiation of red and white fibers in muscle from fetal, neonatal and infant rhesus monkeys, *J. Histochem. Cytochem.*, 15, 93, 1967.
63. **Karpati, G. and Engel, W. K.**, Neuronal trophic function — a new aspect demonstrated histochemically in developing soleus muscle, *Arch. Neurol.*, 17, 542, 1967.
64. **Goldspink, G.**, Succinic dehydrogenase content of individual muscle fibers at different ages and stages of growth, *Life Sci.*, 8, 791, 1969.
65. **Davies, A. S.**, Postnatal changes in the histochemical fibre types of porcine skeletal muscle, *J. Anat.*, 113, 213, 1972.
66. **Hammarberg, C.**, Histochemical staining patterns of muscle fibres in the gastrocnemius, soleus and anterior tibialis muscles of the adult cat, as viewed in serial sections for lipids and succinic dehydrogenase, *Acta Neurol. Scand.*, 50, 272, 1974.
67. **Collin-Saltin, A.-S.**, Enzyme histochemistry on skeletal muscle of human foetus, *J. Neurol. Sci.*, 39, 169, 1978.
68. **Bell, R. D., MacDougall, J. D., Billeter, R., and Howard, H.**, Muscle fiber-types and morphometric analysis of skeletal muscle in six year old children, *Med. Sci. Sport Exercise*, 12, 28, 1980.
69. **deDuve, C., Wattiaux, R., and Baudhuin, P.**, Distribution of enzymes between subcellular fractions in animal tissues, *Adv. Enzymol.*, 24, 291, 1962.
70. **Novikoff, A. B., Woo-Yung, S., and Drucker, J.**, Mitochondrial localization of oxidative enzymes: staining results with two tetrazolium salts, *J. Biophys. Biochem. Cytol.*, 9, 47, 1961.
71. **Ho, K. W., Heusner, W. W., Van Huss, J., and Van Huss, W. D.**, Postnatal muscle fibre histochemistry in the rat, *J. Embryol. Exp. Morphol.*, 76, 37, 1983.
72. **Cooper, C. C., Cassens, R. G., Kastenschmidt, L. L., and Briskey, E. J.**, Histochemical characterization of muscle differentiation, *Dev. Biol.*, 23, 169, 1970.
73. **Pearse, A. G. E.**, *Histochemistry: Theoretical and Applied*, J. & A. Churchill, London, 1960, 851.
74. **Bianchi, C. P.**, Pharmacology of excitation-contraction coupling in muscle — introduction — statement of problem, *Fed. Proc.*, 28, 1624, 1969.
75. **Winegrad, S.**, The intracellular site of calcium activation of contraction in frog skeletal muscle, *J. Gen. Physiol.*, 55, 77, 1970.
76. **Ezerman, E. B. and Ishikawa, H.**, Differentiation of the sarcoplasmic reticulum and T-system in developing chick skeletal muscle in vitro, *J. Cell Biol.*, 35, 405, 1967.
77. **Walker, S. M. and Schrodt, G. R.**, Triads in skeletal muscle fibers of 19-day fetal rats, *J. Cell Biol.*, 37, 564, 1968.
78. **Schiaffino, S. and Margreth, A.**, Co-ordinated development of the sarcoplasmic reticulum and T-system during postnatal differentiation of rat skeletal muscle, *J. Cell Biol.*, 41, 855, 1969.
79. **Edge, M. B.**, Development of apposed sarcoplasmic reticulum at T-system and sarcolemma and change in orientation of triads in rat skeletal muscle, *Dev. Biol.*, 23, 634, 1970.
80. **Ebashi, S.**, Excitation-contraction coupling, *Annu. Rev. Physiol.*, 38, 293, 1976.
81. **Endo, M.**, Calcium release from the sarcoplasmic reticulum, *Physiol. Rev.*, 57, 71, 1977.
82. **Ebashi, S.**, Calcium binding activity of vesicular relaxing factor, *J. Biochem.*, 50, 236, 1961.
83. **Ebashi, S. and Endo, M.**, Calcium ion and muscle contraction, *Prog. Biophys. Mol. Biol.*, 18, 123, 1968.
84. **Winegrad, S.**, Intracellular calcium movements of frog skeletal muscle during recovery from tetanus, *J. Gen. Physiol.*, 51, 65, 1968.

85. **Hamoir, G. and Focant, B.**, Proteinic differences between the sarcoplasmic reticulum of the superfast swimbladder and the fast skeletal muscles of the toadfish, *Opsanus tau, Mol. Physiol.*, 1, 353, 1981.
86. **Gillis, J. M., Piront, A., and Gossein-Rey, C.**, Parvalbumins. Distribution and physical state inside the muscle cell, *Biochem. Biophys. Acta*, 585, 444, 1979.
87. **Potter, J. D., Johnson, J. D., and Mandel, F.**, Fluorescence stopped flow measurements of Ca^{2+} and Mg^{2+} binding to parvalbumin, *Fed. Proc.* 37(Abstr.), 1608, 1978.
88. **Campbell, K. P., Franzini-Armstrong, C., and Shamoo, A. E.**, Further characterization of light and heavy sarcoplasmic reticulum vesicles. Identification of the "sarcoplasmic reticulum feet" associated with heavy sarcoplasmic reticulum vesicles, *Biochem. Biophys. Acta*, 602, 97, 1980.
89. **Jorgensen, A. O., Kalnins, V., and MacLennan, D. H.**, Localization of sarcoplasmic reticulum proteins in rat skeletal muscle by immunofluorescence, *J. Cell Biol.*, 80, 372, 1979.
90. **Meissner, G.**, Isolation and characterization of two types of sarcoplasmic reticulum vesicles, *Biochem. Biophys. Acta*, 389, 51, 1975.
91. **Pellegrino, C. and Franzini, C.**, An electron microscope study of denervation atrophy in red and white skeletal muscle fibers, *J. Cell Biol.*, 17, 327, 1963.
92. **Page, S. G.**, A comparison of the fine structure of frog slow and fast twitch muscle fibres, *J. Cell Biol.*, 26, 477, 1965.
93. **Page, S. G.**, Structure and some contractile properties of fast and slow muscles of the chicken, *J. Physiol.*, 205, 131, 1969.
94. **Eisenberg, B. R. and Kuda, A. M.**, Stereological analysis of mammalian skeletal muscle. II. White vastus muscle of the adult guinea pig, *J. Ultrastruct. Res.*, 51, 176, 1975.
95. **Heilmann, C. and Pette, D.**, Molecular transformations in sarcoplasmic reticulum of fast-twitch muscle by electrostimulation, *Eur. J. Biochem.*, 93, 437, 1979.
96. **Eisenberg, B. R., Kuda, A. M., and Peter, J. B.**, Stereological analysis of mammalian skeletal muscle. I. Soleus muscle of the adult guinea pig, *J. Cell Biol.*, 60, 732, 1974.
97. **Martonosi, A. and Halpin, R. A.**, Sarcoplasmic reticulum. X. The protein composition of sarcoplasmic reticulum membranes, *Arch. Biochem. Biophys.*, 144, 66, 1971.
98. **MacLennan, D. H., Seeman, P., Iles, G. H., and Yip, C. C.**, Resolution of enzymes of biological transport. II. Membrane formation by the adenosine triphosphatase of sarcoplasmic reticulum, *J. Biol. Chem.*, 246, 2702, 1971.
99. **Meissner, G., Conner, G. E., and Fleischer, S.**, Isolation of sarcoplasmic reticulum by zonal centrifugation and purification of Ca^{2+}-pump and Ca^{2+}-binding proteins, *Biochem. Biophys. Acta*, 298, 246, 1973.
100. **Sarzala, M. G., Pilarska, M., Zubrzycka, E., and Michalak, M.**, Changes in the structure, composition and function of sarcoplasmic reticulum membrane during development, *Eur. J. Biochem.*, 57, 25, 1975.
101. **Martonosi, A., Boland, R., and Halpin, R. A.**, The biosynthesis of sarcoplasmic reticulum membranes and the mechanism of calcium transport, *Cold Spring Harbor Symp. Quant. Biol.*, 37, 455, 1973.
102. **Boland, R., Martonosi, A., and Tillack, T. W.**, Developmental changes in the composition and function of sarcoplasmic reticulum, *J. Biol. Chem.*, 249, 612, 1974.
103. **Tillack, T. W., Boland, R., and Martonosi, A.**, Ultrastructure of developing sarcoplasmic reticulum, *J. Biol. Chem.*, 249, 624, 1974.
104. **Jilka, R. L., Martonosi, A. N., and Tillack, T. W.**, Effect of the purified [Mg^{2+} + Ca^{2+}]-activated ATPase of sarcoplasmic reticulum upon the passive Ca^{2+} permeability and ultrastructure of phospholipid vesicles, *J. Biol. Chem.*, 250, 7511, 1975.
105. **Scales, D. and Inesi, G.**, Assembly of ATPase protein in sarcoplasmic reticulum membranes, *Biophys. J.*, 16, 735, 1976.
106. **Martonosi, A.**, Membrane transport during development in animals, *Biochem. Biophys. Acta*, 415, 311, 1975.
107. **Duggan, P. F. and Martonosi, A.**, Sarcoplasmic reticulum. IX. The permeability of sarcoplasmic reticulum membranes, *J. Gen. Physiol.*, 56, 147, 1970.
108. **MacLennan, D. H. and Wong, P. T. S.**, Isolation of a calcium sequestrin protein from the sarcoplasmic reticulum, *Proc. Natl. Acad. Sci. U. S. A.* 68, 1231, 1971.
109. **Michalak, M. and MacLennan, D. H.**, Assembly of the sarcoplasmic reticulum. VI. Biosynthesis of the high-affinity calcium binding protein in rat skeletal muscle cell cultures, *J. Biol. Chem.*, 255, 1327, 1980.
110. **Martonosi, A., Roufa, D., Boland, R., Reyes, E., and Tillack, T. W.**, Development of sarcoplasmic reticulum in cultured chicken muscle, *J. Biol. Chem.*, 252, 318, 1977.
111. **Ha, D. B., Boland, R., and Martonosi, A.**, Synthesis of the calcium transport ATPase of sarcoplasmic reticulum and other muscle proteins during development of muscle cells *in vivo* and *in vitro*, *Biochem. Biophys. Acta*, 585, 165, 1979.
112. **Holland, P. C. and MacLennan, D. H.**, Assembly of sarcoplasmic reticulum. II. Biosynthesis of the adenosine triphosphatase in rat skeletal muscle cell culture, *J. Biol. Chem.*, 251, 2030, 1976.
113. **Zubrzycka, E. and MacLennan, D. H.**, Assembly of the sarcoplasmic reticulum. Biosynthesis of calsequestrin in rat skeletal muscle cell cultures, *J. Biol. Chem.*, 251, 7733, 1976.

114. **Kushmerick, M. J.**, Energetics of muscle contraction, in *Handbook of Physiology*, Vol. 10, Peachey, L. D., Adrian, R. H., and Geiger, S. R., Eds., American Physiological Society, Bethesda, MD, 1983, 189.
115. **Wendt, I. R. and Gibbs, C. L.**, Energy production of mammalian fast-twitch and slow-twitch muscles during development, *Am. J. Physiol.*, 226, 642, 1974.
116. **Close, R.**, Force:velocity properties of mouse muscles, *Nature (London)*, 206, 718, 1965.

SKELETAL MUSCLE GROWTH, HYPERTROPHY, REPAIR, AND REGENERATION

G. J. Lang and A. R. Luff

INTRODUCTION

During normal growth, both *in utero* and postnatally, individual muscles increase in size in approximate proportion to the increasing size of the organism. Following myogenesis, this growth is achieved largely by increases in length and cross-sectional area of individual muscle fibers.[1] The extraordinary adaptability (or plasticity) of the muscle fiber not only allows it to increase in size, but also to alter its biochemical characteristics according to the specific demands placed on the muscle. A fiber may increase its cross-sectional area considerably (hypertrophy), above that required for normal growth, in response to a demand for increased force production and, if subjected to an endurance training regime, will tend toward an oxidative metabolism. In addition, skeletal muscles possess considerable capacity to repair fibers cut or crushed and to regenerate muscle fibers following the effect of disease, toxins, or transplantation.

POSTNATAL MUSCLE GROWTH

Muscle fiber number is virtually fixed at birth even in an animal such as the rat[2,3] which is born at a relatively immature stage of development. Consequently, the increase in mass of a skeletal muscle of up to 50 times during postnatal development is due almost exclusively to a massive increase in the size of individual muscle fibers. The number of myofibrils increases from about 75 in a newly formed fiber to over 1000 in an adult fiber[4] (Figure 1). However, the range of myofibril size (0.4- to 1.25-μm diameter) changes very little during development. This increase in number has been attributed to myofibril splitting and a possible mechanism has been proposed by Goldspink.[5]

Muscle fibers also increase in length during the normal course of development. This is mainly due to an increase in sarcomere number[6] which occurs at the ends of the myofibrils[7] as there is little change in sarcomere length. Sarcomere number can be readily altered by experimental procedures in which the muscle is maintained at long or short lengths[5] (Figure 2). This seems to involve an autoregulatory mechanism within the muscle fiber.

HYPERTROPHY

Skeletal muscle exhibits a degree of specialization by which it conforms to various functional requirements. There is a substantial matching of the morphological, metabolic, biochemical, and contractile properties of muscle to its function. Therefore, it is not surprising to find that when the functional demands placed on a muscle are altered, the muscle adapts to enable it to cope better with those altered demands. Muscle hypertrophy is one such adaptation. Hypertrophy is defined as the enlargement of an organ due to an increase in the size of its constituent cells. In the adult muscle, hypertrophy occurs exclusively as a result of an increase in cross-sectional area as anatomical factors set a limit on muscle length. Although it has been reported that increases in fiber number can account for a substantial proportion of the growth in response to overload,[8-10] it is by no means certain that hyperplasia of muscle cells occurs. The majority of evidence favors hypertrophy over hyperplasia as the major cause of increased muscle size through overload. For a more detailed treatise on the subject, the reader is referred to the review by Saltin and Gollnick.[11]

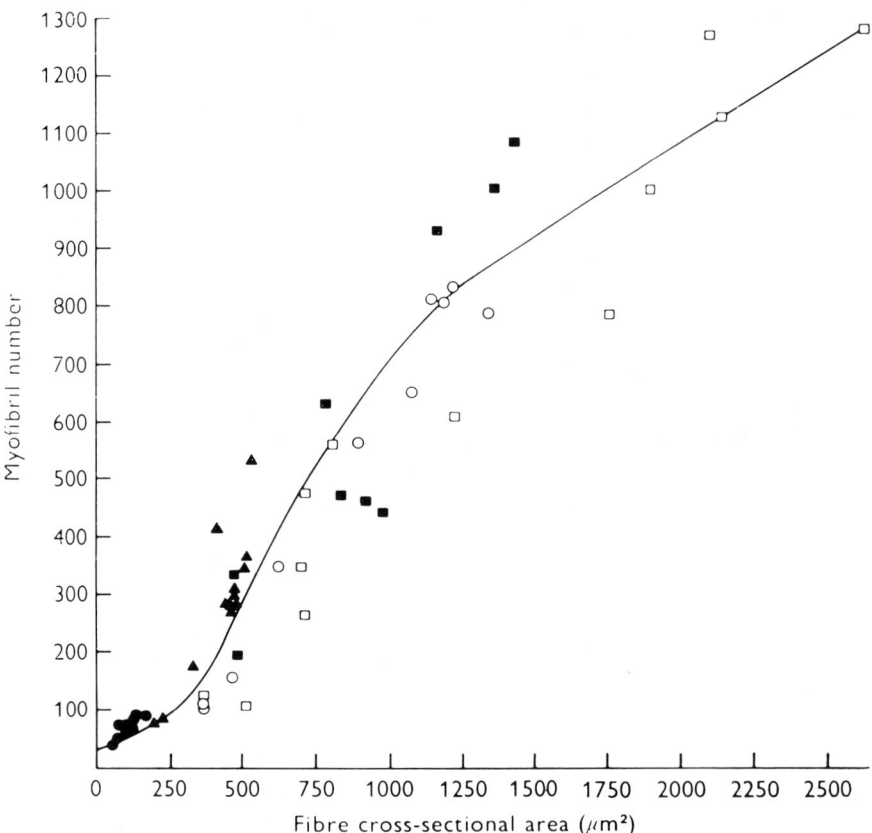

FIGURE 1. The relationship between the number of myofibrils and the size and age of muscle fibers. The total number of myofibrils within individual muscle fibers is plotted against the fiber cross-sectional area. The fibers in which the myofibrils were measured were taken from mice of the following body weights and ages: ●, 2 g (2 d); ▲, 10 g (2.5 weeks); ○, 20 g (6 weeks); ■, 30 g, (6 months); □, 40 g (1 year). (From Goldspink, G., *J. Cell Sci.*, 6, 593, 1970. With permission.)

Two distinct types of growth can be recognized: (1) that which occurs during the normal development of the organism and (2) that which occurs in response to increased work demands placed on the muscle. These are termed developmental and work-induced growth, respectively. The reason for the distinction is that developmental growth has an absolute requirement for growth hormone[12] as well as a dependence upon other pituitary hormones,[13] insulin,[14] and adequate diet.[15] Work-induced growth, on the other hand, occurs in both hypophysectomized[16] and severely diabetic[17] animals. It seems that activity may also have a protective effect on the negative nitrogen balance seen in fasted animals.[18] The amount and type of muscle activity appear to be the primary determinants of skeletal muscle mass and can override both nutritional and endocrine factors to produce muscle hypertrophy. These observations point strongly to an internal cellular mechanism for the induction of skeletal muscle hypertrophy, a mechanism which enables each cell to function independently with regard to the synthesis and degradation of its cellular proteins.

Work-Induced Growth

The anatomical aspects of work-induced growth were probably first described in the classical study by Morpurgo of Sienna in 1897 (see Denny-Brown[19]). He used treadmill running to induce a 54% enlargement in the sartorius muscle of the dog, which occurred as a result of sarcoplasmic proliferation with no contribution by the myofibrillar component.

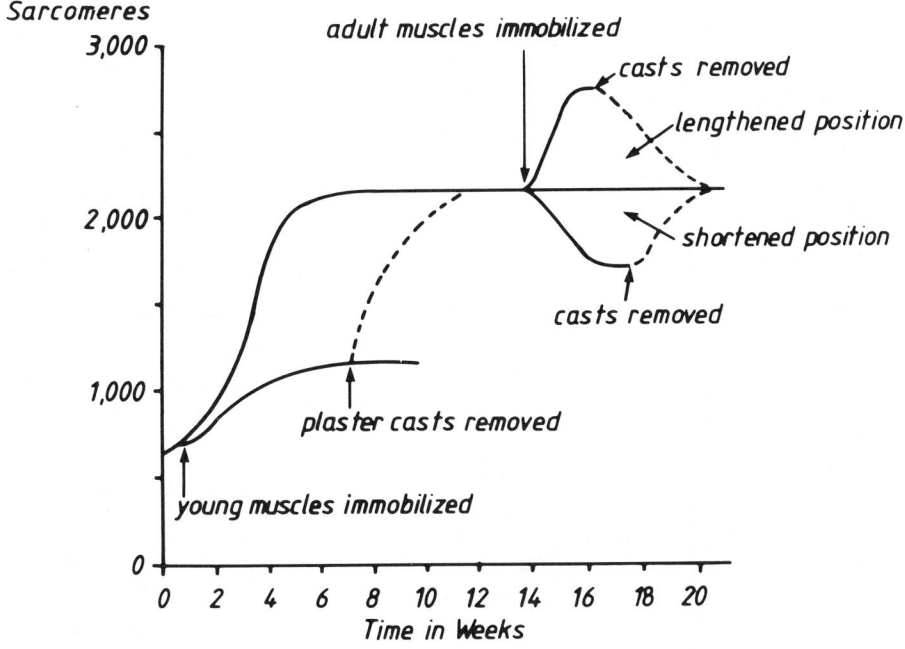

FIGURE 2. Increase in sarcomere number in mouse soleus with age and effect of immobilization of muscle at different ages in different ways. Recovery from effects of plaster cast immobilization is shown by dashed lines. (From Goldspink, G., in *Handbook of Physiology, Section 10: Skeletal Muscle*, Peachey, L. D., Ed., American Physiology Society, Bethesda, MD, 1983, chap. 18. With permission.)

Table 1
PERCENTAGE CHANGES IN SELECTED PARAMETERS FOLLOWING 6 WEEKS OF HIGH-RESISTANCE TRAINING (STRENGTH TRAINING) OR 6 WEEKS OF BICYCLE ERGOMETER TRAINING (ENDURANCE TRAINING)

	Myofibrillar volume density (%)	Mitochondrial volume density (%)	Muscle mass (%)	Peak torque (%)
Effect of strength training	No change	9.6	11	15
Effect of endurance training	6.4	38.6	No change	Not given

Note: Strength training stimulates myofibrillar proliferation resulting in increased muscle mass and torque, whereas endurance training is specific for mitochondrial protein synthesis at the expense of myofibrillar proteins.

Adapted from Howald.[21]

Studies in which hypertrophy is induced by high-resistance training or chronic overload generally show myofibrillar proliferation with a proportional increase in the ability of the muscle to produce tension.[20,21] Thus, the method of inducing hypertrophy is significant in terms of the adaptation which occurs. Most research shows that endurance training causes either no change or an overall decrease in mean fiber area[22,23] (Table 1). Skeletal muscle hypertrophy may be induced noninvasively by the use of a high-resistance training program or invasively by incapacitating synergist muscles resulting in compensatory hypertrophy. Other models of hypertrophy exist,[24-26] but their relevance to the process of work-induced growth is questionable. Whether or not compensatory hypertrophy and high-resistance training stimulate muscle growth via the same mechanism has been a contentious issue.[27] Indications are that a common trigger is responsible in both cases.[28,29] Skeletal muscle which

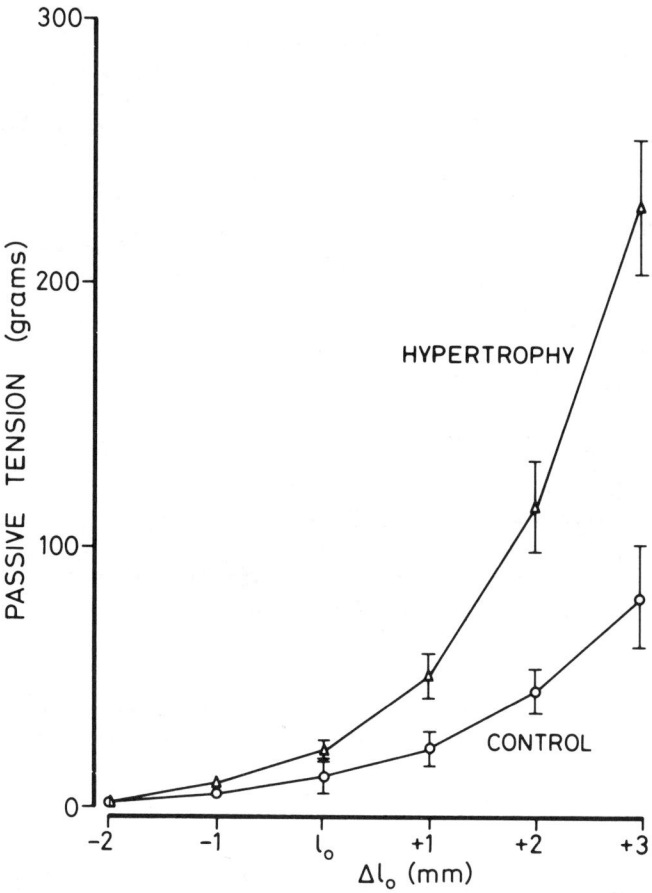

FIGURE 3. Passive force-length relation for normal rat plantaris muscle and one which has undergone compensatory hypertrophy for 4 weeks following removal of the synergistic gastrocnemii. Force is expressed as grams weight and muscle length is expressed in millimeters in greater or less than tetanic optimum length (L_o). (Lang, G. J., unpublished observations.)

has undergone hypertrophy exhibits changes not only in size, but in various contractile, metabolic, structural, histological, and biochemical properties.

Structural Effects

In a muscle which has undergone compensatory hypertrophy, one obvious effect is the increase in tendon bulk. This has been noticed particularly in the plantaris and soleus muscles of the rat following the elimination of the gastrocnemius muscle. Al Zaid et al.[30] found that the collagen content of the compensatory hypertrophied soleus muscle increases from 7.6 to 8.5 $\mu g\ mg^{-1}$ wet weight of tissue ($p < 0.0001$). The advantage of this connective tissue in supporting the increased contractile demands is obvious. In recent experiments,[139] it was found that the passive tension maintained by the hypertrophied plantaris muscle of the rat is three to four times greater than that of the control muscle at lengths greater than optimum (Figure 3). Increased connective tissue in parallel with the muscle fibers could account for this observation. Activities of enzymes involved in the synthesis of connective tissue (e.g., protocollagen proline hydroxylase) have also been shown to increase.[31] The relevance of these changes is accentuated by the observation that increased collagen is necessary for the

Table 2
SUMMARY OF PUBLISHED RESULTS INDICATING THE DEGREE OF HYPERTROPHY (INCREASE IN MUSCLE MASS) IN RESPONSE TO DIFFERENT TYPES OF TRAINING REGIMES

	Type of activity[a]			
	Weight lifting	Compensatory hypertrophy	Sprint training	Endurance training
Degree of hypertrophy	6—36%	53—88%	6%	11—14% (Refs. 135,136)
				No change (Ref. 138)
				Atrophy (Refs. 22,23)
	(Refs. 8,130,131)	(Refs. 44,82,133,134)	(Ref. 23)	
Hypertrophy of fiber types	FG > FOG > SO	SO > FOG > FG	FOG & FG > SO	SO > FOG & FG
	(Refs. 8,15,37,87,89)	(Refs. 44,133,134)	(Refs. 38,132)	(Refs. 135,136,137)

Note: The relative response of the different muscle fiber types is indicated in the lower part of the table.

[a] FG, fast glycolytic or IIB; FOG, fast oxidative glycolytic or IIA; SO, slow oxidative or I.

differentiation of cells in culture.[32] Relative proportions of the different isozymes of connectin change with training.[33] It has been suggested that alterations in the connective tissue content of muscle are of particular importance in the maintenance of posture[34] and that it is mechanical stress which is the signal for connective tissue proliferation.[35]

Histochemical Fiber Types

A plethora of literature exists which indicates the nonuniform responses of muscle fibers to hypertrophy, even within a given muscle. Table 2 is a simplified summary of published results. High-resistance training has often been shown to be specific for fast-twitch fibers;[36,37] endurance training has been shown to stimulate slow-twitch fiber hypertrophy.[22,38] This is not surprising in light of the glycogen depletion patterns observed with differing intensities of effort, i.e., fast-twitch depletes faster than slow-twitch with jumping, and slow-twitch depletes faster than fast-twitch with running.[39] These observations may be explained on the basis of the size principle of motoneuron recruitment[40] which states essentially that the excitability (and therefore incorporation) of motoneurons is an inverse function of their size. At high levels of activation, however, there is evidence that the small S (slow-twitch) units are not incorporated or are recruited after the large fast-fatiguable (FF) and fast-fatigue-resistant (FR) (fast-twitch) units.[41] This reversal of recruitment order could account for most observations of differential fiber hypertrophy in different training regimes. Other reports of differential responses of the various fiber types within a given muscle can be found.[15,42-46]

Protein Synthesis and Degradation

As the size of a skeletal muscle is ultimately dependent on the balance between the rates of protein synthesis and degradation, much attention has focused on these aspects in studies of skeletal muscle hypertrophy. Increased protein synthesis, decreased protein degradation, or both will lead to the accumulation of muscle protein and indeed every combination has been demonstrated experimentally.[13,47-49] In the study of Laurent and Millward,[49] increases in both protein synthesis and protein degradation resulted in two thirds of the muscle protein being "wasted" and one third being accumulated. A differential response of the fiber types was also reported, whereby fast-twitch fibers hypertrophied as a result of protein synthesis and slow-twitch fibers hypertrophied as a result of decreased degradation.[46] Since the turnover of protein in slow muscle is significantly faster than in fast muscle,[50] decreased degradation may represent a more viable strategy for hypertrophy in these muscles.

Many independent studies have confirmed the contribution of increased protein synthesis to skeletal muscle hypertrophy.[51] Others have shown that the capacity of the genetic apparatus is increased due to higher concentrations of the nucleic acids DNA and RNA[52,53] and also of DNA-dependent RNA polymerase.[54,55] However, much of this nucleic acid is located externally to the muscle cell in the connective tissue, interstitial cells, and satellite cells.[56-58] The fate of these cells is controversial. They may be incorporated into the muscle fibers as occurs in postnatal growth;[59,60] they may form new fibers[61] or they may function in laying down the collagen matrix for increasing the connective tissue content of the muscle. It is significant that muscles made to contract under load accumulate amino acids rapidly, whereas those that undergo unloaded shortening fail to increase their accumulation rate above control values.[13] This is another indication of the critical nature of tension as the trigger for skeletal muscle hypertrophy.

Stretch has been proposed as the major stimulus for postnatal skeletal muscle growth.[62,63] However, as stretch results in the development of tension (whether active or passive), it is likely to be stimulating positive protein balance via the same mechanisms as active contraction against load. If this were the case, then simple anatomical considerations, combined with a knowledge of the length-tension curves and the functional length range, generally would show that the ability of a muscle to develop tension by active contraction is far greater than that which can be developed by passive stretching. Thus, contraction under load must be viewed as the major stimulus to skeletal muscle hypertrophy.

The ultimate signal to the genetic apparatus for increased protein synthesis must be chemical in nature. However, the development of tension seems to be the critical factor in skeletal muscle hypertrophy. The mechanochemical transduction mechanism is as yet unidentified. The potency of mechanical tension as a stimulus is quite strong. In the classic studies by Muller and Hettinger,[64,65] muscle hypertrophy occurred as a result of one maximal 6-s contraction per day. Further investigation demonstrated that the threshold for a training effect is in the vicinity of two thirds maximal voluntary strength. Below this value, only maintenance of muscle bulk is observed. The candidates for the intracellular trigger are many, including the inositol lipids,[66] creatine,[67] ADP and AMP,[68] cAMP,[69] somatomedins produced in the muscle cell,[70] and the polyamines spermine and spermidine (reviewed by Tabor and Tabor[71]).

Contractile Properties

A number of investigators have studied the contractile adaptations of skeletal muscle to hypertrophy. Maximum tetanic tension increases in parallel with muscle size;[72,73] in addition, some qualitative adaptations have been shown to occur. Gonyea and Bonde-Petersen[74] demonstrated a general slowing of the contractile properties following 10 to 61 weeks of weight lifting in cats. In the flexor carpi radialis, tetanic tension increased by 41% and was accompanied by a 34% increase in the time to peak of the isometric twitch and a 32% decrease in the rate of tetanic tension development. This type of response is in general agreement with that found by Vrbova[75] and Jewell and Zaimis,[76] but in contrast to those of many other investigators who showed no change in these parameters.[44,72,77-79] Binkhorst and van't Hof[78] did show a decrease in the extrapolated maximum speed of shortening of the plantaris muscle following compensatory hypertrophy, although the alteration in fiber angle due to the increased size of the muscle is partially responsible.

Biochemical and Metabolic Adaptations

Much attention has been focused on certain biochemical correlates of muscle function, particularly myosin, its light chains and catalytic function.[80-84] Myosin ATPase activity has

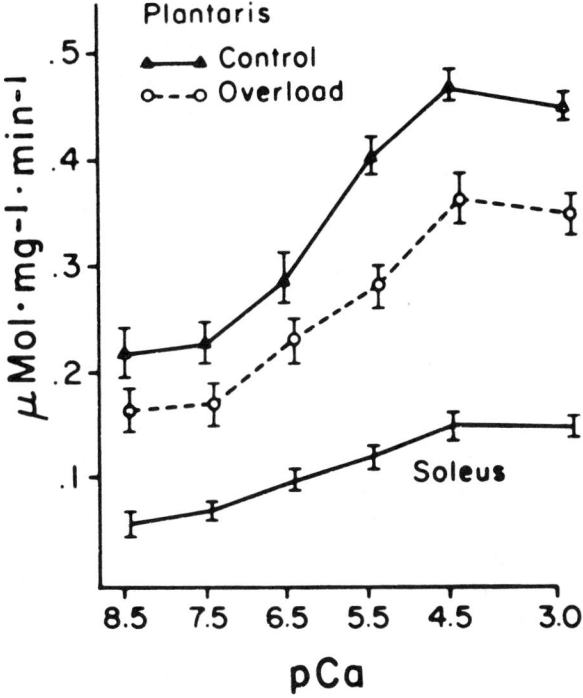

FIGURE 4. Myofibril ATPase activity of normal soleus, normal plantaris, and overloaded (hypertrophied) plantaris muscle expressed as a function of free Ca^{2+} concentration. Values are reported as means ± SE for eight samples representing each muscle. ATPase of overloaded muscles was lower than controls at each level of calcium ($p < 0.05$). (From Baldwin, K. M., Cheadle, W. G., Martinez, O. M., and Cooke, D. A., *J. Appl. Physiol.*, 42, 312, 1977. With permission.)

been correlated with speed of muscle shortening.[85] Baldwin et al.[80] and Noble et al.[82] showed a decrease in the ATPase activity of myosin from hypertrophied muscle (Figure 4). However, much of the contractile data suggests no change in the intrinsic contractile properties of the muscle. In a study conducted on the actin binding properties of myosin from hypertrophied muscle, Noble and Lang[84] inferred an increase in the myosin cross bridge engagement time to actin. This is typical of slow muscle myosin and accounts for the greater energetic efficiency of slow muscle in isometric contractions.[86] There is also a shift in the myosin light chain pattern toward that of a slow muscle.[81] As high-resistance training is conducted at low velocities of muscle shortening, a shift to the synthesis of slow muscle myosin would be of considerable practical advantage on an energetic basis. The biochemical data support the concept that there is general slowing of muscle properties, an observation often not borne out by the contractile data.

The regulatory enzymes of the energy yielding metabolic pathways are remarkably refractory to stimulation by high-resistance training, unlike their relatively easy induction by endurance training. Costill et al.[87] showed no changes in phosphorylase, phosphofructokinase, or lactate dehydrogenase following strength training. This is coincidental with the observation that these same enzymes in weight lifters and shot-putters are within the normal range for sedentary individuals.[88] Mitochondrial enzymes are not perturbed by high-resistance training[87,89] and actually show a decrease in concentration due to their dilution by the myofibrillar proteins.[80] Thus, the signals to the muscle in high-resistance training are fairly exclusive for the contractile proteins, having little effect on the metabolic profile of skeletal muscle.

Proposed Mechanisms

Considerable interest exists in identifying the factors that contribute to the growth and hypertrophy of mammalian skeletal muscle. The trigger which eventually leads to the accumulation of skeletal muscle protein has, thus far, completely eluded physiologists. It is well established that increases in nucleic acids and their related enzymes contribute to the enhanced synthetic capabilities of the hypertrophying muscle, but what triggers this? We know tension is of ultimate importance and the existence of a mechanochemical transducer has already been postulated. Where does it reside? The most likely location is in a structure which supports mechanical tension during contraction such as the cross-bridge or the Z-line. Exercise for strength causes a more pronounced accumulation of protein metabolites than other exercises.[90] Some amino acids (particularly leucine, isoluecine, and valine) are known to stimulate protein synthesis and decrease degradation.[91] If they were also a product of high-resistance training, then they could act as the trigger for skeletal muscle hypertrophy. Their most likely mechanism of action would be via ornithine decarboxylase which has the shortest half-life of any enzyme known,[71] thus optimizing its function as a regulatory enzyme. An increase in the activity of ornithine decarboxylase will result in elevated levels of spermine and spermidine which are potent stimulators of the genetic apparatus and are required for the completion of the larger protein molecules. This system may function as a biological supercompensating mechanism (such as the callous response), to strengthen the proteosynthetic stimulus in the face of increased protein degradation which occurs at intense workloads.[92]

DEGENERATION AND REPAIR

Although this section is primarily concerned with muscle regeneration, it is necessary to consider muscle degeneration because the specific cause of the original degeneration determines or at least has a substantial influence on the quality and extent of the subsequent regeneration. The ability of skeletal muscle to regenerate, or at least to functionally recover from destructive lesions, has been known from work in the 19th century. A classic form of skeletal muscle degeneration is known as "Zenker's waxy degeneration" and a detailed description is provided by Field.[93] This was originally described in detail by Zenker and is typically associated with typhoid fever; however, it also appears to be a feature of other acute and chronic infections. In this type of degeneration, the fibers are destroyed, but the sarcolemma and endomysium remain intact, with the result that eventual regeneration may be complete.

More recently, experiments with the local anesthetic bupivacaine on muscle in experimental animals have produced a rapid degeneration and subsequent regeneration, mainly, it seems, because the sarcolemma and blood vessels remain intact.[94,95] Although this process may not be identical to Zenker's degeneration, it certainly shows similarities.

A variety of methods have been used for inflicting injury on skeletal muscle in order to evaluate its capacity to recover. These include local freezing, heat, ischemia, crushing, and cutting (some of which are reviewed in detail by Field[93]). The extent to which a muscle may recover from any one of these abuses generally seems to be determined by whether or not the sarcolemma is destroyed and by the degree of disorganization caused by the original injury.

It appears that significant recovery follows the crushing of a muscle,[96] although this was not quantified. The original crush destroyed the endomysial sheaths and sarcolemmae; however, the epimysium remained intact along with some fine nerve fibers. Presumably, this provided sufficient "skeleton" on which regeneration could occur. A further experimental procedure was used by Le Gros Clarke and Wadja[97] in which part of a muscle was made ischemic. As a result, the sarcolemma and endomysial membranes remained intact. Fibroblasts grew between the dead fibers providing a tube of endomysial connective tissue.

Sprouting occurred from the stumps of surviving fibers. The sprouts consisted of plasmodial outgrowths of granular sarcoplasm and parallel rows of longitudinally arranged nuclei. Longitudinal striations appeared, followed eventually by cross-striations. Le Gros Clarke and Wadja[97] estimated that the longitudinal growth rate of the regenerating fibers is 1 to 1.5 mm/d. An excellent description of this process is given by Le Gros Clarke.[96]

Whether regeneration is "continuous" or "discontinuous" is one of those esoteric arguments which has preoccupied workers in the field for many years. "Continuous" regeneration occurs by sarcoplasmic outgrowths or buds from surviving muscle fiber stumps as described above by Le Gros Clarke.[96] "Discontinuous" regeneration results from new myoblasts which appear within the sarcolemmal sheath fusing to form a new fiber. The problem has been reexamined more recently by Ali,[98] who concluded that the type of regeneration depends on the severity of the original damage and that in any given situation it is probable that both processes occur. Few would now disagree with this conclusion.

TRANSPLANTATION

Regeneration of muscle following transplantation has received considerable attention over the past 20 years, probably for two main reasons: (1) the emerging interest in the plasticity of muscle and its neurotrophic control, and (2) the growing sophistication in reconstructive surgical techniques which require the survival of muscle implants. Attempts to transplant muscle date back to the 19th century. Results from such experiments were extremely variable and gave rise to conflicting views on the viability of muscle grafts. This early work has been reviewed concisely by Thompson.[99]

The resurgence of interest in muscle transplantation occurred with the work of Carlson in the late 1960s and 1970s following the success of Studitsky's group[100] in the early 1960s (see Carlson[101] for a detailed review of these experiments). Essentially, Studitsky introduced the idea of bringing the transplanted muscle to a "plastic condition" by trauma or prior denervation. This appeared to significantly improve the recovery and regeneration of the transplanted muscle. Carlson[102] demonstrated that when the ankle extensors of the rat were removed, minced into small fragments, and replaced in their original bed (orthotopic transplantation), sufficient regeneration occurred to produce a small strip of histologically identifiable muscle tissue. Subsequently, Carlson and Gutmann[103] showed that such a preparation developed only about 5 to 10% of the isometric tetanic force of an appropriate control muscle (Table 3) and contained an enormous amount of connective tissue. The physiological characteristics of the regenerating muscle were initially similar to those of a slow-twitch muscle, although they speeded up as recovery proceeded.

Minced muscle transplants, while amply demonstrating the regenerative capacity of muscle, give rise to a "muscle" with a functional capacity of only a small fraction of the original. Free muscle grafting in which a whole or part of a muscle is transplanted and the tendons are reattached, but without necessarily providing a blood supply or innervation, was tried unsuccessfully several times in the early part of this century. However, Bosova[104] obtained some success with free muscle grafts by denervating the muscle 10 to 15 d prior to transplantation. With a normal free muscle transplant, the fibers in the center of the muscle rapidly become ischemic, leaving a surviving rim of fibers. After about 1 week, a gradient of regenerating fibers toward the center became established.[105] With prior denervation, the central ischemic core disappeared more rapidly and regenerating fibers had the same degree of maturity across the muscle. By the end of the first week, a uniform population of maturing fibers existed in the extensor digitorum longus (EDL) and soleus muscles of the rat.[105] These regenerating fibers were formed *de novo* by a secondary myogenesis, initiated by satellite cells which survived the necrosis.[106]

While there seems to be little doubt that prior denervation significantly improves the

Table 3
FORCE (mN) DEVELOPED BY THE NORMAL LATERAL GASTROCNEMIUS MUSCLE OF THE RAT AND BY REGENERATED MUSCLES

"Age" (d)	Normal muscle (mN)		Minced, regenerated muscle (mN)	
9	75.4 ± 33.3	(5)	6.5 ± 0.8	(4)
40	2320 ± 567	(3)	76.5 ± 10.2	(5)
100	5070 ± 530	(4)	242 ± 102	(4)

Note: Regenerated muscles had been removed, chopped into small fragments and replaced in their original bed. "Age" refers to true age from birth for normal muscles and to postoperative time for regenerated muscles. Animals were operated at the age of 1 month. The degree of regeneration of a minced muscle is relatively poor compared with normal muscle.

Modified from Carlson and Gutmann.[103]

success of a free muscle graft in the short term (a few days or weeks), it is questionable whether it is of long-term (several months) benefit. Carlson[107] found that prior denervation of rat EDL led to a very high survival rate of fibers at 4 d, but that by 60 d, it did not offer any improvement. A similar pattern of results was obtained with cat EDL;[108,109] after 170 d predenervation was not advantageous and all transplanted muscles developed about 24% of the tension of the control muscles. A further study with cat EDL up to 518 d[110] demonstrated that the transplanted muscle achieved 50% of the mass of the control which essentially confirmed the earlier studies. However, experiments with rat tibialis anterior[111] indicated that an improvement in the recovery of a free muscle graft was still evident at 3 months. In addition, it was also demonstrated that crushing the nerve 2 weeks prior to transplantation, combined with prior denervation of the muscle, produced the best recovery. These results should be viewed cautiously for two reasons. Firstly, in most investigations, there is considerable variability in the degree of recovery within any one experimental group. Secondly, the extent of recovery seems to vary as much between investigators as it does between the various experimental protocols used in any particular laboratory.

The success of whole muscle transplants is significantly improved if the muscle nerve remains intact. After 60 d, rat EDL with the nerve intact recovered to 82% of control tension compared with 61% for a standard free graft. In addition, muscle contraction could be elicited via nerve stimulation after 8 d instead of 3 weeks. Interestingly, Carlson et al.[112] found that a nerve-intact graft with a high sciatic crush was not impaired in its recovery, suggesting that the presence of the nerve is required, but that the early restoration of neuromuscular transmission is not essential.

Further detailed studies of the long- and short-term functional recovery of transplanted, nerve-intact cat EDL muscles[113,114] showed that the muscle could recover up to 64% of the tension of the control and that muscle mass and fiber size were similar to those of the control (Table 4). However, when the nerve was cut and reanastomosed, tension was only 34% of the control[113] (Table 4). During the early stages of recovery, the dynamic characteristics of the muscle tended to be slowed,[115] but after about 6 months, they became near normal and about 95% of fibers were identified histochemically as fast-twitch.[116] In general, the recovery of cat muscles was still not as good as that seen in equivalent experiments with rat muscles and this was attributed to the larger core of necrotic fibers.[113]

Where free muscle grafting has been used in patients, predenervation has resulted in an improved and satisfactory survival of the transplanted muscle.[99] However, the opportunity to use nerve-intact grafting is unlikely to arise in the clinical situation.

Table 4
CAT EDL[a] MUSCLES WHICH HAD BEEN ORTHOTOPICALLY TRANSPLANTED AND WHERE THE NERVE HAD BEEN CUT AND RESUTURED TO THE MUSCLE NA[b] AND WHERE THE NERVE HAD BEEN LEFT INTACT

	Control muscle (12)	Nerve-anastomosed grafts (9)	Nerve-intact grafts (13)
Muscle mass (g)	2.82 ± 0.18	1.87 ± 0.25	3.22 ± 0.32
Muscle cross-sectional area (cm^2)	0.92 ± 0.03	0.55 ± 0.08	0.85 ± 0.08
Maximum isometric tetanic tension (N)	25.2 ± 1.0	8.5 ± 1.2	16.0 ± 1.8

Note: Each group includes animals with recovery times of 120 and 240 d. All values for nerve-intact grafts were significantly different from NA; in addition the tetanic tension of the nerve-intact grafts was significantly different from control.

[a] EDL, extensor digitorum longus.
[b] NA, nerve anastomosed.

Modified from Faulkner et al.[113]

Table 5
MOTOR UNIT SIZE AND NUMBER IN RAT EDL MUSCLES WHICH HAD BEEN ORTHOTOPICALLY TRANSPLANTED AND REINNERVATION ALLOWED TO OCCUR SPONTANEOUSLY (STANDARD AUTOGRAFTS) AND WITH THE NERVE INTACT

	Control muscle	Standard autografts	Nerve-intact autografts
Average motor unit maximum isometric tetanic force (mN)	38.0 ± 3.6 (86)	30.2 ± 5.4 (45)	32.1 ± 4.3 (36)
Estimated number of motor units per muscle	47 ± 7	21 ± 4	42 ± 4

Note: Recovery time was about 80 d for standard and about 52 d for nerve-intact autografts.

Functional Aspects

The motor unit pattern in rat EDL muscles following free grafting and nerve-intact grafting has been determined by Cote and Faulkner.[117] They also showed, as have others (see above), better recovery of tension in nerve-intact grafts, although tension was still less than normal. In free grafted muscles, motor unit number was about half normal, whereas nerve-intact grafts had close to normal numbers. In both types, the average motor unit tension was less than normal (Table 5). They did not find any qualitative difference in the units. Gorniak et al.,[118] using chronic electromyographic (EMG) recording, demonstrated the functional involvement of orthotopically implanted cat EDL muscles. In addition, an exercise-induced increase in muscle mass has been shown for a free grafted soleus muscles in rats.[119] Quick and Rogers[120] presented physiological and morphological evidence that muscle spindles and spindle-like activity reappeared in rat EDL muscles following orthotopic implantation and bupivicaine treatment. However, the spindles and their activity were somewhat aberrant.

Cellular Processes

Typically, a free muscle graft retains only a thin rim of surviving muscle fibers.[105] Deeper fibers are inevitably ischemic; their cytoplasmic contents are removed by macrophages over subsequent days. Within the first 24 h of the regeneration of minced muscle, Snow[121] observed small, viable undifferentiated cells scattered through the mess of degenerating fibers. These cells appeared to be similar to the satellite cells (small cells lying beneath the basal lamina of fibers) of normal muscle. Subsequently, Snow[122] labeled satellite cells of uninjured rat muscle with tritiated thymidine, transplanted the muscle to littermates, and found that 4 to 6 d later, the nucleii of regenerating myotubes were labeled. This provided an elegant demonstration that satellite cells in young rats can differentiate into multinucleated myotubes. Jones[123] has cultured cells from regenerating muscle and found them to be similar to embryonic cells in growth rates and fusion characteristics; in addition, the embryonic cells would fuse with the regenerating cells. This supports the view that myogenic cells from regenerating muscle are identical to embryonic myoblasts.

In transplantation studies, doubt has been cast on some results because of the presence of surviving intact fibers. Several investigators have therefore treated their transplanted muscles with bupivicaine and hyaluronidase in order to provide a uniform population of regenerating fibers.[124] Such fibers regenerate and generally appear normal. However, Ontell et al.[125] (Figure 5) noticed that in orthotopically transplanted, bupivicaine-treated EDL muscles of mice the regenerating myotubes would branch and recombine along the length of the muscle to form a complex syncytium. They subsequently showed that these branched myotubes persist as branched fibers.[106]

It would be reasonable to expect that the sequence of protein synthesis which occurs during development recurs during regeneration. The presence of both embryonic myosin light chains[126] and fetal myosin heavy chains[127] has been detected in rat muscles within a few days of injury. By 2 weeks, both rat EDL and soleus contained myosin light chains typical of fast-twitch muscle. This is a further indication that the changes in early regeneration (and early development) can proceed in the absence of the nerve.

The use of denervation 2 to 3 weeks prior to free muscle grafting as a method of plasticizing the muscle has been discussed above. Results of Hess and Rosner[128] lend validity to this procedure. They showed that while satellite cells are relatively rare in normal muscle, they are abundant following denervation. The increase occurs by nuclei budding off existing fibers and remaining beneath the basal lamina of the parent fiber. Filaments form in the bud and it takes on the appearance of a myoblast. It seems likely that this forms the basis of the "priming" effect produced by prior denervation of a free muscle graft.

It would seem that some care is necessary in the identification of cell types in the early stages of regeneration. Trupin and Hsu[129] have found that invading macrophages can assume a fusiform shape and adopt a sublaminar position, while lacking phagocytic features. They argue that such cells could be mistaken for satellite and myogenic cells.

The recovery of a minced muscle in either its own bed or that of another muscle provides a useful model for the study of regeneration, even though the functional recovery is relatively poor. Free muscle transplantation is an improvement and prior denervation is also beneficial. The best recovery described so far occurs when the nerve to the muscle is left intact during transplantation.

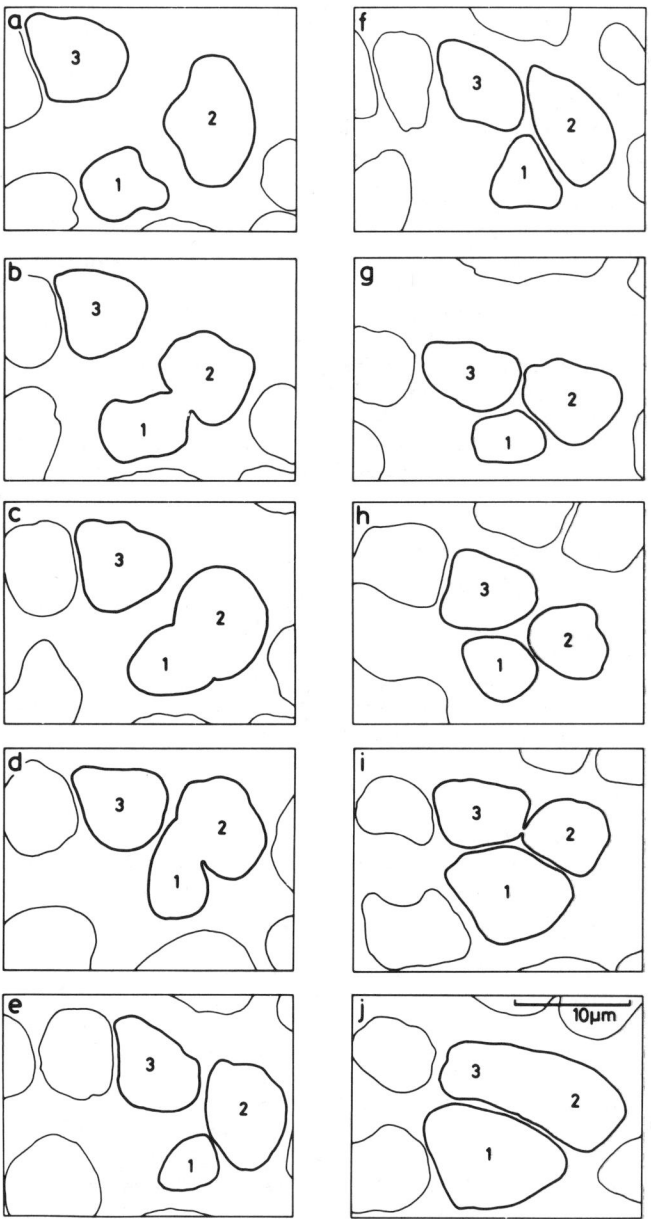

FIGURE 5. Line drawings from electron micrographs of Ontell et al.[125] of serial sections of regenerating muscle to show myotubes branching and interconnecting along the length of a muscle. In (a) three discrete myotubes are seen; in (b) myotubes 1 and 2 commence to fuse, but have dissociated by (e). At (i) a connection appears between myotubes 2 and 3, leading to complete fusion in (j). Distance between successive sections is in micrometer: 30, 15, 15, 15, 60, 80, 140, 15, and 15. Calibration bar in (j) is approximate.

REFERENCES

1. **Goldspink, G.,** Post-embryonic growth and differentiation of striated muscle, in *The Structure and Function of Muscles,* Vol. 1, Bourne, G. H., Ed., Academic Press, New York, 1972, chap. 5.
2. **Goldspink, G. and Rowe, R. W. D.,** Studies on post-embryonic growth and development of skeletal muscle. II. Some physiological and structural changes that are associated with the growth and development of skeletal muscle fibers, *Proc. R. Irish Acad. Sect. B,* 66, 85, 1968.
3. **Ontell, M. and Dunn, R. F.,** Neonatal muscle growth: a quantitative study, *Am. J. Anat.,* 152, 539, 1978.
4. **Goldspink, G.,** The proliferation of myofibrils during muscle fibre growth, *J. Cell Sci.,* 6, 593, 1970.
5. **Goldspink, G.,** Alterations in myofibril size and structure during growth, exercise, and changes in environmental temperature, in *Handbook of Physiology, Section 10: Skeletal Muscle,* Peachey, L. D., Ed., American Physiological Society, Bethesda, MD, 1983, chap. 18.
6. **Williams, P. and Goldspink, G.,** Longitudinal growth of striated muscle fibres, *J. Cell Sci.,* 9, 751, 1971.
7. **Griffin, G., Goldspink, G., and Williams, P. E.,** Region of longitudinal growth in striated muscle fibres, *Nature (London) New Biol.,* 232, 28, 1971.
8. **Gonyea, W., Ericson, G. C., and Bonde-Petersen, F.,** Skeletal muscle fibre splitting induced by weight-lifting exercise in cats, *Acta Physiol. Scand.,* 99, 105, 1977.
9. **Hall-Craggs, E. C. B.,** The longitudinal division of fibres in overloaded rat skeletal muscle, *J. Anat.,* 107, 459, 1970.
10. **Reitsma, W.,** Some structural changes in skeletal muscles of the rat after intensive training, *Acta Morphol. Neerl. Scand.,* 7, 229, 1970.
11. **Saltin, B. and Gollnick, P. D.,** Skeletal muscle adaptability: significance for metabolism and performance, in *Handbook of Physiology, Section 10: Skeletal Muscle,* Peachey, L. D., Ed., American Physiological Society, Bethesda, MD, 1983, chap. 19.
12. **Smith, P. E.,** Hypertrophy and replacement therapy in the rat, *Am. J. Anat.,* 45, 205, 1930.
13. **Goldberg, A. L., Etlinger, J. D., Goldspink, D. F., and Jablecki, C.,** Mechanism of work induced hypertrophy of skeletal muscle, *Med. Sci. Sports,* 7, 185, 1975.
14. **Narahara, H. T. and Holloszy, J. O.,** The actions of insulin, trypsin and electrical stimulation on amino acid transport in muscle, *J. Biol. Chem.,* 249, 5435, 1974.
15. **Goldspink, G. and Ward, P. S.,** Changes in rodent muscle fibre types during post-natal growth, undernutrition and exercise, *J. Physiol.,* 296, 453, 1979.
16. **Goldberg, A. L.,** Work induced growth of skeletal muscle in normal and hypophysectomized animals, *Am. J. Physiol.,* 312, 1193, 1967.
17. **Goldberg, A. L.,** Role of insulin in work induced growth of skeletal muscle, *Endocrinology,* 83, 1071, 1968.
18. **Li, J. B. and Goldberg, A. L.,** Effects of food deprivation on protein synthesis and degradation in rat skeletal muscle, *Am. J. Physiol.,* 231, 441, 1976.
19. **Denny-Brown, D.,** Experimental studies pertaining to hypertrophy, regeneration and degradation, in *Proc. Assoc. Research in Nervous and Mental Disease,* Vol. 38, Adams, R. D., Eaton, L. M., and Shy, G. M., Eds., Williams & Wilkins, Baltimore, 1960, 147.
20. **Roy, R. R., Baldwin, K. M., Martin, T. P., Chimarust, S. P., and Edgerton, V. R.,** Biochemical and physiological changes in overloaded rat fast and slow-twitch ankle extensors, *J. Appl. Physiol.,* 59, 639, 1985.
21. **Howald, H.,** Malleability of the motor system: training for maximizing power output, *J. Exp. Biol.,* 115, 365, 1985.
22. **Jansson, E. and Kaijser, L.,** Muscle adaptation to extreme endurance training in man,, *Acta Physiol. Scand.,* 100, 315, 1977.
23. **Parsons, D., Musch, T. I.,, Moore, R. L., Haidet, G. C., and Ordway, G. A.,** Dynamic exercise training in foxhounds. II. Analysis of skeletal muscle, *J. Appl. Physiol.,* 59, 190, 1985.
24. **Ashmore, C. R. and Summers, P. J.,** Stretch induced growth of chicken wing muscles: myofibrillar proliferation, *Am. J. Physiol.,* 51, C93, 1981.
25. **Stewart, D. M., Sola, O. M., and Martin, A. W.,** Hypertrophy as a response to denervation in skeletal muscle, *Z. Vgl. Physiol.,* 76, 146, 1972.
26. **Martin, W. D. and Romond, E. H.,** Effects of chronic rotation and hypergravity on muscle fibers of soleus and plantaris muscles of the rat, *Exp. Neurol.,* 49, 758, 1975.
27. **Gutmann, E. and Hajek, I.,** Differential reaction of muscle to exercise use in compensatory hypertrophy and increased phasic activity, *Physiol. Biochem.,* 20, 205, 1971.
28. **Michel, R., Noble, E., and Gardiner, P.,** Rat plantaris EMG during locomotion before and after ablation of synergists, *Med. Sci. Sports Exercise,* 17, 234, 1985.
29. **Lang, G. J. and Luff, A. R.,** Skeletal muscle hypertrophy: stretch or overload, *Proc. Aust. Phsyiol. Pharmacol. Soc.,* 15, 149P, 1984.

30. **Al Zaid, N., Goldspink, G., and Williams, P. E.,** Increase in collagen concentration in overloaded skeletal muscles, *J. Physiol.,* 310, 55P, 1981.
31. **Turto, H., Lindy, S., and Halme, J.,** Protocollagen proline hydroxylase activity in work induced hypertrophy of rat muscle, *Am. J. Physiol.,* 226, 63, 1974.
32. **Hauschka, S. D. and Konigsberg, I. R.,** The influence of collagen on the development of muscle clones, *Proc. Natl. Acad. Sci. U.S.A.,* 55, 119, 1966.
33. **Yamaguchi, M., Nakayama, Y., and Nishikawa, J.,** Studies of exercise and an elastic protein "connectin" in hindlimb muscle of growing rat, *Jpn. J. Physiol.,* 35, 21, 1985.
34. **Kovanen, V., Suominen, H., and Heikkinen, E.,** Connective tissue of fast and slow skeletal muscles of the rat-effects of endurance training, *Acta Physiol. Scand.,* 108, 173, 1980.
35. **Blanchard, O., Cohen-Solal, L., Tardieu, C., Allain, J. C., Tarbary, C., and Le Lous, M.,** Tendon adaptation to different long term stresses and collagen reticulation in soleus muscle, *Connect. Tissue Res.,* 13, 261, 1985.
36. **Gollnick, P. D., Armstrong, R. B., Saubert, C. W., Phiel, K., and Saltin, B.,** Enzyme activity and fiber composition in skeletal muscle of untrained and trained men, *J. Appl. Physiol.,* 33, 312, 1972.
37. **Thorstensson, A., Hulton, B., Von Dobeln, W., and Karlsson, J.,** Effect of strength training on enzyme activities and fibre characteristics in human skeletal muscle, *Acta Physiol. Scand.,* 96, 392, 1976.
38. **Saltin, B., Nazar, K., Costill, D. L., Stein, E., Jansson, E., Essen, B.,, and Gollnick, P. D.,** The nature of the training response; peripheral and central adaptations to one legged exercise, *Acta Physiol. Scand.,* 96, 289, 1976.
39. **Gillespie, C. A., Simpson, D. R., and Edgerton, V. R.,** Motor unit recruitment as reflected by muscle fiber glycogen loss in a prosimian (bushbaby) after running and jumping, *J. Neurosurg. Psychiatry,* 37, 817, 1974.
40. **Hennemann, E.,** Relation between size of neurons and their susceptibility to discharge, *Science,* 26, 1345, 1957.
41. **Desmedt, J. E. and Godaux, E.,** Ballistic contraction in man: characteristic recruitment pattern of single motor units of the tibialis anterior muscle, *J. Physiol.,* 264, 673, 1977.
42. **Herbison, G. J., Mazher Jaweed, M., and Ditunno, J. F.,** Response of type I fibers to weight lifting in rat plantaris, *Arch. Phys. Med. Rehab.,* 62, 342, 1981.
43. **Christman, J. V.,** The Effects of Vertical Jump Training on Rat Skeletal Muscle, Master's thesis, University of California, Los Angeles, 1978.
44. **Walsh, J. V., Burke, R. E., Rymer, W. Z., and Tsairis, P.,** Effect of compensatory hypertrophy studied in individual motor units in medial gastrocnemius of the cat, *J. Neurophysiol.,* 41, 496, 1978.
45. **Henriksson-Larson, K., Friden, J., and Wretling, M. L.,** Distribution of fibre sizes in human skeletal muscle. An enzyme histochemical study in m. tibialis anterior, *Acta Physiol. Scand.,* 123, 171, 1985.
46. **Watt, P. W., Kelly, F. J., Goldspink, D. F., and Goldspink, G.,** Exercise induced morphological and biochemical changes in skeletal muscles of the rat, *J. Appl. Physiol.,* 53, 1144, 1982.
47. **Goldspink, D. F.,** The influence of activity on muscle size and protein turnover, *J. Physiol.,* 264, 283, 1977.
48. **Dohm, G. L., Kasperek, G. J., Tapscott, E. B., and Beecher, G. R.,** Effects of exercise on synthesis and degradation of muscle protein, *Biochem. J.,* 188, 255, 1980.
49. **Laurent, G. J. and Millward, D. J.,** Protein turnover during skeletal muscle hypertrophy, *Fed. Proc.,* 39, 42, 1980.
50. **Goldberg, A. L.,** Protein synthesis in tonic and phasic skeletal muscles, *Nature (London),* 216, 1219, 1967.
51. **Poortmans, J. R.,** Effects of long lasting physical exercise and training on protein metabolism, in *Metabolic Adaptation to Prolonged Physical Exercise,* Howald, H. and Poortmans, J. R., Eds., Birkhauser, Basel, 1975, 212.
52. **Meerson, F. Z.,** The myocardium in hyperfunction, hypertrophy and heart failure, *Circ. Res.,* 25(Suppl. 2), 1969.
53. **Milward, D. J., Garlick, P. J., James, W. P. T., Nnanyelugo, D. O., and Ryatt, J. S.,** Relationship between protein synthesis and RNA content in skeletal muscle, *Nature (London),* 241, 204, 1973.
54. **Rogozkin, V. and Feldkoren, B.,** The effect of retabolil and training on activity of RNA polymerase in skeletal muscles, *Med. Sci. Sports,* 11, 345, 1979.
55. **Sobel, B. E. and Kaufman, S.,** Enhanced RNA polymerase activity in skeletal muscle undergoing hypertrophy, *Arch. Biochem. Biophys.,* 137, 469, 1970.
56. **Jablecki, C. K., Heuser, J. E., and Kaufman, S.,** Autoradiographic localization of new RNA synthesis in hypertrophying skeletal muscle, *J. Cell Biol.,* 57, 743, 1973.
57. **Goss, R. J.,** Hypertrophy versus hyperplasia, *Science,* 153, 1615, 1966.
58. **Enesco, M.,** Increase in the number of nuclei in various striated muscles of the growing rat, *Anat. Rec.,* 139, 225, 1961.

59. Schiaffino, S., Pierobon Bormioli, S., and Aloisi, M., The fate of newly formed satellite cells during compensatory muscle hypertrophy, *Virchows Arch.*, 21, 113, 1976.
60. Moss, F. P. and LeBlond, C. P., Nature of dividing nuclei in skeletal muscle of growing rats, *J. Cell Biol.*, 44, 459, 1970.
61. Solleo, A., Anastasi, G., LaSpada, G., Falzea, G., and Denaro, M. G., New muscle fibre production during compensatory hypertrophy, *Med. Sci. Sports Exercise*, 12, 268, 1980.
62. Gutmann, E., Schiaffino, S., and Hanzlikova, V., Mechanisms of compensatory hypertrophy in skeletal muscle of the rat, *Exp. Neurol.*, 31, 451, 1971.
63. Schiaffino, S. and Hanzlikova, V., On the mechanism of compensatory hypertrophy on skeletal muscles, *Experientia*, 26, 152, 1970.
64. Muller, E. A. and Hettinger, T., Veber unterschiede der trainingsgeschwindigkeit atrophierter und normalen muskeln, *Arbeitsphysiologie*, 15, 223, 1953.
65. Muller, E. A., The regulation of muscular strength, *J. Assoc. Phys. Ment. Rehab.*, 11, 41, 1957.
66. Jablecki, C., Dienstag, J., and Kaufman, S., [^{3}H] inositol incorporation into phosphatidyl-inositol in work induced growth of rat muscle, *Am. J. Physiol.*, 232, E324, 1977.
67. Ingwall, J. S., Morales, P., and Stockdale, P. E., Creatine and control of myosin synthesis in differentiating skeletal muscle, *Proc. Natl. Acad. Sci. U.S.A.*, 69, 2250, 1972.
68. Buse, M. G. and Reid, S. S., A possible regulation of protein turnover in muscle, *J. Clin. Invest.*, 56, 1250, 1975.
69. Sheppard, J. R., Difference in the cAMP levels in normal and transformed cells, *Nature (London), New Biol.*, 236, 14, 1972.
70. Sturek, M., Lambs, D. R., and Snyder, A. C., Somatomedin-like activity and muscle hypertrophy, *I.R.C.S. Med. Sci.*, 9, 760, 1981.
71. Tabor, C. W. and Tabor, H., 1,4-Diaminobutane (putrescine), spermidine, and spermine, *Annu. Rev. Biochem.*, 45, 285, 1976.
72. Roy, R. R., Baldwin, K. M., Martin, T. P., Chimarusti, S. P., and Edgerton, V,. R., Biochemical and physiological changes in overloaded rat fast and slow-twitch ankle extensors, *J. Appl. Physiol.*, 59, 639, 1985.
73. Binkhorst, R. A., The effect of training on some isometric contraction characteristics of a fast muscle, *Pflugers Arch.*, 309, 193, 1969.
74. Gonyea, W. and Bonde-Petersen, S. F., Alterations in muscle contractile properties and fiber composition after weight lifting exercise in cats, *Exp. Neurol.*, 59, 75, 1978.
75. Vrbova, G., The effect of motoneurone activity on the speed of contraction of striated muscle, *J. Physiol.*, 169, 513, 1963.
76. Jewell, P. A. and Zaimis, E. J., Changes at the neuromuscular junction of red and white muscle fibres in the cat induced by disuse atrophy, *J. Physiol.*, 124, 429, 1954.
77. Freeman, P. L. and Luff, A. R., Contractile properties of hindlimb muscles in rat during surgical overload, *Am. J. Physiol.*, 242, C259, 1982.
78. Binkhorst, R. A. and van't Hof, M. A., Force-velocity relationship and contraction time of the rat fast plantaris muscle due to compensatory hypertrophy, *Pflugers Arch.*, 342, 145, 1973.
79. van Linge, B., The response of muscle to strenuous exercise, *J. Bone Jt. Surg.*, 44B, 711, 1962.
80. Baldwin, K. M., Cheadle, W. G., Martinez, O. M., and Cooke, D. A., Effect of functional overload on enzyme levels in different types of skeletal muscle, *J. Appl. Physiol.*, 42, 312, 1977.
81. Baldwin, K. M., Valdez, V., Herrick, R. E., MacIntosh, A. M., and Roy, R. R., Biochemical properties of overloaded fast-twitch skeletal muscle, *J. Appl. Physiol.*, 52, 467, 1982.
82. Noble, E. G., Dabrowski, B. L., and Ianuzzo, C. D., Myosin transformation in hypertrophied rat muscle, *Pflugers Arch.*, 396, 260, 1983.
83. Rapp, G. and Weicker, H., Comparative studies of fast muscle myosin light chains after different training programmes, *Int. J. Sports Med.*, 3, 58, 1982.
84. Noble, E. G. and Lang, G. J., Changes in the actin-activated myosin ATPase of hypertrophied rat plantaris muscle, *Med. Sci. Sports Exercise*, 1984.
85. Barany, M., ATPase activity of myosin correlated with speed of muscle shortening, *J. Gen. Physiol.*, 50, 197, 1967.
86. Awan, M. Z. and Goldspink, G., Energy utilization by mammalian fast and slow muscle in doing external work, *Biochim. Biophys. Acta*, 216, 229, 1970.
87. Costill, D. L., Coyle, E. F., Fink, W. F., Lesmes, G. R., and Witzmann, F. A., Adaptations in skeletal muscle following strength training, *J. Appl. Physiol.*, 46, 96, 1979.
88. Costill, D. L., Daniels, J., Evans, W., Fink, W., Krahenbuhl, G., and Saltin, B., Skeletal muscle enzymes and fiber composition in male and female track athletes, *J. Appl. Physiol.*, 40, 149, 1976.
89. MacDougall, J. D., Ward, G. R., Sale, D. G., and Sutton, J. R., Biochemical adaptation of human skeletal muscle to high resistance training and immobilization, *J. Appl. Physiol.*, 43, 700, 1977.

90. **Marakova, A. F.**, Biochemical changes in muscles of animals during experimental training of various kinds, *Ukr. Biokhim. Zh.*, 30, 903, 1958.
91. **Fulks, R. M., Li, J. B., and Goldberg, A. L.**, Effects of insulin and amino acids on protein turnover in rat diaphragm, *J. Biol. Chem.*, 250, 290, 1975.
92. **Kasperek, G. J. and Snider, R. D.**, Increased protein degradation after eccentric exercise, *Eur. J. Appl. Physiol.*, 54, 30, 1985.
93. **Field, E. J.**, Muscle regeneration and repair, in *Structure and Function of Muscle*, Vol. 3, Bourne, G. H., Ed., Academic Press, New York, 1960, 139.
94. **Benoit, P. W. and Belt, W. D.**, Destruction and regeneration of skeletal muscle after treatment with a local anaesthetic, bupivacaine (Marcaine), *J. Anat.*, 107, 547, 1970.
95. **Hall-Craggs, E. C. B.**, Rapid degeneration and regeneration of a whole skeletal muscle following treatment with bupivacaine (Marcaine), *Exp. Neurol.*, 43, 349, 1974.
96. **LeGros Clark, W. E.**, An experimental study of the regeneration of mammalian striped muscle, *J. Anat.*, 80, 24, 1946.
97. **LeGros Clark, W. E. and Wajda, H. S.**, The growth and maturation of regenerating striated muscle fibres, *J. Anat.*, 81, 56, 1947.
98. **Ali, M. A.**, Myotube formation in skeletal muscle regeneration, *J. Anat.*, 128, 553, 1979.
99. **Thompson, N.**, Autogenous free grafts of skeletal muscles, *Plast. Reconstr. Surg.*, 48, 11, 1971.
100. **Studitsky, A. N.**, Free auto- and homografts of muscle tissue in experiments on animals, *Ann. N.Y. Acad. Sci.*, 120, 789, 1964.
101. **Carlson, B. M.**, A review of muscle transplantation in mammals, *Physiol. Bohemoslov.*, 27, 387, 1978.
102. **Carlson, B. M.**, Regeneration of the rat gastrocnemius muscle from sibling and non-sibling muscle fragments, *Am. J. Anat.*, 128, 21, 1970.
103. **Carlson, B. M. and Gutmann, E.**, Development of contractile properties of minced muscle regenerates in the rat, *Exp. Neurol.*, 36, 239, 1972.
104. **Bosova, N. N.**, Free autoplastic transplantation of whole muscles, *Bull. Exp. Biol. Med.*, 53, 88, 1962.
105. **Carlson, B. M. and Gutmann, E.**, Regeneration in free grafts of normal and denervated muscles in the rat: morphology and histochemistry, *Anat. Rec.*, 183, 47, 1975.
106. **Bourke, D. L. and Ontell, M.**, Branched myofibers in long-term whole muscle transplants: a quantitative study, *Anat. Rec.*, 209, 281, 1984.
107. **Carlson, B. M.**, A quantitative study of muscle fiber survival and regeneration in normal, predenervated and Marcaine treated free muscle grafts in the rats, *Exp. Neurol.*, 52, 421, 1976.
108. **Faulkner, J. A., Maxwell, L. C., Mufti, S. A., and Carlson, B. M.**, Skeletal muscle fiber regeneration following heterotrophic autotransplantation in cats, *Life Sci.*, 19, 289, 1976.
109. **Mufti, S. A., Carlson, B. M., Maxwell, L. C., and Faulkner, J. A.**, The free autografting of entire limb muscles in the cat: morphology, *Anat. Rec.*, 188, 417, 1977.
110. **Maxwell, L. C., Faulkner, J. A., White, T. P., and Hansen-Smith, F. M.**, Growth of regenerating skeletal muscle fibers in cats, *Anat. Rec.*, 209, 153, 1984.
111. **Hall-Craggs, E. C. B. and Brand, P.**, Effect of previous nerve injury on the regeneration of free autogenous muscle grafts, *Exp. Neurol.*, 57, 275, 1977.
112. **Carlson, B. M., Foster, A. H., Bader, D. M., Hnik, P., and Vejsada, R.**, Restoration of full mass in nerve-intact muscle grafts after delayed reinnervation, *Experientia*, 39, 171, 1983.
113. **Faulkner, J. A., Markley, J. M., McCully, K. K., Watters, C. R., and White, T. P.**, Characteristics of cat skeletal muscles grafted with intact nerves or with anastomosed nerves, *Exp. Neurol.*, 80, 682, 1983.
114. **Maxwell, L. C.**, Muscle fiber regeneration in nerve-intact and free skeletal muscle autografts in cats, *Am. J. Physiol.*, 246, C96, 1984.
115. **Faulkner, J. A., Niemeyer, J. H., Maxwell, L. C., and White, T. P.**, Contractile properties of transplanted extensor digitorum longus muscles of cats, *Am. J. Physiol.*, 238, C120, 1980.
116. **Maxwell, L. C., Faulkner, J. A., Mufti, S. A., and Turowski, A. M.**, Free autografting of entire limb muscles in the cat: histochemistry and biochemistry, *J. Appl. Physiol.*, 44, 431, 1978.
117. **Cote, C. and Faulkner, J. A.**, Motor unit function in skeletal muscle autografts of rats, *Exp. Neurol.*, 84, 292, 1984.
118. **Gorniak, G. C., Gans, C., and Faulkner, J. A.**, Muscle fiber regeneration after transplantation: prediction of structure and physiology from electromyograms, *Science*, 204, 1085, 1979.
119. **White, T. P., Villanacci, J. F., Morales, P. G., Segal, S. S., and Essig, D. A.**, Exercise induced adaptations of rat soleus muscle grafts, *J. Appl. Physiol.*, 56, 1325, 1984.
120. **Quick, D. C. and Rogers, S. L.**, Stretch receptors in regenerated rat muscle, *Neuroscience*, 10, 851, 1983.
121. **Snow, M. H.**, Myogenic cell formation in regenerating rat skeletal muscle. I. A fine structural study, *Anat. Rec.*, 188, 181, 1977.
122. **Snow, M. H.**, An autoradiographic study of satellite cell differentiation into regenerating myotubes following transplantation of muscles in young rats, *Cell Tissue Res.*, 186, 535, 1978.

123. **Jones, P. H.,** In vitro comparison of embryonic myoblasts and myogenic cells isolated from regenerating adult rat skeletal muscle, *Exp. Cell Res.*, 139, 401, 1982.
124. **Carlson, B. M. and Gutmann, E.,** Free grafting of the extensor digitorum longus muscle in the rat after Marcaine pretreatment, *Exp. Neurol.*, 53, 82, 1976.
125. **Ontell, M., Hughes, D., and Bourke, D.,** Secondary myogenesis of normal muscle produces abnormal myotubes, *Anat. Rec.*, 204, 199, 1982.
126. **Carraro, U., Dalla-Libera, L. D., and Catani, C.,** Myosin light and heavy chains in muscle regenerating in absence of the nerve: transient appearance of the embryonic light chain, *Exp. Neurol.*, 79, 106, 1983.
127. **Sartore, S., Gorza, L., and Schiaffino, S.,** Fetal myosin heavy chains in regenerating muscle, *Nature (London)*, 298, 294, 1982.
128. **Hess, A. and Rosner, S.,** The satellite cell bud and myoblast in denervated mammalian muscle fibers, *Am. J. Anat.*, 129, 21, 1970.
129. **Trupin, G. L. and Hsu, L.,** The identification of myogenic cells in regenerating skeletal muscle. II. Early mammalian regenerates, *Dev. Biol.*, 68, 72, 1979.
130. **Ho, K. W., Roy, R. R., Tweedle, C. D., Heusner, W. W., Van Huss, W. D., and Carrow, R. E.,** Skeletal muscle fiber splitting with weight-lifting exercise in rats, *Am. J. Anat.*, 157, 433, 1980.
131. **Goldspink, G. and Howells, K. F.,** Work induced hypertrophy in exercised normal muscles of different ages and the reversibility of hypertrophy after cessation of exercise, *J. Physiol.*, 239, 179, 1974.
132. **Coyle, E. F., Feiring, D. C., Rotkis, T. C., Cote, R. W., Roby, F. B., Lee, W., and Wilmore, J. H.,** Specificity of power improvements through slow and fast isokinetic training, *J. Appl. Physiol.*, 51, 1437, 1981.
133. **Iannuzzo, C. D., Chen, V., Armstrong, R. B., Dabrowski, B., and Noble, E.,** An experimental model to study chronically hypertrophied skeletal muscle, *Adv. Physiol. Sci.*, 24, 279, 1975.
134. **Lang, G. J. and Luff, A. R.,** Effects of overload on the histochemistry and morphometry of rat plantaris muscle, *Proc. Aust. Physiol. Pharmacol. Soc.*, 16, 271P, 1985.
135. **Carrow, R. E., Brown, R. E., and Van Huss, W. D.,** Fiber size and capillary to fiber ratios in skeletal muscle of exercised rats, *Anat. Rec.*, 159, 33, 1967.
136. **Muller, W.,** Temporal progress of muscle adaptation to endurance training in hind limb muscles of young rats: a histochemical and morphometrical study, *Cell Tissue Res.*, 156, 61, 1974.
137. **Gollnick, P. D., Armstrong, R. B., Saltin, B., Sanbert, C. W., Sembrowich, W. L., and Shepherd, R. E.,** Effect of training on enzyme activity and fiber composition of human skeletal muscle, *J. Appl. Physiol.*, 34, 107, 1973.
138. **Edgerton, V. R., Gerchman, L., and Carrow, R.,** Histochemical changes in rat skeletal muscle after exercise, *Exp. Neurol.*, 24, 110, 1969.
139. **Lang, G. J.,** unpublished observations.

DEVELOPMENT OF BLOOD AND HEMATOPOIESIS

DEVELOPMENTAL CHANGES IN BLOOD AS A WHOLE

N. T. Shahidi and W. B. Ershler

HEMOPOIESIS IN THE FETUS

In view of rapid growth during fetal life and change in oxygen availability after birth, qualitative and quantitative changes in hemopoiesis are particulary dramatic *in utero* and in the newborn period. While developmental changes involve all hemopoietic cell lines, changes in erythropoiesis are most pronounced.

Intrauterine hematopoiesis is marked by several overlapping functional and anatomic stages, and traditionally it is divided into three periods: mesoblastic (connective tissue), hepatic, and myeloid. In the mesoblastic period, blood-forming islands can be first detected in the second or third week after conception in the yolk sac in the area of vasculosa.[1] Blood islands in the yolk sac differentiate in two directions. Peripheral cells in the islands form the walls of the first blood vessels, whereas the centrally located cells become the primitive blood cells or hematocytoblasts.[2,3]

There are indications that this process of differentiation requires intimate association between mesodermal and endodermal tissues.[4,5] The newly formed erythroid cells enter circulation after being hemoglobinized in about the third week of gestation when the heart starts pumping blood. These newly formed cells in the yolk are quite large and the vast majority of them retain their nuclei. The mean corpuscular volume (MCV) and the mean cell diameter exceed 180 fl and 10.5 μm, respectively. These large cells probably synthesize the primitive globin chains epsilon (ϵ) and zeta (ζ). Since only minute amounts of these primitive hemoglobins are available, their structure and function have not been fully determined. The amino acid sequence of the ζ-chain shows considerable homology to α-chain. By the sixth week of gestation, the mesoblastic erythropoiesis begins to decline and by the end of the first trimester it can no longer be detected.[1,6]

The hepatic phase of hematopoiesis begins in the fifth week of gestation[7,8] and gradually increases so that the liver is the main anatomic site of hematopoiesis from the third to the sixth fetal month and continues its hemopoietic activity into the first month of postnatal life. Although a small number of granulocytes and megakaryocytes are present in the liver, the predominant hemopoietic component is erythroid progenitors. During the peak of hepatic hemopoietic activity, erythroid precursors represent at least 50% of the total nucleated cells of this organ.[9] Such an erythroid preponderance in the liver is probably under local control, for hemopoiesis in the bone marrow involves all three lines equally.

As erythropoiesis enters the hepatic phase, alpha and gamma globin chains are synthesized in various combinations. The above globin chains result in the production of Gower I and II hemoglobins ($zeta_2$ $epsilon_2$ and $alpha_2$ $epsilon_2$), hemoglobin Portland ($zeta_2$ $gamma_2$), and fetal hemoglobin ($alpha_2$ $gamma_2$). The Gower hemoglobins and hemoglobin Portland are detectable only in the first 3 months of gestation. Near the end of the first month of gestation, hemoglobins Gower I and II comprise 42 and 24% of the total hemoglobin, respectively, with fetal hemoglobin making up the remainder.

While scattered foci of megakaryopoiesis are present in human fetal liver,[1-7] granulopoiesis is notably absent. Recent *in vitro* studies, however, have revealed the presence of myeloid progeneitor cells.[10]

During the hepatic phase, there is a steady decrease in the size of the erythrocytes with a sharp drop from 180 to 130 fl between 12 and 24 weeks of gestation. Subsequently, the erythrocyte size declines more gradually to reach the mean cord blood value, 120 fl. Concommitantly, the number of erythrocytes, the hemoglobin concentration, and the volume of packed red cells increase steadily. The mean red cell count at 12 weeks of gestation is 1.5

$\times$ 10⁶ per cubic millimeter, whereas it is 3.5 $\times$ 10⁶ at 24 weeks of gestation. Similarly, by the 12th gestational week, the mean hemoglobin concentration and hematocrit are 9 g% and 33%, respectively, whereas by the 24th week of gestation, the hemoglobin has risen to approximately 14 g/dl and the hematocrit has risen to 40%.

The above data are primarily based on blood samples obtained from aborted fetuses or under sampling conditions which may have altered hemalogical values. A recent study using an electronic counting technique and fetal blood sample obtained *in utero* has yielded somewhat lower values for red cell count, hemoglobin, and hematocrit. Table 1 summarizes the hemoglobin concentration, hematocrit, and red cell values in 163 normal fetuses during pregnancy. Nucleated red blood cells comprise approximately 12% of erythroid cells in the peripheral blood at 18 to 20 weeks of gestation, declining gradually to about 4% by 26 to 30 weeks.[11] At term usually no more than 0.01 to 0.02% of circulating erythrocytes are nucleated. The reticulocyte count shows a similar pattern in that at 12 weeks of gestation, they approximate 40% of erythroid cells, declining to about 10% by 20 weeks and to approximately 5% at term.

The activity of hepatic hematopoiesis begins to decline at about 20 weeks postconception although it persists until shortly after birth. During the hepatic phase of erythropoiesis, there is evidence for significant erythropoiesis in the spleen. The relative contribution of splenic erythropoiesis, however, is probably small. In addition, other sites such as thymus and kidney may also contain foci of erythropoiesis.

Hematopoetic activity in the bone marrow begins during the third and fourth fetal months and progresses steadily. During the last trimester of gestation, the bone marrow is the main site of hematopoiesis. Marrow cellularity reaches its peak at about 8 gestational months, after which the cellularity remains constant, although the total volume occupied by hemopoietic tissue continues to increase.[12] In contrast to the hepatic phase in which erythroids are the predominant cells, the bone marrow contains erythroids, myeloids, and lymphocytes in equal distribution.

CONTROL OF ERYTHROPOIESIS DURING FETAL LIFE

The control of erythropoiesis in fetal life is not well understood. During the rapid growth of the fetus, there is an increased demand for erythropoiesis involving committment of stem cell compartments and maturation of terminally differentiated cells. Animal studies have revealed that a much higher proportion of pluripotent stem cells is in the cell cycle in the fetus than in adults.[13-15] It has been shown that the erythroid cells of the human fetus are responsive to erythropoietin. Erythropoietin has been detected in the cord blood of full-term and premature newborns.[16,17] Increased levels are found in hypoxic human newborns whether the hypoxia is due to fetal anemia, placental dysfunction, or maternal hypoxemia.[18] Animal studies have revealed that the control of fetal erythropoiesis is different from that seen in postnatal life. For instance, hypertransfusion of pregnant mice does not decrease fetal erythropoiesis.[19] Administration of large doses of erythropoietin to pregnant rats fails to stimulate fetal erythropoiesis.[20] It is not clear whether increased rate of erythropoiesis in the fetus is due to increased concentration of erythropoietin or to increased sensitivity of the cells to this hormone. In this context it is of interest to note that the cell cycle time of the erythroid progenitors (burst forming unit-erythroid, BFU-e, and colony forming unit-erythroid, CFU-e) in the fetus is significantly shorter than that in adults.[21,22] Various studies suggest that the major site of erythropoietin production in the fetus is in the liver.[23-25]

WHITE BLOOD CELLS AND PLATELETS PRODUCTION DURING FETAL LIFE

In contrast to erythropoiesis, myelopoiesis or lymphopoiesis is not detectable in the yolk sac. Myelopoiesis can be observed in fetal connective tissue as early as 7 weeks of gestation

Table 1
HEMOGLOBIN AND RED CELL VALUES IN 163 LIVE FETUSES OF 18 TO 30 WEEKS OF GESTATION

Weeks of gestation	RBC (× 10^{12}/l)	Hb (g/100 ml)	Ht (%)	MCV (fl)	MCH (pg)	MPHC (g/100 ml)	Red cell distribution with
18—20 (n = 25)	2.66 ± 0.29	11.47 ± 0.78	35.86 ± 3.29	133.92 ± 8.83	43.14 ± 2.71	32 ± 2.38	20.64 ± 2.28
21—22 (n = 55)	2.96 ± 0.26	12.28 ± 0.89	38.53 ± 3.21	130.06 ± 6.17	41.39 ± 3.32	31.73 ± 2.78	20.15 ± 1.92
23—25 (n = 61)	3.06 ± 0.26	12.40 ± 0.77	38.59 ± 2.41	126.19 ± 6.23	40.48 ± 2.88	32.14 ± 3.2	19.29 ± 1.62
26—30 (n = 22)	3.52 ± 0.32	13.35 ± 1.17	41.54 ± 3.31	118.17 ± 5.75	37.94 ± 3.67	32.15 ± 3.55	18.35 ± 1.67

Note: Abbreviation used: RBC, red blood count; Hb, hemoglobin concentration; Ht, hematocrit; MCV, mean corpuscular volume; MCH, MPHC.

From Forntier, F., Dallon, F., Galacteros, F., Bardakjian, J., Rainanut, M., and Benzard, E., *Pediatr. Res.*, 20, 342, 1986. With permission.

and begins in the bone marrow about 10 to 11 weeks of gestation. Between the 10th and 20th week of gestation, granulocytes and their precursors represent 30 to 40% of the cellular elements found in the bone marrow.[26] The number of circulating granulocytes during the first half of gestation is quite limited and does not exceed 1000 per cubic millimeter.[9,27] During the last trimester of pregnancy, the granulocyte count increases rapidly and at birth levels of 7000 per cubic millimeter in the absence of any apparent infection have been observed. Table 2 summarizes the white blood cell and differential counts in 163 normal fetuses. The studies were performed recently on blood samples obtained from live fetuses *in utero*.[11]

Lymphopoiesis has been detected in the fetal liver, lymph plexus, and thymus at the seventh week of gestation. Lymphocytes can be detected in the bone marrow and spleen at about the 20th week of gestation.[28] Circulating lymphocytes appear in fetal blood by the fifth week. Subsequently, their number increases rapidly and by the 20th week of gestation may reach 10,000 per cubic millimeter.[27] During the second half of gestation, the number of lymphocytes declines gradually and at birth it averages 3000 per cubic millimeter. T-lymphocytes are usually found at the seventh week of gestation. B-lymphocytes bearing sIgM develop between 8 and 10 weeks of gestation, initially within the fetal liver and from preexisting pools of pre-B-cells. Beginning at 10 to 12 weeks, subpopulations of these cells express different surface immunoglobin classes (isotypes) in addition to sIgM. The precise order in which isotype diversity emerges is difficult to ascertain. Usually cells bearing IgG are detected earlier than those having IgD or IgA. Usually by the 16th week of gestation, more than 90% of all lymphocytes in both the thymus and the blood have T or B characteristics.[29,30] Monocytes are usually seems between the fourth and fifth fetal month.[27,31]

Megakaryocytes can be detected in the yolk sac as early as the fifth week of gestation.[27,31] Subsequently, they could also be found in the liver until the end of the gestation period. After the third fetal month, megakaryocytes are almost always present in the bone marrow.[6] Circulating platelets can be detected in the blood by the 11th gestational week.[32] Beginning the third trimester, the number of circulating platelets is similar to that found in adults.[12] A most recent study[11] on blood samples from 163 live fetuses aged from 18 to 30 weeks confirms earlier reports (see Table 2).

HUMORAL IMMUNE SYSTEM

IgM production from fetal lymphopid tissue has been detected by 10 weeks.[33] IgG and IgA synthesis begins by 12 and 13 weeks, respectively.[34] IgG when passively transferred from the maternal circulation has been found in the fetus as early as the 38th day of gestation. The concentration of IgG is approximately 100 mg/dl until the 17th week of gestation at which time it increases proportionately to gestational age. At term the cord blood IgG concentration is usually greater than maternal IgG levels. After birth the half-life of maternal IgG in the circulation is about 25 d. IgA is not usually detectable until several days after birth, although its synthesis is initiated as early as 30 weeks of gestation. During severe perinatal infection, however, the fetus is able to produce a significant amount of IgM and IgA. The degree of IgM response is related to the time of infection, its severity, and the maturity of the infant. Since IgM does not cross the placenta, the presence of specific IgM antibodies demonstrates the presence of an intrauterine infection.

HEMOSTATIS IN THE FETUS

Studies on the platelets of a small number of fetuses at about 20 weeks of gestation have revealed a significant defect of aggregation in response to epinephrine and collagen when compared to platelets from adults.[35] Clotting of fetal blood has been noticed from about 11 to 12 weeks of gestation.[36] Fibrinogen synthesis has been noticed in cultures of

Table 2
PLATELETS, WBC, AND DIFFERENTIAL COUNTS IN 163 FETUSES OF 18 TO 30 WEEKS OF GESTATION

Weeks of gestation	Platelets (10^9/l)	WBC ($\times 10^9$/l)	Lymphocytes (%)	Neutrophils (%)	Eosinophils (%)	Monocytes (%)	Normoblasts (%)
18—20 (n = 25)	242.1 ± 34.48	4.20 ± 0.83	80 ± 9	5 ± 2	1.5 ± 2.5	1.5 ± 2	12 ± 8
21—22 (n = 55)	258.2 ± 53.65	4.19 ± 0.84	81 ± 7	5.5 ± 3.5	1 ± 1	1.5 ± 1.5	11 ± 7
23—25 (n = 61)	259.43 ± 42.45	3.95 ± 0.69	82 ± 6	7.5 ± 4.5	2 ± 2	1.5 ± 1.5	7 ± 4
26—30 (n = 22)	253.54 ± 36.6	4.44 ± 0.85	84 ± 6	8.5 ± 2.5	2 ± 1	1.5 ± 1	4 ± 3.5

Note: WBC, white blood cells.

Modified from Forntier, F., Dallon, F., Galacteros, F., Bardakjian, J., Rainanut, M., and Benzard, E., *Pediatr. Res.*, 20, 342, 1986.

Table 3
COAGULATION FACTORS OF 103 NORMAL FETUSES OF 19 TO 27 WEEKS OF GESTATION

Fetuses (weeks of gestation)	VIII C (%)	VIII RAg (%)	IX C (%)	V C (%)	II C (%)
19—21 (n = 51)	40 ± 12	59 ± 12.5	9 ± 2.5	39 ± 11	13 ± 4
22—24 (n = 44)	39 ± 13.5	64 ± 13	9 ± 3	40.5 ± 5	14 ± 2
25—27 (n = 44)	42.5 ± 12	63 ± 13	12 ± 4	39 ± 9	14 ± 3.5
Mothers[a]	160 ± 80	190 ± 110	90 ± 20	85 ± 10	95 ± 15

[a] The mean values for the respective mothers of the fetuses are also shown.

Modified from Forntier, F., Dallon, F., Galacteros, F., Bardakjian, J., Rainanut, M., and Benzard, E., *Pediatr. Res.*, 20, 342, 1986.

fetal liver as early as 5.5 weeks. The fibrinogen concentration in the fetuses of 12 to 14 weeks gestation is usually under 100 mg/dl. After approximately 6 months of gestation, mean fibrinogen levels are within the adult range. A lower limit of normal has been estimated at about 150 mg/dl at birth. Since clotting techniques for determination of fibrinogen have resulted in somewhat lower levels in the fetuses and neonates than those based on protein determination, it has been suggested that fetal fibrinogen may be structurally different from that in adults.[37] Indeed, it has been shown by chromatographic methods that fetal fibrinogen digests exhibit different peptides as compared to adult fibrinogen. It has further been found that the fetal fibrinogen is less susceptible to the action of thrombin, particularly at alkaline pH, than adult fibrinogen.[38,39] The difference between adult and fetal fibrinogen has been shown to lie in the A chain.[40]

The most reliable samples for assay of various coagulation factors in early fetal life are obtained by fetoscoptic sampling of pure fetal blood.[11,41-43] The concentration of various coagulation factors in 103 normal fetuses[11] is shown in Table 3. In this study, Factor VIII and Factor V activities were 40% and Factor II and IX were 10% of the normal adult's value. All vitamin K-dependent factors remain low in fetal life and decline further unless vitamin K is administered to the mother prior to the delivery or to the infant after birth.

As determined by euglobulin lysis time, there is significant fibrinolytic activity in the fetuses as early as 16 weeks of gestation. The activity of plasminogen activator remains elevated until birth when it falls rapidly to reach adult levels within 4 h and is associated with a steady prolongation of euglobulin lysis time. There is a progressive increase in the concentration plasminogen from 20% between 12 and 24 weeks of gestation to 43% at term. Inhibitors of plasminogen activation are above adult levels throughout the second half of fetal life and gradually decline to the adult range within 6 months after birth.

HEMOPOIESIS AT BIRTH

The hematologic values, primarily hemoglobin, hematocrit, and red cell count, at birth are affected by various factors such as gestational age, the time of clamping the umbilical cord vessels, and the site of the sampling. There is a significant discrepancy between the capillary samples obtained by skin prick as compared to a sample of blood collected by venipuncture. For instance, it has been shown[48] that during the first hour of life the hemoglobin concentration of capillary blood averaged 20.3 g/dl, while that from the jugular vein averaged 16.7 g/dl. In some newborns, the discrepancy was as high as 8 g/dl. Other investigators[49,50] have reported a difference of 5 to 10% between the capillary and venous samples obtained a few hours after birth. The difference between the capillary and the venous hemoglobin concentration persists until the fifth to sixth day of life. Gestational age has been found to contribute to the variability in the cord blood hemoglobin levels by some investigators,[51] particularly in infants born prior to 32 weeks of age.[52] However, in a subsequent study[53] using an electronic cell counter, no significant difference in hemoglobin, hematocrit, and red cell counts was observed in capillary samples taken on the first day of life from infants with gestational age ranging from 24 weeks to term. Postmature infants do not exhibit increased hemoglobin concentration.[54]

The time of sampling also plays an important role in hematologic values at birth. The hemoglobin concentration increases significantly during the first few hours after birth. The shrinking plasma volume and placental transfusion are responsible for this increase.

At birth approximately 25 to 30% of fetal blood volume is contained in the placental and cord vessels. However, at the end of the first minute, about one half of this is transfused into the newborn.[55,56] Delayed clamping of the cord therefore results in a higher red cell mass and hematocrit concentration. At 72 h of age, infants with immediate cord clamping have a red cell mass of 31 ml/kg of body weight, compared to 49 ml/kg in those with delayed cord clamping.[55] In a similar study, the hematocrit and red cell mass were, respectively, 15 and 32% greater in infants with delayed cord clamping.[57] The total blood volume of the infant rapidly adjusts after birth, decreasing in plasma volume, while red cell volume remains essentially unchanged.[57] In one study,[58] it was found that the hemoglobin concentration increased from 16.6 to 19.1 g/dl during the period of 8 h. At the same time, the plasma protein concentration increased from 6.5 to 7 g/dl. The magnitude of the increase in hemoglobin concentration resulting from plasma shrinkage can be ascertained in infants whose cord has been clamped immediately after birth.[59] It has been estimated that about 25% of the placental transfusion takes place within 15 s of birth and 50% takes place by the end of the first minute,[57,60] and even more rapid placental transfusion occurs in infants whose mothers have received ergotamin derivatives at the onset of the third stage of labor.[61] It has been shown that the hydrostatic pressure produced by placing the infant below the mother accelerates placental transfusion. Infants held above placenta level may bleed into it.[62] Consequently, it has been suggested to keep the infant born by Caesarean section 20 cm below the placenta for approximately 30 s before clamping the cord.[63]

In addition to the above factors, significant blood exchange between mother and fetus can further influence blood values at birth. It has been estimated that in as many as 50% of pregnancies some fetal cells pass into maternal circulation at some time during gestation or shortly before birth. In 10% of pregnancies, these losses range from 0.5 to 40 ml, while in 1% of pregnancies, losses as great as 100 ml have been observed.[64] Inversely, the passage of maternal blood into the fetus may cause plethora with hemoglobin concentrations as high as 26 g/dl.[65] The occurrence of such a phenomenon has been estimated at approximately 1%.[66] In addition to the above physiologic factors, various pathological conditions such as nutritional deficiencies and viral and protozoal infections in the mother can alter hematologic values in the infant.

Considering the above limitations, normal cord blood hemoglobin levels have been reported to range from 14 to 20 g/dl with a mean value of 16.8 g/dl.[63] It has been found that the hemoglobin levels in first-born infants are about 0.5 g higher than in second and subsequent births.[67-69] It has been suggested[70] that a hemoglobin concentration of 13.5 g/dl should be considered the lowest normal value.

The number of circulating erythrocytes, as the hemoglobin concentration, shows significant variation at birth. Several studies[53,71,72] have reported values between 4.6×10^6 to 5.2×10^6 per cubic millimeter. The red cell count also rises during the first few hours of life, but declines subsequently to cord blood levels by the end of the first week.

The erythrocytes at birth are significantly larger than those of normal adults and exhibit greater variability in their size. The average diameter of erythrocytes has been estimated to be between 8 and 8.5 μm at birth.[73-75] The diameter of the erythrocytes decreases usually

Table 4
HEMATOLOGICAL VALUES OF 63 CORD BLOOD SAMPLES AT BIRTH AND THE VENOUS BLOOD SAMPLES FROM THE RESPECTIVE MOTHERS OBTAINED 1 DAY AFTER DELIVERY

	Newborns (cord samples at delivery)	Respective mothers at the day of the blood sampling
WBC ($\times 10^9$/l)	11.11 ± 4.42	8.76 ± 0.48
RBC ($\times 10^{12}$/l)	3.56 ± 0.76	3.86 ± 0.05
Hb (g/100 ml)	13.29 ± 1.6	12.1 ± 0.17
Ht (%)	41.2 ± 6.02	36 ± 0.5
MCV (fl)	118.81 ± 14	93.9 ± 0.8
MCH (pg)	38.25 ± 9.83	31 ± 0.1
MCHC (g/100 ml)	34.99 ± 11.4	33.6 ± 0.2
Red cell distribution width	20.53 ± 1.4	13.6 ± 0.8
Platelets ($\times 10^9$/l)	299.5 ± 58.4	244 ± 10

Modified from Forntier, F., Dallon, F., Galacteros, F., Bardakjian, J., Rainanut, M., and Benzard, E., *Pediatr. Res.*, 20, 342, 1986.

to a normal adult size of 7.5 μm by approximately 6 months of age. Electronic counting methods yield similar values.[76] Relatively higher values have been found in premature infants.[77] The mean corpuscular hemoglobin is also elevated at birth (33.5 to 41.4 pg), whereas the mean corpuscular hemoglobin concentration is within the normal adult range (30 to 35%). It has been found[78,79] that the erythrocytes with greater volume rarely contain fetal hemoglobin, while the small cells (diameter of less the 6 μm) have a higher proportion of fetal hemoglobin. This finding is consistent with the knowledge that the largest and the youngest erythrocytes synthesize less fetal hemoglobin at the time of birth. The average value for MCV at birth in full-term infants has been estimated at 106.4 ± 5.7 fl.[76] Significantly higher values (115 ± 5 fl) have been measured in preterm infants.[77] In a recent study,[11] hematological values of 63 cord blood samples, obtained at birth, were compared to the venous blood samples from the respective mothers, obtained 1 d after delivery (Table 4).

The reticulocyte count at birth ranges from 4 to 7% with a mean of about 5%.[80,81] Infants born prematurely have a higher reticulocyte count. Values between 6 and 10% are frequently observed in these infants. The percentage of reticulocytes increases slightly during the first 2 to 3 postnatal days after which it tends to fall abruptly to values of about 1% by the 7th day of life. The presence of reticulocytosis beyond this period is indicative of a hemolytic process of blood loss. Nucleated red blood cells are invariably present in the blood at the time of birth and during the first day of life. The number of nucleated red cells in a full-term infant is approximately 500 per cubic millimeter. Toward the end of the 2nd day, not more than 30 nucleated red cells are usually present and by the 4th day of life, they are usually absent in the peripheral blood of the term infants. The number of nucleated red cells may be significantly higher in the premature infant. Levels as high as 1500 nucleated red cells per cubic millimeter have been observed in premature infants. The nucleated red cells disappear from the circulation within a few days regardless of gestational age. It would be unusual to see nucleated red cells in the peripheral blood smear after the first week of life. The presence of a large number of these cells beyond 7 d of life may be suggestive of a hemolytic process, hemorrhage, or hypoxia.

In addition to the above quantitative differences between the erythrocytes of the newborn infant and those of the adult, interference contrast microscopy studies of the red cell membrane have also shown morphological differences. Approximately 2% of the erythrocytes in adults appear to have cytoplasmic vacuoles which by this technique appear as pits and craters on the surface of the red cells. By contrast, approximately 50% of the red cells in premature infants and 25% in term infants have these abnormalities.[82] Similar abnormalities have been observed using scanning electron microscopy.[83,84]

Since these red cell defects have been observed in patients without spleen or nonfunctional spleen, such as in patients with sickle cell anemia,[85] their presence in the newborn infant may suggest reticuloendothelial or splenic dysfunction.

ERYTHROCYTE METABOLISM

There is now ample evidence that erythrocytes produced by the fetus are basically different from red cells produced in older infants and children.[86-88] The red cells of the newborn infants consume greater quantities of glucose than do cells from adults.[89,90] This may be due to the younger red cell population in the infant. The erythrocytes of premature infants, however, appear to consume less glucose than would be expected from the mean red cell age.[91] This apparent decrease in glucose consumption can be overcome by increasing the phosphate concentration.[92] Galactose consumption in newborn erythrocytes is significantly greater than in adult erythrocytes,[89] and the activity of galactokinase is approximately three times greater than in adult erythrocytes.[93] In the glycolytic pathway, numerous reports[94-97] have indicated that the activities of several enzymes such as glyceraldehyde, 3-phosphate dehydrogenase, phosphoglyceratekinase, phosphoglucose isomerase, and enolase are higher in the red cells of the newborn infant than would be anticipated from the young red cell age. The activity of the rate-limiting enzyme hexokinase, however, is similar to that of adult cells. The activity of another rate-limiting enzyme, phosphofructokinase, is about 70% of the activity present in normal adult erythrocytes.[98] Decreased phosphofructokinase activity of fetal erythrocytes has been ascribed to an accelerated decay of an unstable enzyme.[99] Qualitative differences in the isoenzymes have also been described in the red cells from newborn infants as compared to adult red cells.[100-103]

In the pentose phosphate pathway, the activity of glucose 6-phosphate dehydrogenase and 6-phosphogluconate dehydrogenase is higher in erythrocytes of the newborn as compared to those from adults.[104-107] Despite the high activity of the above enzymes, the red cells from newborn infants are more susceptible to oxidant-induced injury.[108] This is evidenced by reduced glutathione instability, development of Heinz bodies, and methemoglobinemia. In addition to the above NADPH generating enzymes, glutathione peroxidase and glutathione reductase play a major role in peroxide detoxification. It has been shown that the levels of glutathione peroxidase are inappropriately low for the metabolic age of erythrocytes of the newborn.[108-111] The levels of catalase and methemoglobin reductase are likewise reduced. Another factor which may further contribute to increased susceptibility of erythrocytes of the newborn to oxidant is decreased levels of membrane SH groups.[112]

The measurement of ATP in erythrocytes of the newborn has yielded conflicting results. Although some investigators have reported higher than normal adult levels of ATP of erythrocytes in the first several days of life,[113-115] others have been unable to confirm these findings.[116,117,118] Additionally, it has been found that, unlike erythrocytes of the adult, those from newborn infants cannot maintain their ATP levels during short periods of *in vitro* incubation[90] and that the incorporation of orthophosphate into ATP of erythrocytes of the newborn is much slower than that observed in adults.[119,120]

Lower levels of several nonglycolytic enzymes such as adenylate kinase, acetylcholinesterase and carbonic anhydrase have been reported.[121-125] The activity of type I and type II isoenzymes of carbonic anhydrase is 15 and 25% of normal adult values, respectively. The difference in carbonic anhydrase activity between fetal and adult erythrocytes has been exploited to produce selective hemolysis of the adult cells in an adult and fetal erythrocyte mixture. Red cells from adults lyse more rapidly when incubated with ammonium chloride.[126] The isoenzyme pattern of carbonic anydrase changes with gestational age and consequently it has been suggested that the determination of isoenzyme pattern may distinguish the small infants from those born prematurely.[127]

RED CELL MEMBRANE

Although structural analyses of red cell membrane proteins[12,129] have yielded no difference between red cells of adults and newborns, there are several functional differences distinguishing the red cell membrane of the newborn from that of the adult. These include increased K^+ leakage on storage,[130] a decreased level of ouabain-sensitive ATPase,[141] increased insulin binding sites,[132] and decreased digoxin receptor sites.[133]

One of the interesting features of the membrane of the erythrocyte in the newborn infant is the degree of lateral mobility which is not observed in the erythrocytes from adults. It has been demonstrated that the fetal erythrocyte can take up a variety of substances such as ferritin labeled anti-A antibodies[134] and concanavalin A[135] by endocytosis, whereas the erythrocytes from adults are unable to do so. It has been shown that the area in the membrane, where invagination and endocytic vesicles occur, lacks spectrin.[136] In addition, a large population of erythrocytes from preterm newborn infants is less filtrable than that of the normal adult.[137-139] This decreased deformibility of erythrocytes of the newborn may play a role in reduced red cell life span.[138]

Investigation dealing with developmental changes in membrane antigens has primarily involved the ABO, Lewis, P, and Ii antigens.[140] The A and B antigens are not fully developed at birth and reach adult values by the age of 2 to 4 years. The antigens of the Lewis group (Le^a, Le^b) are expressed weakly at birth, developing after a few weeks with Le^a appearing before Le^b.

The most pronounced developmental change in membrane antigens between neonatal and adult red cells occurs in the Ii system. In contrast to red cells of adults, those from newborns react strongly with anti-I and weakly with anti-I antibodies.[141]

The red cells of the newborn infant demonstrate increased mechanical fragility,[142,143] but the osmotic fragility is normal to slightly decreased. A small percentage of cells, however, may be more fragile.[142,144,145]

DIPHOSPHOGLYCERATE (2,3DPG) AND OXYGEN DELIVERY OF ERYTHROCYTES

It is well known that the oxygen dissociation curve of the fetal blood is left shifted. The P50 is approximately 6 to 8 mmHg lower than that in cells of adults. Such a difference in the P50 is the result of decreased affinity of 2,3DPG for hemoglobin F. Purified solutions of hemoglobin A and F exhibit similar oxygen dissociation curves. Such a differential binding of 2,3DPG may not solely be responsible for the preferential binding of the oxygen by the fetal cells *in vivo*. Other mechanisms such as difference in pH gradient across the placenta may also play an important role. Indeed, it has been shown that infants who have been transfused *in utero* with red cells of adults suffer no obvious deleterious effect.[146] Similarly, the fetuses of mothers with high oxygen affinity hemoglobin appear to be normal.[147,148] After birth the oxygen dissociation curve with increased synthesis of β-chain hemoglobin shifts progressively to the right.

BONE MARROW AND ERYTHROPOIETIN IN THE NEWBORN PERIOD

The bone marrow at birth is quite active. Nevertheless, the cellularity of the bone marrow decreases during the first week of life and then slowly attains normal adult levels between the first and third month of life.[149] Various estimates[149,150] indicate that on the first day of life the erythroid elements account for approximately 36% of all nucleated cells in the marrow, but by the end of the first week, the erythroid elements decline to approximately 10% of the total nucleated cells. The majority of the cells contained in the bone marrow are of myeloid origin. The erythroid to myeloid ratio drops from approximately 1.5:1 at birth to less than 1:6 by the first week of life. Normal adult values are usually established by approximately the second month of life.

The presence of erythropoietin activity in the plasma has been demonstrated at birth.[151,152] Higher levels have been found in term infants as compared to those born prematurely.[153,154] Erythropoietin activity disappears after the first day of life to reappear in the second and the third months.[152] It has been demonstrated that the fetus is able to react to hypoxia by the 32nd week of gestation with increased erythropoietin production. Elevated levels have been demonstrated in infants with anemia and hypoxia caused by congenital heart disease.[155,156]

BLOOD VOLUME AND ERYTHROCYTE SURVIVAL

In normal full-term infants during the first day of life, the mean blood volume is about 85 ml/kg 157 which is slightly higher than the mean adult value of 77 ml/kg. It should be pointed out, however, that the delayed clamping of the cord results in a higher value. The mean red cell volume is about 42 ml/kg. In premature infants, the blood volume per kilogram of body weight tends to be somewhat higher than in full-term newborns. This is largely due to the increased plasma volume.[158,159]

Erythrocyte survival has been measured in premature and full-term infants using a variety of techniques. The methods include differential agglutination, disappearance of fetal hemoglobin containing erythrocytes transfused into adults, and ^{51}Cr and [^{32}P] DFP tagged erythrocytes. Because of inherent technical difficulties, and because of the high proportion of young red cells in the cord blood, the above techniques have yielded variable results. In general, however, the erythrocytes from full-term and premature infants have a significantly shorter survival when compared to those from adults, ranging from 45 to 70 d.[160,162]

THE WHITE BLOOD CELLS

The number of polymorphonuclear neutrophils increases in both full-term and premature infants in the first 24 h of life. It stays relatively constant during the first few days and then begins to fall by the end of the first week. The site of sampling plays an important role. In general the white blood cell count obtained from the cord blood is approximately 70% of that obtained simultaneously from capillary samples.[163] The time of sampling is also very important, for the total neutrophil count rises during the first 12 h of life and then slowly declines to reach a plateau by the 4th day of life. The premature infants have slightly lower total white blood cell counts. The mean neutrophil count during the first 12 h of life is approximately 12,000 per cubic millimeter in infants born after 37 weeks of gestation as compared with the value of 8000 per cubic millimeter in infants with gestational ages of 32 to 37 weeks and 6000 per cubic millimeter in those born prior to the 32nd week of gestation.[164] Usually by the fourth and fifth days of life, these differences are no longer apparent. After approximately the first week of life in term infants, the lymphocytes become the predominant cells. This relative lymphocytosis is maintained until the fourth year of life. It is not uncommon to see immature white blood cells in the blood of newborn infants. These immature forms include promyelocytes and occasionally blast forms, although these are more often observed during infection. Because of significant variation in the neutrophil count immediately after birth, the total white blood cell count and differential are of limited clinical value for the diagnosis of neonatal infection.[165]

In neonates the specific response to chemotactic stimulus and random mobility of neutrophils is reduced.[166] It has been suggested that the basic defects resulting in these functional abnormalities are of membrane origin or defective microtubule assembly.[167] Neonatal neutrophils seem to have normal recognition, but exhibit defective opsinization. The reasons for inadequate opsinization are not clear; however, there is paucity of specific IgM antibodies.[168] It has been further shown that serum from neonates is deficient in several complement components.[168,169] Neutrophils from newborn infants under stress such as those with infection fail to adequately phagocytize.[170,171] The neutrophils from newborns reduce the nitroblue tetrazolium (NBT) with variable efficiency despite increased oxygen consumption.[172] Chemoluminescence, which is also an assay of neutrophil function, is similarly lower in newborn infants.[171]

PLATELETS

Normal newborn infants have platelet counts in the same range as normal older children and adults.[173,174] However, some reports have indicated that 14 to 15% of infants have platelet counts less than 150,000 per cubic millimeter during the first month of life.[175,176] In a study of 298 premature infants, approximately 2% had platelet counts of less than 100,000 per cubic millimeter. The platelet count has been found to be higher in the umbilical vein than in the umbilical artery[177] and tends to rise to the 300,000- to 400,000-cubic millimeter range during the first 3 to 4 weeks of postnatal life in both full-term and premature infants. Neonatal platelet counts of less than 100,000 per cubic millimeter should be regarded as abnormal even in premature infants.

THE HEMATOLOGIC PICTURE DURING THE FIRST MONTH OF LIFE

After birth there is a steady decline in the hemoglobin, hematocrit, and red cell count. The nadir erythropoiesis is reached at about 7 to 9 weeks. At that time, the hemoglobin concentration usually ranges between 9.5 to 11 g/dl and hematocrit ranges from 30 to 33%. The red cell count is approximately 3.4×10^6 per cubic millimeter.[178,179] Reticulocyte count is usually less than 1% after the first week of life. Concomitantly, the number of erythroid precursors in the bone marrow declines dramatically during the first week[180] and erythropoietin activity in the plasma becomes undetectable during this period. Shortened red cell life span, decreased bone marrow erythropoietic activity, and increased plasma volume due to rapid growth all contribute to the decline in hemoglobin concentration and the red cell mass. Such a physiological anemia, however, is compensated for by the reduced oxygen affinity of hemoglobin A which gradually replaces hemoglobin F. The newborn is able to respond to increased oxygen demand and hypoxic infants do not show signs of decreased erythropoietic activity.

The decline in the erythropoietic activity in premature infants during the first few weeks of life tends to occur earlier and more rapidly. Nadir counts for these infants are about 6.5 to 9 g/dl of hemoglobin.[178,181,182] Similarly, in the absence of iron deficiency, the premature infant is able to respond appropriately to hypoxemia by increasing their erythropoietic activity.

ERYTHROPOIESIS IN INFANCY AND CHILDHOOD

After having reached the nadir concentration by about 3 months after birth, the hemoglobin concentration begins to rise after 6 months to a mean level of approximately 12.5 g/dl in normal infants with adequate iron supply.[183] This level is maintained until about 2 years of age. It subsequently increases in both boys and girls.[184] At puberty, however, when the hemoglobin concentration reaches a plateau in girls, it continues to rise in boys.[184] Such a difference has been shown to be due to endogenous production of androgens.[185] Increased hemoglobin concentration in adult males is associated with an increase in total bone marrow erythroid activity. Indeed, after puberty the capacity for bone marrow erythroid cells to produce hemoglobin is far greater than in the prepubertal period.[186]

In addition to the alterations in hemoglobin concentration, there are significant changes in red cell indices in children in whom iron deficiency and thalassemia syndromes have been ruled out.[187] The MCV continues to fall during the first year of life with a mean level of 77 ± 7 fl in infants aged 10 to 17 months, rising slightly to 81 ± 5 fl in children between 2 and 7 years of age.

HEMATOPOIESIS AND AGING

It is the common belief that the total number of bone marrow stem cells (colony forming units, CFU) does not decline with age.[188-191] In aging rodent studies, however, the proliferative potential of CFU has been reported to decline,[192-196] although the relevance of such findings has been questioned.[197-199] For example, investigation of colony formation by marrow from young and old mice serially transplanted (every 12 d) into lethally irradiated young hosts revealed a decline in colony formation with each transfer that was greater with old marrow. Nevertheless, in heterochronic bone marrow transplant experiments, marrow cells from old donors function as well as those from young donors in reconstituting lethally irradiated recipient mice.[198] It has been shown, however, that the erythropoietic response after bleeding, which is initially brisk for both young and old mice, falls off more quickly in the old.[199] Also, long-term bone marrow cultures established from young donors maintained CFU-S (colony forming units-spleen) longer than those established from the old.[200] These observations are consistent with the hypothesis that diploid cells have only a limited capacity for *in vitro* proliferation and that this capacity decreases over the life span.[201] It can therefore be concluded that marrow stem cells are sustained in number throughout the life span, but most likely are reduced in proliferative capacity. This reduction is most notable in serial cell transfer experiments when the cells have far exceeded the life span of the normal host.

HEMATOLOGIC CONSEQUENCES OF AGING

Red Cells

Physical changes within the red cell coincide with aging; these changes are subtle, however, and may be of no clinical significance. Accordingly, mean cellular volume[202,203] and cell diameter[204] are slightly increased, as is osmotic fragility.[205] The survival of red cells, however, is not different in the aged.[206,207] In contrast, hemoglobin concentration and hematocrit do change significantly with age. In several studies, it has been shown that the hemoglobin concentration gradually falls.[208-212] The mechanisms for such a decline have not been completely determined, but it is noted that serum iron[213] and B_{12},[214] as well as plasma and red cell folate levels,[214,215] also decline with age. Age-associated decreased androgenic activity may also partially explain the decline in hemoglobin concentration. Despite the measurable decline in hemoglobin concentration with age, normal, healthy old people do not exhibit clinical manifestations of anemia.

Granulocytes

Minor changes in the granulocyte series are also observed with aging. The absolute number of granulocytes has been shown to fall slightly in older populations;[216,217] however, some studies show no difference. Granulocyte function, however, might be more significantly age-reduced. For example, the granulocyte response to bacteria pyrogen was found to be much less in elderly adults (over 70 years) when compared to young adults.[218] Other assessments of granulocyte reserve[219] have also shown some decline with age; however, granulocyte phagocytosis and killing are considered intact throughout the life span.[220]

Platelets

Platelets are quantitatively and morphologically unchanged throughout the adult years.[221] Occasional reports of age-associated platelet dysfunction have appeared;[222,223] however, the clinical significance of any of these observations has yet to be established.

Lymphocytes

Perhaps of all the formed elements in the bloodstream, the cell most affected by age is the lymphocyte. There has been some controversy regarding the nature of the age-related lymphocyte alterations. Several investigators have reported quantitative deficiencies in total lymphocyte pools[224,225] or T-cell fractions,[226] whereas others have not been able to confirm these changes.[227] Nevertheless, there is a striking decline in lymphocyte function in all mammalian species that can be first noticed shortly after sexual maturation with gradual progression thereafter throughout the adult years. Whereas quantitative changes are unimpressive, qualitative changes in lymphocyte function are primarily responsible for immune senescence.[228] The age-related lymphocyte dysfunction has been shown to correlate with thymus gland involution, and accordingly T-cell changes are most noticeable. For example, the proliferative response to T-cell mitogens (such as phytohemagglutinin)[229-231] or alloantigens (such as in a mixed lymphocyte reaction)[232] declines with age, as do lymphokine production[233,234] and the cytotoxic response after *in vitro* sensitization.[235] Clinically, such reductions in T-cell function are associated with reduced cell-mediated immunity, which may account for greater rates of infection, neoplasia, and other diseases. B-cell function, such as response to B-cell mitogens or antibody response, is less prominently involved.[236] In fact, the reduced antibody responses that have been reported with age[237,238] are generally in response to T-dependent antigens and may accordingly be attributed to deficient T-cell help.[239] With age, there is also increased autoantibody[240,241] and paraprotein[242,243] production of unknown clinical significance. Their occurrences are also observed in experimental animals and their incidence with age may be increased by further inhibiting T-cell immunity such as by thymectomy.[244] Therefore the age-associated occurrence of autoantibody or monocolonal protein may reflect disordered T-cell immunoregulation and not an intrinsic defect in B-cell function at all. There have been reports, however, that responses to certain T-independent antigens are also reduced with age, and it is now clear that B-cell function is not totally spared from the aging process.[245] Immunoglobulin levels peak during young adulthood, decline during the middle years, and rise again during the seventh and eight decades.[246] It has become apparent therefore that age-associated lymphocyte changes are complex. Total lymphocyte numbers are only mildly altered; however, shifts in lymphocyte subsets have been reported, and related functional disturbance resulting from these quantitative changes may be important. Age-related lymphocyte dysfunction, however, is well described, with the most profound age decline being in T-cell immunity. The resulting immune-deficient state may be causally related to age-associated diseases such as cancer and infections, as well as to seemingly unrelated degenerative conditions such as atherosclerosis, dementia, and diabetes.

BONE MARROW APPEARANCE AND AGING

Another change of practical importance in advanced years is that of bone marrow cellularity. Whereas hematopoietic tissue comprises nearly 80% of potential space in the anterior iliac crest early in life, this value falls to about 50% by the age of 30 years and is stable thereafter until age 60 years. After 60 years, however, anterior crest bone marrow cellularity falls to below 30% by the eighth decade.[247] Relative cellular composition of marrow from elderly shows no difference when compared to the young.[248,249]

REFERENCES

1. **Gilmour, J. R.**, Normal haemopoiesis in intrauterine and neonatal life, *J. Pathol.*, 52, 25, 1961.
2. **Maximox, A. A.**, Relation of blood cells to connective cells, tissue and endothelium, *Physiol. Rev.*, 4, 533, 1924.
3. **Bloom, W. and Bartelmez, G. W.**, Hematopoiesis in young human embryos, *Am. J. Anat.*, 67, 21, 1940.
4. **Wilt, F. H.**, The beginnings of erythropoiesis in the yolk sac of the chick embryo, *Ann. N.Y. Acad. Sci.*, 241, 99, 1976.
5. **Rifkind, R. A., Chui, D., and Epler, H.**, An ultrastructural study of early morphogenetic events during the establishment of fetal hepatic erythropoiesis, *J. Cell Biol.*, 40, 343, 1969.
6. **Wintrobe, M. M., Ed.**, *Clinical Hematology*, 7th ed., Lea & Febiger, Philadelphia, 1975, 53.
7. **Zamboni, L.**, Electron microscopic studies of blood embryogenesis in humans. II. The hemopoietic activity in the fetal liver, *J. Ultrastruct. Res.*, 12, 525, 1965.
8. **Djaldetti, M., Ovadia, J., Bessler, H., Fishman, P., and Halbrecht, I.**, Ultrastructural study of the erythropoietic events in human embryonic livers, *Biol. Neonate*, 26, 367, 1975.
9. **Thomm, D. B. and Yaffey, J. M.**, Human foetal haematopoiesis. I. Cellular composition of foetal blood, *Br. J. Haematol.*, 8, 290, 1962.
10. **Porcellini, A., Manna, A., Manna, M., Talvi, N., Delfini, C., Moretti, L., and Rizzoli, V.**, Ontogeny of granulocyte-macrophage progenitor cells in human fetus, *Int. J. Cell Cloning*, 1, 92, 1983.
11. **Forntier, F., Dallon, F., Galacteros, F., Bardakjian, J., Rainanut, M., and Benzard, E.**, Hematological values of 163 normal fetuses between 18 and 30 weeks of gestation, *Pediatr. Res.*, 20, 342, 1986.
12. **Kalpaktsoglou, P. K. and Emery, J. L.**, The effect of birth on the haemopoietic tissue of the human bone marrow, *Br. J. Haematol.*, 11, 453, 1965.
13. **Becker, A. J., McCulloch, E. A., Similnovitch, L., and Till, J. E.**, The effect of differing demands for blood cells production on DNA synthesis by hemopoietic colony forming cells of mice, *Blood*, 26, 296, 1965.
14. **Kubanek, B., Renricca, N., Porcellini, A., Howard, D., and Stohlman, F.**, The pattern of stem cell repopulation in heavily irradiated mice receiving transplants of fetal liver, *Blood*, 35, 65, 1970.
15. **Schofield, R.**, A comparative study of the repopulating potential of grafts from various haemopoietic sources: CFU repopulation, *Cell Tissue Kinet.*, 3, 119, 1970.
16. **Halvorsen, S.**, Plasma erythropoietin levels in the cord blood and in blood during the first weeks of life, *Acta Paediatr.*, 52, 425, 1963.
17. **Harlvorsen, S. and Finne, P. H.**, Erythropoietin production in the human fetus and newborn, *Ann. N.Y. Acad. Sci.*, 149, 576, 1968.
18. **Finne, P. H.**, Erythropoietin levels in cord blood as an indicator of intrauterine hypoxia, *Acta Paediatr.*, 55, 478, 1966.
19. **Jacobson, L. O., Goldwasser, E., Gurney, C. W., Fired, W., and Plzak, L.**, Studies of erythropoietin: the hormone regulating red cell production, *Ann. N.Y. Acad. Sci.*, 77, 551, 1959.
20. **Matoth, Y., and Zaizov, R.**, Regulation of erythropoiesis in the fetal rat, *Isr. J. Med. Sci.*, 7, 839, 1971.
21. **Hassan, M. W., Lutton, J. D., Levere, R. D., Rieder, R. F., and Cederqvist, L. L.**, In vitro culture of erythroid colonies from human fetal liver and umbilical cord blood, *Br. J. Haematol.*, 41, 477, 1979.
22. **Stamtoyannopoulos, G., Rosenblum, B. B., Papavannopoulou, T., Brice, M., Nakamoto, B., and Shepard, T. H.**, Hb F and Hb A production in erythroid cultures from human fetuses and neonates, *Blood*, 54, 440, 1979.
23. **Zajani, E. D., Peterson, E. N., Gordon, A. S., and Wasserman, L. R.**, Erythropoietin production in the fetus: role of kidney and maternal anaemia, *J. Lab. Clin. Med.*, 83, 281, 1974.
24. **Lucarelli, G., Porcellini, A., Carnevali, C., Carmena, A., and Stohlman, F.**, Fetal and neonatal erythropoiesis, *Ann. N.Y. Acad. Sci.*, 149, 544, 1968.
25. **Peschle, C. and Condorelli, M.**, Regulation of fetal and adult erythropoiesis, in *Congenital Disorders of Erythropoiesis*, CIBA Foundation Symp. 37, Elsevier, Excerpta Medica, North-Holland, Amsterdam, 1976, 25.
26. **Kelemen, E., Calvo, W., and Fliedner, T. M.**, *Atlas of Human Hemapoietic Development*, Springer-Verlag, Berlin, 1979.

27. **Playfair, J. H. L., Wofendale, M. R., and Kay, H. E. M.,** The leukocytes of peripheral blood in the human foetus, *Br. J. Haematol.,* 9, 336, 1963.
28. **Kyriazis, A. A. and Esterley, J. R.,** Development of lymphoid tissues in the human embryo and early fetus, *Arch. Pathol.,* 90, 348, 1970.
29. **Prindull, G.,** Maturation of cellular and humoral immunity during human embryonic development, *Acta Paediatr. Scand.,* 63, 607, 1970.
30. **Wylran, L., Carr, M. C., and Fudenberg, H. H.,** Effect of serum on human rosette forming cells in fetuses and adult blood, *Clin. Immunol. Immunopathol.,* 1, 408, 1973.
31. **Knoll, W.,** Der Gang der Erythropoeses beim menschlichen Embryo, *Acta Haematol.,* 2, 369, 1949.
32. **Bleyer, W. A., Hakami, N., and Shepard, T. H.,** The development of hemostasis in the human fetus and newborn infant, *J. Pediatr.,* 79, 838, 1971.
33. **Gitlin, D. and Blasucci, A.,** Development of γG, γA, γM, β_{1C}/β_{1A}, C'1 esterase inhibitor, ceruloplasmin, transferrin, hemopexin, haptoglobin, fibrinogen, plasminogen, α_1-antitrypsin, orosmucoid, β-lipoprotein, α_2-macroglobin, and prealbumin in the human conceptus, *J. Clin. Invest.,* 48, 1433, 1969.
34. **Boxer, L. A.,** Immunological function and leukocyte disorders in newborn infants, *Clin. Haematol.,* 7, 123, 1978.
35. **Pandolfi, M., Astedt, B., Cronberg, L., and Nilsson, I. M.,** Failure of fetal platelets to aggregate in response to adrenaline and collagen, *Proc. Soc. Exp. Biol. Med.,* 141, 1081, 1972.
36. **Zillicus, H., Ottelin, A. M., and Matsson, T.,** Blood clotting and fibrinolysis in human fetuses, *Biol. Neonat.,* 10, 108, 1966.
37. **Kunzer, W.,** Fetales fibrinogen, *Klin. Wochenschr.,* 39, 536, 1961.
38. **Witt, I., Muller, H., and Kunzer, W.,** Evidence for the existence of foetal fibrinogen, *Thromb. Diath. Haemorrh.,* 22, 101, 1969.
39. **Witt, I. and Muller, H.,** Phosphorus and hexose content of human fetal fibrinogen, *Biochem. Biophys. Acta,* 221, 402, 1970.
40. **Witt, I. and Tesch, R.,** Molecular characterization of human foetal fibrinogen, *Thromb. Haemostasis Abstr.,* 42, 79, 1979.
41. **Mibashan, R. S., Rodeck, C. H., Thumpston, J. K., Edwards, R. J., Singer, J. D., White, J. M., and Campbell, S.,** Plasma assay for fetal factors VIIIC and IX for prenatal diagnosis of haemophilia, *Lancet,* ii, 1309, 1979.
42. **Mibashan, R. S., Peake, I. R., Rodeck, C. H., Thumpston, J. K., Furlong, R. A., Gorer, R., Bains, L., and Bloom, A. L.,** Dual diagnosis of prenatal haemophilia A by measurement of fetal factor VIIIC and VIIIC antigen (VIIICAg), *Lancet,* ii, 994, 1980.
43. **Holmberg, L., Endrickson, P., Ekelund, H., and Astedt, B.,** Coagulation in the human fetus. Comparison with term newborn infants, *J. Pediatr.,* 85, 860, 1974.
44. **Ekelund, H., Hedner, U., and Nilsson, I. M.,** Fibrinolysis in human foetuses, *Acta Paediatr. Scand.,* 59, 369, 1970.
45. **Ekelund, H. and Finnstrom, O.,** Fibrinolysis in pre-term infants and in infants small for gestational age, *Acta Paediatr.,* 61, 185, 1972.
46. **Ekelund, H., Hedner, U., and Nilsson, I. M.,** Fibrinolysis in newborns, *Acta Paediatr. Scand.,* 59, 33, 1970.
47. **Ekelund, H.,** Fibrinolysis in the first year of life, *Acta Paediatr. Scand.,* 61, 5, 1972.
48. **Octtinger, L., Jr. and Mills, W. B.,** Simultaneous capillary and venous hemoglobin determinations in newborn infant, *J. Pediatr.,* 35, 362, 1949.
49. **Vahlquist, B.,** Das Serumeisen. Eine padiatrischklische and experimenteele Studie, *Acta Paediatr.,* 28, 1, 1941.
50. **Mollison, P. L.,** *Blood Transfusion in Clinical Medicine,* 3rd ed., Blackwell Scientific, Oxford, 1961, 614.
51. **Walker, J. L. and Turnbull, E. P. N.,** Haemoglobin and red cells in the human fetus and their relation to the oxygen content of the blood in the vessels of the umbilical cord, *Lancet,* ii, 312, 1953.
52. **Schulman, I.,** Characteristics of the blood in foetal life, in Walker, J. and Turnbull, A. C., Eds., *Oxygen Supply to the Human Fetus,* Charles C Thomas, Springfield, IL, 1959, 43.
53. **Zaizov, R., and Matoth, Y.,** Red cell values on the first postnatal day as a function of gestational age, *Am. J. Hematol.,* 1, 276, 1976.
54. **Rooth, G. and Sjostedt, S.,** Haemoglobulin in cord blood in normal and prolonged pregnancy, *Arch. Dis. Child.,* 32, 91, 1957.
55. **Usher, R., Shephard, M., and Lind, J.,** The blood volume of the newborn infant and placental transfusion, *Acta Paediatr.,* 52, 497, 1963.
56. **Yao, A. C., Lind, J., Thsala, R., and Michellson, K.,** Placental transfusion in the premature infant with observations on clinical course and outcome, *Acta Paediatr.,* 58, 561, 1969.
57. **McCue, L., Garner, F. B., Hurt, W. G., et al.,** Placental transfusion, *J. Pediatr.,* 72, 15, 1968.
58. **Gairdner, D., Ed.,** *Recent Advances in Paediatrics,* Little, Brown, Boston, 1958, chap. 2.

59. **Gairdner, D., Marks, J., Roscoe, J. D., and Brettell, R. O.**, The fluid shift from the vascular compartment immediately after birth, *Arch. Dis. Child.*, 33, 489, 1958.
60. **Yao, A. C., Moinian, M., and Lind, J.**, Distribution of blood between infant and placenta after birth, *Lancet*, 2, 871, 1969b.
61. **Yao, A. C., Hirvensalo, M., and Lind, J.**, Placental transfusion rate and uterine contraction, *Lancet*, 1, 380, 1968.
62. **Gunther, M.**, The transfer of blood between baby and placenta in the minutes after birth, *Lancet*, 1, 1277, 1957.
63. **Oski, F. A. and Naiman, J. L.**, *Hematologic Problems in the Newborn*, W. B. Saunders, Philadelphia, 1982.
64. **Cohen, F., Zuelzer, W. W., Gustafson, D. C., et al.**, Mechanisms of isoimmunization. I. The transplacental passage of fetal erythrocytes in homospecific pregnancies, *Blood*, 23, 621, 1964.
65. **Michael, A. F., Jr. and Mauer, A. M.**, Maternal-fetal transfusion as a cause of plethora in the neonatal period, *Pediatrics*, 28, 458, 1961.
66. **Andrews, B. F. and Thompson, J. W.**, Materno-fetal transfusion. A common phenomenon, *Pediatrics*, 29, 500, 1962.
67. **Guest, G. M. and Brown, E. W.**, Erythrocytes and hemoglobin of the blood, infancy and childhood, *Am. J. Dis. Child.*, 93, 486, 1957.
68. **Reinhardt, M. C. and Marti, H. R.**, Haematological data of African newborns and their mothers in Abidjan, *Helv. Paediatr. Acta*, 33(Suppl. 41), 85, 1978.
69. **Lind, T., Gerrard, J., Sheridan, T. S., and Walker, W.**, Effect of maternal parity and infant sex upon the haematological values of cord blood, *Acta Paediatr.*, 66, 33, 1977.
70. **Sisson, T. R. C.**, Blood hemoglobin levels in the neonatal period, *Q. Rev. Pediatr.*, 13, 124, 1958.
71. **Wegelius, R.**, On changes in peripheral blood picture of newborn infant immediately after birth, *Acta Paediatr.*, 35, 1, 1948.
72. **Lippman, H. S.**, Morphological and quantitative study of blood corpuscles in newborn period, *Am. J. Dis. Child.*, 27, 472, 1924.
73. **Saragea, T.**, Le diametre des hematies des l'homme aux differents ages de la vie, *C. R. Soc. Biol.*, 86, 312, 1922.
74. **Heissen, A. and Schalloer, R.**, Uber die Grossenverhaltnisse der roten Blutkorperchen bein Neugeborenen und Saugling, *Z. Kinderheilkd.*, 46, 105, 1928.
75. **Faxen, N.**, Red blood picture in healthy infants, *Acta Paediatr.*, 19, 1, 1937.
76. **Matoth, Y., Zaizov, R., and Varasano, I.**, Postnatal changes in some red cell parameters, *Acta Paediatr. Scand.*, 60, 317, 1971.
77. **Stockman, J. A., III and Oski, F. A.**, RBC values in low-birth-weight infants during the first seven weeks of life, *Am. J. Dis. Child.*, 134, 945, 1980.
78. **Breathnach, C. S.**, Red cell diameters in human cord and neonatal blood, *Q. J. Exp. Physiol.*, 47, 148, 1962.
79. **Komazawa, M., Garcia, A. M., and Oski, F. A.**, The relation of red cell size to fetal hemoglobin concentration in the term infant, *J. Pediatr.*, 85, 114, 1974.
80. **Seip, M.**, The reticulocyte level and the erythrocyte production judged from reticulocyte studies, in newborn infants during the first week of life, *Acta Paediatr.*, 44, 355, 1955.
81. **Zinkham, W. H.**, Peripheral blood and bilirubin values in normal full-term primaquine sensitive Negro infants: effects of vitamin K, *Pediatrics*, 31, 983, 1963.
82. **Holroyde, C. P., Oski, F. A., and Gardner, F. H.**, The "pocked" erythrocyte, *N. Engl. J. Med.*, 281, 516, 1969.
83. **Preston, F. E. and Shahani, R. T.**, Surface ultramicroscopy of neonatal erythrocytes, *Lancet*, 1, 1177, 1970.
84. **David, H., Gross, J., Verlings, I., et al.**, Ultrastrukturelle besonderheiten von roten blutzenne nuegeborener, *Folia Haematol.*, 100, 261, 1973.
85. **Pearson, H. A., McIntosh, S., Rooks, Y., et al.**, Interference phase microscopic enumeration of pitted RBC and splenic hypofunction in sickle cell anemia, *Pediatr. Rev.*, 12, 471, 1978.
86. **Hollan, S. R.**, On foetal-type erythropoiesis, *Haematologica*, 6, 185, 1972.
87. **Oski, F. A. and Komazawa, M.**, Metabolism of the erythrocytes of the newborn infant, *Semin. Hematol.*, 12, 49, 1975.
88. **Stockman, J. A., III, Clark, D. A., et al.**, Erythrocytes of the human neonate, *Curr. Top. Hematol.*, 1, 193, 1978.
89. **Lachhein, L., Grube, E., Johnigk, C., et al.**, Der Verbrauch an Glucose, Galaktose, Ribose and Inosin von Erwachsenen — und Nabelschnur-Erythrocyten, *Klin. Wochenschr.*, 39, 875, 1961.
90. **Oski, F. A. and Naiman, J. L.**, Red cell metabolism in the premature infant. I. Adenosine triphosphate levels, adenosine triphosphate stability, and glucose consumption, *Pediatrics*, 36, 104, 1965.

91. **Oski, F. A. and Smith, C.,** Red cell metabolism in the premature infant. III. Apparent inappropriate glucose consumption for cell age, *Pediatrics,* 41, 473, 1968.
92. **Bentley, H. P., Jr., Alford, C. A., Jr., and Diseker, M.,** Erythrocyte glucose consumption in the neonate, *J. Lab. Clin. Med.,* 76, 311, 1970.
93. **Ng, W. G., Donnell, G. N., and Bergren, W. R.,** Galactokinase activity in human erythrocytes of individuals at different ages, *J. Lab. Clin. Med.,* 66, 115, 1965.
94. **Gross, R. R., Schroeder, E. A. R., and Brounstein, S. A.,** Energy metabolism in the erythrocytes of premature infants compared to full term newborn infants and adults, *Blood,* 21, 755, 1963.
95. **Konrad, P. M., Valentine, W. N., and Paglia, D. E.,** Enzymatic activities and glutathione content of erythrocytes in the newborn: comparison with red cells of older normal subjects and those with comparable reticulocytosis, *Acta Haematol.,* 48, 193, 1972.
96. **Gahr, M., Meves, H., and Schroter, W.,** Fetal properties in red blood cells of newborn infants, *Pediatr. Res.,* 13, 1231, 1979.
97. **Travis, S. F., Kumar, S. P., Paez, P. C., et al.,** Red cell metabolic alternations in postnatal life in term infants: glycolytic enzymes and glucose-6-phosphate dehydrogenase, *Pediatr. Res.,* 14, 1349, 1980.
98. **Oski, F. A.,** Red cell metabolism in the newborn infant. V. Glycolytic intermediates and glycolytic enzymes, *Pediatrics,* 44, 84, 1969.
99. **Travis, S. F. and Garvin, J. H., Jr.,** In vivo lability of red cell phosphofructokinase in term infants: the possible molecular basis of the relative PFK deficiency in neonatal red cells, *Pediatr. Res.,* 11, 1159, 1977.
100. **Holmes, E. W., Jr., Malone, J. I., Winegrad, A., and Oski, F. A.,** Hexokinase isoenzymes in human erythrocytes. Association of Type II with fetal hemoglobin, *Science,* 156, 646, 1967.
101. **Chen, S. H., Anderson, J. E., et al.,** Lysozyme patterns in erythrocytes from human fetuses, *Am. J. Hematol.,* 2, 23, 1977.
102. **Gahr, M.,** Isoelectric focusing of hexokinase and glucose-6-phosphate dehydrogenase isoenzymes in erythrocytes of newborn infants and adults, *Br. J. Haematol.,* 4, 529, 1980.
103. **Witt, I. and Witz, D.,** Reinigung and characktersierung von Phosphopyruvate-hydraise (= Enolase; EC 4.2.1.11) aus neugeboren und erwachsenen Erythrozten, *Hoppe Seylers Z. Physiol. Chem.,* 351, 1232, 1970.
104. **Gross, R. T. and Hurwitz, R. E.,** The pentose phosphate pathway in human erythrocytes; relationship between the age of the subject and enzyme activity, *Pediatrics,* 22, 453, 1958.
105. **Zinkham, W. H.,** An in-vitro abnormality of glutathione metabolism in erythrocytes from normal newborns; mechanism and clinical significance, *Pediatrics,* 23, 18, 1959.
106. **Fois, A., Biagoli, M. L., and Contu, L.,** Comportamento della glucosio-6-fos fato deidrogenase'e del glutathione redotta negli eritrociti dei neonate a termine e dei prematuri, *Haematologica,* 46, 178, 1961.
107. **Witt, I., Muller, H., and Kunzer, W.,** Vergleichende biochemische Untersuchungen an erykthrocyten aus neugeborenen und erwachsenen Blut, *Klin. Wochenschr.,* 45, 262, 1967.
108. **Stocks, J., Offerman, E. L., Modell, C. B., and Dormandy, T. L.,** The susceptibility to autoxidation of human red cell lipids in health and disease, *Br. J. Haematol.,* 23, 713, 1972.
109. **Whaun, J. M. and Oski, F. A.,** Relation of red blood cell glutathione peroxidase to neonatal jaundice, *J. Pediatr.,* 76, 555, 1970.
110. **Bracci, R.,** II. Deficit de glutatione-perossidasi eritocitaria nel neonate con malattia emolitica, *Minerva Pediatr.,* 20, 2692, 1968.
111. **Bracci, R., Corvaglia, E., Princi, P., et al.,** The role of GSHK-peroxidase deficiency in the increased susceptibility to Heinz body formation in the erythrocytes of newborn infants, *Ital. J. Biochem.,* 18, 100, 1969.
112. **Schroter, W. and Bodemann, H.,** Experimentally induced cation leaks of the red cell membrane, *Biol. Neonate,* 15, 291, 1970.
113. **Gross, R. T., Schroeder, E. A. R., and Brounstein, S. A.,** Energy metabolism in the erythrocytes of premature infants compared to full term newborn infants and adults, *Blood,* 21, 755, 1963.
114. **DeLuca, C., Stevenson, J. H., and Kaplan, E.,** Simultaneous multiple-column chromatography: its application to the separation of the adenine nucleotides of human erythrocytes, *Ann. Biochem.,* 4, 39, 1962.
115. **Oski, F. A. and Naiman, J. L.,** Red cell metabolism in the newborn infant. I. Adenosine triphosphate levels, adenosine triphosphate stability and glucose consumption, *Pediatrics,* 36, 104, 1965.
116. **Zipursky, A., LaRue, T., and Israels, L. G.,** The in vitro metabolism of erythrocytes from newborn infants, *Can. J. Biochem.,* 38, 727, 1960.
117. **Greenwalt, T. J. and Ayers, V. E.,** The phosphate partition of the erythrocytes of normal newborn infants and of infants with hemolytic disease, *J. Clin. Invest.,* 35, 1404, 1956.
118. **Caruso, P., Conti, F., and Londrillo, A.,** Diagramma delle attivita Enzimaticke endoeritocitarie nel neonato, nel lattante, nel bambino, *Minerva Pediatr.,* 15, 1136, 1963.
119. **Zipursky, A., La Ruse, T., and Israels, L. G.,** The in vitro metabolism of erythrocytes from newborn infants, *Can. J. Biochem. Physiol.,* 38, 727, 1960.

120. **Greenwalt, J. T., Ayers, V. E., and Morell, S. A.**, Phosphate partition in the erythrocytes of normal newborn infants and infants with erythroblastosis fetalis. III. P 32 uptake and incorporation, *Blood*, 19, 468, 1962.
121. **Rudolph, N.**, Adenylate kinase activity in red cells of newborn infants, Program, Society for Pediatric Research, 1969, 160.
122. **Jones, P. E. H. and McCance, R. A.**, Enzyme activities in the blood of infants and adults, *Biochem. J.*, 45, 464, 1949.
123. **Burman, D.**, Red cell cholinesterase activity in infancy and childhood, *Arch. Dis. Child.*, 32, 362, 1961.
124. **Wehinger, H.**, Zur Natur und ontogenetischem Entwicklung von Carbonanhydrase-Isoenzymem in menschlichen Erythrozyten, *Blut*, 27, 172, 1973.
125. **Sell, J. E. and Petering, H. G.**, Carbonic anhydrases from human neonatal erythrocytes, *J. Lab. Clin. Med.*, 84, 369, 1974.
126. **Boyer, S. H., Noyes, A. N., et al.**, Enrichment of erythrocytes of fetal origin from adult-fetal blood mixtures via selective hemolysis of adult blood cells; an aid to antenatal diagnosis of hemoglobinopathies, *Blood*, 47, 883, 1976.
127. **Mondrukp, M., Anker, N., et al.**, Carbonic anhydrase isozyme determination: an aid to the diagnosis of "small-for-date" infants, *Clin. Chem. Acta*, 100, 107, 1980.
128. **Shapiro, D. L.**, Spectrin phosphorylation in intact neonatal and adult erythrocyte membranes, *Arch. Biochem. Biophys.*, 193, 264, 1979.
129. **Shapiro, D. L. and Pasqualini, P.**, Erythrocyte membrane proteins of premature and full-term newborn infants, *Pediatr. Res.*, 12, 176, 1978.
130. **Blum, S. F. and Oski, F. A.**, Red cell metabolism in the newborn infant. IV. Transmembrane potassium flux, *Pediatrics*, 43, 396, 1969.
131. **Whaun, J. and Oski, F. A.**, Red cell stromal adenosine triphosphatase (ATPase) of newborn infants, *Pediatr. Res.*, 3, 105, 1969.
132. **Kappy, M. S. and Plotnick, L.**, Studies of insulin binding in children using human erythrocytes in small amounts of blood, *Diabetes*, 28, 1001, 1979.
133. **Kearin, M., Kelly, J. G., et al.**, Digoxin "receptors" in neonates: an explanation of less sensitivity to digoxin than in adults, *Clin. Pharmacol. Ther.*, 28, 346, 1980.
134. **Haberman, S., Blanton, P., and Martiln, J.**, Some observations of the ABO antigen sites of the erthrocyte membranes of adults and newborn infants, *J. Immunol.*, 98, 150, 1967.
135. **Schekman, R. and Singer, S. J.**, Clustering and endocytosis of membrane receptors can be induced in mature erythrocytes of neonatal but not adult humans, *Proc. Natl. Acad. Sci. U.S.A.*, 73, 4075, 1976.
136. **Tokuyasu, K. T., Schekman, R., and Singer, S. J.**, Domains of receptor mobility and endocytosis in the membranes of neonatal human erythrocytes and reticulocytes are deficient in spectrin, *J. Cell Biol.*, 80, 481, 1979.
137. **Tillman, W., Wagner, D., and Schröter, W.**, Verminderte flexibiltat der erythrozykten von neurgeborenen, *Blut*, 34, 281, 1977.
138. Erythrocyte deformability in the fetus, preterm and term neonate, *Pediatr. Res.*, 20, 93, 1986.
139. **Buchan, P. C.**, Evaluation and modification of whole blood filtration in the measurement of erythrocyte deformability in pregnancy and the newborn, *Br. J. Haematol.*, 45, 97, 1980.
140. **Mollison, P. L.**, *Blood Transfusion in Clinical Medicine*, 5th ed., Blackwell Scientific, Oxford, 1972.
141. **Marsh, W. L.**, Anti-i: a cold antibody defining the Ii relationship in human red cells, *Br. J. Haematol.*, 7, 200, 1961.
142. **Sjolin, S.**, The resistance of red cells in vitro. A study of the osmotic properties, the mechanical resistance and the strong behavior of red cells of fetuses, children and adults, *Acta Pediatr.*, 43, 1, 1954.
143. **Goldbloom, R. B., Fischer, E., Reinhold, J., et al.**, Studies on the mechanical fragility of erythrocytes. I. Normal values for infants and children, *Blood*, 8, 165, 1953.
144. **Danon, Y., Kleinman, A., and Danon, D.**, The osmotic fragility and density distribution of erythrocytes in the newborn, *Acta Haematol.*, 43, 242, 1970.
145. **Luzzatto, L., Esan, G. L. F., and Oglemudia, S. E.**, The osmotic fragility of red cells in newborns and infants, *Acta Haematol.*, 43, 248, 1970.
146. **Novy, M. J., Frigoletto, F. D., Easterday, C. L., Umansky, I., and Nelson, N. M.**, Changes in cord blood oxygen affinity after intrauterine transfusions for erythroblastosis, *N. Engl. J. Med.*, 285, 589, 1971.
147. **Parer, J. T.**, Reversed relationship of oxygen affinity in maternal and fetal blood, *Am. J. Obstet. Gynecol.*, 108, 323, 1970.
148. **Charache, S., Jacobson, R., Brimhall, B., Murphy, E. A., Hathaway, P., Winslow, R., Jones, R., Rath, C., and Simkovich, J.**, Hb Potomac (101 Glu replaced by ASp): speculation on placental oxygen transport in carriers of high affinity hemoglobins, *Blood*, 51, 331, 1978.
149. **Gairdner, D., Marks, J., and Roscoe, J. D.**, Blood formation in infancy: normal erythropoiesis, *Arch. Dis. Child.*, 27, 124, 1952a.
150. **Shapiro, L. M. and Bassen, F. A.**, Sternal marrow changes during first week of life, *Am. J. Med. Sci.*, 202, 341, 1941.

151. **Halvorsen, S.**, Plasma erythropoietin levels in cord blood and in blood during the first weeks of life, *Acta Paediatr.*, 52, 425, 1963.
152. **Mann, D. L., Sites, M. D., Donati, R. M., et al.**, Erythropoietic stimulating activity during the first ninety days of life, *Proc. Soc. Exp. Biol. Med.*, 118, 212, 1965.
153. **Finne, P. H.**, Erythropoietin levels in cord blood as an indicator of intrauterine hypoxia, *Acta Paediatr. Scand.*, 55, 478, 1966.
154. **Stoutenborough, K. A., Sutherland, J. M., Meineke, H. A., et al.**, Erythropoietin levels in cord blood of control infants and infants with respiratory distress syndrome, *Acta Paediatr. Scand.*, 58, 121, 1969.
155. **Althoff, H., Dahm, P., and Werner, H.**, Presence of erythropoietin in umbilical cord blood, *Arch. Kinderheilkd.*, 157, 238, 1957.
156. **Jones, B. and Klingberg, W. G.**, A study of erythropoietin in 2 types of hemolytic anemia — erythroblastosis fetalis and sickle cells anemia, *J. Pediatr.*, 56, 752, 1960.
157. **Mollison, P. L., Veall, N., and Cutbush, M.**, Red cell and plasma volume in newborn infants, *Arch. Dis. Child.*, 25, 252, 1950.
158. **Schulman, I., Smith, C. H., and Stern, G. S.**, Studies on the anemia of prematurity, *Am. J. Dis. Child.*, 88, 567, 1954.
159. **Usher, R. and Lind, J.**, Blood volume of the newborn premature infant, *Acta Paediatr. Scand.*, 54, 419, 1965.
160. **Pearson, H. A.**, Life-span of the real red blood cell, *J. Pediatr.*, 70, 166, 1967.
161. **Bratteby, L. E., Garby, L., Groth, T., Schneider, W., and Wadman, B.**, Studies on erythro-kinetics in infancy. XIII. The mean life span and the life span frequency function of red blood cells formed during foetal life, *Acta Paediatr. Scand.*, 57, 311, 1968.
162. **Vest, M. F. and Grieder, H. R.**, Erythrocyte survival in the newborn infant, as measured by ^{51}chromium and its relationship to the postnatal serum bilirubin levels, *J. Pediatr.*, 59, 794, 1961.
163. **Dorros, G., Kleiner, G. J., and Romney, S. L.**, Fetal leukocyte pattern in premature rupture of amniotic membranes and in normal and abnormal labor, *Am. J. Obstet. Gynecol.*, 105, 1269, 1969.
164. **Coulombel, L., Dehan, M., Tchernia, G., et al.**, The number of polymorphonuclear leukocytes in relation to gestational age in the newborn, *Acta Paediatr. Scand.*, 68, 709, 1979.
165. **Manro, B. L., Weinberg, A. G., Rosenfeld, C. R., and Browne, R.**, The neonatal blood count in health and disease. I. Reference values for neutrophilic cells, *J. Pediatr.*, 99, 89, 1979.
166. **Vest, M. E.**, Chemotactic function in the human neonate. Humoral and cellular aspects, *Pediatr. Res.*, 5, 487, 1971.
167. **Boxer, L. A., Watanabe, A. M., Rister, M., et al.**, Correction of leukocyte function in Chediak-Higashi syndrome by ascorbate, *N. Engl. J. Med.*, 295, 1041, 1976.
168. **McCracken, G. H., Jr. and Eichenwald, H. F.**, Leukocyte function and the development of opsonic and complement activity in the neonate, *Am. J. Dis. Child.*, 121, 120, 1971.
169. **Feinstein, P. A. and Kaplan, S. R.**, The alternative pathway of complement activation in the neonate, *Pediatr. Res.*, 9, 803, 1975.
170. **Forman, M. L. and Stiehm, E. R.**, Impaired opsonic activity but normal phagocytosis in low-birth weight infants, *N. Engl. J. Med.*, 281, 926, 1969.
171. **Miller, M. E.**, Phagocytic function in the neonate: selected aspects, *Pediatrics*, 64, 709, 1979.
172. **Cocchi, P., Mori, S., and Becattine, A.**, NBT tests in premature infants, *Lancet*, 2, 1425, 1969.
173. **Sell, E. J. and Corrigan, J. J., Jr.**, Platelet counts, fibrinogen concentrations, and factor V and factor VIII levels in healthy infants according to gestational age. I, *Pediatrics*, 82, 1028, 1973.
174. **Alexandre, P., Andre, E., and Streiff, F.**, Etude sur sang capillaire de la coagulation du nouveau-ne sain premature, *Arch. Fr. Pediatr.*, 32, 417, 1975.
175. **Aballi, A. J., Puapondh, Y., and Desposito, F.**, Platelet counts in thriving premature infants, *Pediatrics*, 42, 685, 1968.
176. **Appleyard, W. J. and Brinton, A.**, Venous platelet counts in low birth weight infants, *Biol. Neonate*, 17, 30, 1971.
177. **Foley, M. E., Clayton, J. K., and McNicol, G. P.**, Haemostatic mechanisms in maternal, umbilical vein and umbilical artery blood at the time of delivery, *Br. J. Obstet. Gynaecol.*, 84, 81, 1977.
178. **O'Brien, R. T. and Pearson, H. A.**, Physiologic anemia of the newborn infant, *J. Pediatr.*, 79, 132, 1971.
179. **Gairdner, G., Marks, J., and Roscoe, J. D.**, Blood formation in infancy. II. Normal erythropoieses, *Arch. Dis. Child.*, 27, 214, 1952.
180. **Gairdner, D., Marks, J., and Roscoe, J. D.**, Blood formation in infancy. I. The normal bone marrow, *Arch. Dis. Child.*, 27, 128, 1952.
181. **Gairdner, D., Marks, J., and Roscoe, J. D.**, Blood formation in infancy. IV. The early anaemia of prematurity, *Arch. Dis. Child.*, 30, 203, 1955.
182. **Stockman, J. A.**, Anemia of prematurity, *Semin. Hematol.*, 12, 163, 1975.

183. **Saarinen, U. A. and Siimes, M. A.**, Development changes in red blood cell counts and indices of infants after exclusion of iron deficiency by laboratory criteria and continuous iron supplementation, *J. Pediatr.*, 92, 412, 1978.
184. **Dallman, P. R. and Siimes, M. A.**, Percentile curves for hemoglobin and red cell volume in infancy and childhood, *J. Pediatr.*, 94, 26, 1979.
185. **Shahidi, N. T.**, Androgens and erythropoiesis, medical progress, *N. Engl. J. Med.*, 189, 72, 1973.
186. **Shahidi, N. T. and Clatanoff, D.**, Role of puberty in red cell production in hereditary hemolytic anemias, *Br. J. Haematol.*, 17, 335, 1969.
187. **Koerper, M. A., Mentzer, W. C., Brecher, G., and Dallman, P. R.**, Developmental change in red blood cell volume: implication in screening infants and children for iron deficiency and thalassemia trait, *J. Pediatr.*, 89, 580, 1976.
188. **Chen, M. G.**, Age-related changes in hematopoietic stem-cell properties of a long-lived hybrid mouse, *J. Cell. Physiol.*, 79, 225, 1971.
189. **Coggle, J. E., Gordon, M. Y., Proukakis, C., and Bogg, C. E.**, Age-related changes in the bone marrow and spleen of SAS/14 mice, *Gerontologie*, 21, 1, 1975.
190. **Boggs, D. R., Saxe, D. F., and Boggs, S. S.**, Aging and hematopoiesis. II. The ability of bone marrow cells from young and aged mice to cure and maintain cure in W/W, *Transplantation*, 37, 300, 1984.
191. **Harrison, D. E.**, Proliferative capacity of erythropoietic stem cell lines and aging: an overview, *Mech. Ageing Dev.*, 9, 409, 1979.
192. **Harrison, D. E.**, Normal production of erythrocytes by mouse marrow continuous for 13 months, *Proc. Natl. Acad. Sci. U.S.A.*, 70, 3184, 1973.
193. **Yuhas, J. M. and Storer, J. B.**, The effect of age on two modes of radiation death and on hematopoietic cell survival in the mouse, *Radiat. Res.*, 32, 596, 1967.
194. **David, M. L., Upton, A. C., and Satterfield, L. C.**, Growth and senescence of the bone marrow stem cell pool in RFM/Un mice, *Proc. Soc. Exp. Biol. Med.*, 137, 1452, 1971.
195. **Chen, M. G.**, Impaired Elkind recovery in hematopoietic colony forming cells of aged mice, *Proc. Soc. Exp. Biol. Med.*, 145, 1181, 1974.
196. **Roylance, P. J., Hanna, I. R. A., and Tarbut, R. G.**, Changes with age in the cell proliferation of rat bone marrow, *J. Anat.*, 104, 191, 1969.
197. **Tyan, M. L.**, Age-related changes in mouse CFU-S regulation, in *Immunological Aspects of Aging*, Segre, Q. and Smith, L., Eds., Marcel Dekker, New York, 1983, 57.
198. **Harrison, D. E.**, Normal function of transplanted marrow cell lines from aged mice, *J. Gerontol.*, 30, 279, 1979.
199. **Harrison, D. E.**, Defective erythropoietic responses of aged mice not improved by young marrow, *J. Gerontol.*, 30, 286, 1975.
200. **Mauch, P., Botnick, L. E., Hannon, E. C., Obbagy, J., and Hellman, S.**, Decline in bone marrow proliferative capacity as a function of age, *Blood*, 60, 245, 1982.
201. **Hayflick, L.**, The cell biology of human aging, *N. Engl. J. Med.*, 295, 1302, 1976.
202. **Olbrech, O.**, Blood changes in the aged. I, *Edinburgh Med. J.*, 54, 306, 1947.
203. **Okunu, T.**, Red cell size and age, *Br. Med. J.*, i, 569, 1970.
204. **Spriggs, A. I. and Sladden, R. A.**, The influence of age on red cell diameter, *J. Clin. Pathol.*, 11, 53, 1958.
205. **Detraglia, M., Cook, F. B., Stasiw, D. M., and Cerny, L. C.**, Erythrocyte fragility in aging, *Biochim. Biophys. Acta*, 345, 213, 1974.
206. **Hurdle, A. D. F. and Rosin, A. J.**, Red cell volume and red cell survival in normal aged people, *J. Clin. Pathol.*, 15, 343, 1962.
207. **Woodford, W. E., Webster, D., Dixon, M. P., and MacKenzie, W.**, Red cell longevity in old age, *Gerontol. Clin.*, 4, 183, 1962.
208. **Smith, J. S. and Whitelaw, D. M.**, Hemoglobin values in the elderly, *Can. Med. Assoc. J.*, 105, 816, 1971.
209. **McLennan, W. J., Andrews, G. R., MacLeod, C., and Caird, F. I.**, Anaemia in the elderly, *Q. J. Med.*, 42, 1, 1973.
210. **Myers, A. M., Saunders, C. R. G., and Chalmers, D. G.**, The haemoglobin level of fit elderly people, *Lancet*, ii, 261, 1968.
211. **Jernigan, J. A., Gudat, J. C., Blake, J. L., Bowen, L., and Lezotte, D. C.**, Reference values for blood findings in relatively fit elderly persons, *J. Am. Geriatr. Soc.*, 28, 308, 1980.
212. **Lipschitz, D. A., Mitchell, C. O., and Thompson, C.**, The anemia of senescence, *Am. J. Hematol.*, 11, 47, 1981.
213. **Pirrie, R.**, The influence of age upon serum iron in normal subjects, *J. Clin. Pathol.*, 5, 10, 1952.
214. **Elwood, P. E., Shinton, N. K., Wilson, C. I. D., Sweetnam, P., and Frazer, A. C.**, Haemoglobin, vitamin B_{12} and folate levels in the elderly, *Br. J. Haematol.*, 21, 447, 1971.
215. **Girdwood, R. H., Thompson, A. D., and Williamson, J.**, Folate status in the elderly, *Br. Med. J.*, ii, 670, 1967.

216. **Caird, F. I., Andrews, G. R., and Gallie, T. B.,** The leukocyte count in old age, *Age Ageing,* 1, 239, 1972.
217. **Cruickshank, J. M. and Alexander, M. K.,** The effect of age, sex, parity, haemoglobin level and oral contraception preparations on the normal leukocyte count, *Br. J. Haematol.,* 18, 541, 1970.
218. **Timaffy, M.,** A comparative study of bone marrow function in young and old individuals, *Gerontol. Clin.,* 4, 13, 1962.
219. **Cream, J. J.,** Prednisone-induced granulocytosis, *Br. J. Haematol.,* 15, 259, 1968.
220. **Phair, J. P., Kauffman, C. A., Bjornson, A., Gallagher, J., Adams, L., and Hess, E. V.,** Host defenses in the aged: evaluation of components of the inflammatory and immune responses, *J. Infect. Dis.,* 183, 67, 1978.
221. **Shapleigh, J. B., Mayes, S., and Moore, C. V.,** Hematologic values in the aged, *J. Gerontol.,* 7, 207, 1952.
222. **Bankowski, E., Niewiarowski, S., and Galasinski, W.,** Platelet aggregation by human collagen in relation to its age, *Gerontologia,* 13, 219, 1967.
223. **Banargee, A. K., and Etherington, M.,** Platelet function in old age, *Age Ageing,* 3, 29, 1974.
224. **Diaz-Jouanen, E., Strickland, R. G., and Williams, R. C., Jr.,** Studies of human lymphocytes in the newborn and the aged, *Am. J. Med.,* 58, 620, 1975.
225. **MacKinney, A. A., Jr.,** Effect of aging on the peripheral blood lymphocyte count, *J. Gerontol.,* 33, 213, 1978.
226. **Augener, W., Cohnen, E., Reuter, A., and Brittinger, E.,** Decrease of T lymphocytes during aging, *Lancet,* 1, 164, 1974.
227. **Davey, F. R. and Huntington, S.,** Age-related variation in lymphocyte subpopulations, *Gerontology,* 23, 381, 1977.
228. **Weksler, M. E. and Hutteroth, T. H.,** Impaired lymphocyte function in aged humans, *J. Clin. Invest.,* 53, 99, 1972.
229. **Inkeles, B., Innes, J. B., Kuntz, M. M., Kadish, A. S., and Weksler, M. E.,** Immunological studies of aging. III. Cytokinetic basis for the impaired responses of lymphocytes from aged humans to plant lectins, *J. Exp. Med.,* 145, 1176, 1977.
230. **Del Pozo Perez, M. A., Prieto Valtuena, J., Gonzales Builabert, M. I., and Velasso Alonso, R.,** Effect of age and sex on T and B lymphocytic populations in man, *Biomedicine,* 19, 340, 1973.
231. **Pisciotta, A. V., Westring, D. W., DePrey, C., and Walsh, B.,** Mitogenic effect of phytohemagglutinin at different ages, *Nature (London),* 215, 193, 1967.
232. **Roberts-Thomson, I. C., Whittingham, S., Youngchaiyud, U., and Mackay, I. R.,** Ageing, immune response, and mortality, *Lancet,* 2, 368, 1974.
233. **Gillis, S., Kozak, R., Durantem, and Weksler, M. E.,** Immunologic studies of aging; decreased production and response to T cell growth factor by lymphocytes from aged humans, *J. Clin. Invest.,* 67, 937, 1981.
234. **Miller, R. A. and Studtman, O.,** Enumeration of IL-2 secreting helper T cells by limiting dilution analysis and demonstration of unexpectedly high levels of IL-2 production per responding cell, *J. Immunol.,* 128, 2258, 1982.
235. **McConnachie, P. R., Rachelefsky, G., Steiehm, E. R., and Terasaki, P. I.,** Antibody-dependent lymphocyte killer function and age, *Pediatrics,* 52, 795, 1973.
236. **Callard, R. E., Basten, A., and Waters, L. K.,** Immune function in aged mice. II. B-cell function, *Cell. Immunol.,* 31, 26, 1977.
237. **Ershler, W. B., Moore, A. L., Hacker, M. P., Ninomyia, J. T., Naylor, P. B., and Goldstein, A. L.,** Specific antibody synthesis *in vitro*. II. Age-associated thymosin enhancement of antitetanus antibody synthesis, *Immunopharmacology,* 8, 69, 1984.
238. **Ershler, W. B., Moore, A. L., and Socinski, M. A.,** Influenza and aging: age-related changes and the effects of thymosin on the antibody response to influenza vaccine, *J. Clin. Immunol.,* 4, 445, 1984.
239. **Price, G. G. and Makinodan, T.,** Immunologic deficiencies in senescence. I. Characterization of intrinsic deficiencies, *J. Immunol.,* 108, 403, 1972.
240. **Whittingham, S., Irwin, J., Mackay, I., Marsh, S., and Cowling, D. C.,** Autoantibodies in healthy subjects, *Aust. Ann. Med.,* 18, 130, 1969.
241. **Mackay, I. R.,** Aging and immunologic function in man, *Gerontologia,* 18, 285, 1972.
242. **Axelsson, U., Bachmann, R., and Hallen, J.,** Frequency of pathological proteins (M-components) in 6,995 sera from an adult population, *Acta Med. Scand.,* 179, 235, 1966.
243. **Hallen, J.,** Frequency of "abnormal" serum globulins (M-components) in the aged, *Acta Med. Scand.,* 173, 737, 1963.
244. **Radl, J., Hollander, C. F., Van den Berg, P., and DeGlopper, E.,** Idiopathic paraproteinaemia. I. Studies in an animal model, the aging C57Bl/KaLw Rig mouse, *Clin. Exp. Immunol.,* 33, 395, 1979.
245. **Callard, R. E. and Basten, A.,** Immune function in aged mice. IV. Loss of T cell and B cell function in thymus-dependent antibody responses, *Eur. J. Immunol.,* 8, 552, 1978.

246. **Hallgren, H. M., Buckley, C. E., III, Gilbertson, V. A., and Yunis, E. J.**, Lymphocyte phytohemagglutinin responsiveness, immunoglobulins and autoantibodies in aging humans, *J. Immunol.*, 111, 1101, 1973.
247. **Hartsock, R. J., Smith, E. B., and Petty, C. S.**, Normal variations with aging on the amount of hematopoietic tissue in bone marrow from the anterior iliac crest, *Am. J. Clin. Pathol.*, 43, 326, 1965.
248. **Reich, C., Swirsky, M., and Smith, D.**, Sternal bone marrow in the aged, *J. Lab. Clin. Med.*, 29, 508, 1944.
249. **Shapleigh, J. B., Mayes, S., and Moore, C. U.**, Hematological values in the aged, *J. Gerontol.*, 7, 207, 1952.

ONTOGENY OF HEMOPOIESIS

Mehdi Tavassoli

Embryonic hemopoiesis is a reflection of its evolutionary development. A brief treatment of the evolution of hemopoiesis is therefore useful prior to the exposition of its embryogenesis.

EVOLUTIONARY CONSIDERATIONS

Origin of Hemopoiesis

The least developed invertebrates do not have hemopoiesis as it is now commonly understood. Primitive blood cells, hemocytes, are produced of mesodermal origin,[1,2] in the hemocele during embryonic life. Since the life span of these organisms is relatively short, hemocytes remain with the organism throughout its life span. There is no need for sustained hemopoiesis. Hemocytes are derived from coelomocytes which are, in turn, derived from the epithelial lining of the coelomic cavity.[3]

In somewhat more developed invertebrates, such as worms and insects, these blood cells may undergo division,[4] usually in the primitive circulatory system. However, even these divisions are in the form of amitosis in which the nucleus incorporates tritiated thymidine and divides without the cytoplasmic division being consummated. In even more developed invertebrates, however, division of hemocytes within the circulatory system may take place and be considered the most primitive form of hemopoiesis, if only a very diffuse one.[4-6] Whether the proliferation is through the division of hemocytes or whether a pool of stem cell is present[6,7] is not known.

Segmented worms provide a cornerstone in the course of evolution of hemopoiesis. In less developed invertebrates, the formation of blood cells is a migratory phenomenon. While in segmented worms migratory hemopoiesis is still a dominant form, dense accumulations of cells resembling hemocytes (blood islands or lymphogenous organs) occur and are associated with the lateral vascular channels. The cells may or may not be sessile. These blood islands may be considered the most primitive form of *settlement* for hemopoiesis, to distinguish it from *migratory* hemopoiesis. Variations on this theme provide the "leukocytopoietic" or "lymphogenous" organs of higher invertebrates.

In arthropods, hemocytopoietic organs are still diffuse, but they have taken a more definitive and variegated form:[8-15] stomachal glands in decapods, the gland of Blanchard in the scorpion, and spleens of Kowalevsky in myriopods. The latter structures are diffuse, multiple, and segmentally distributed. Spleens of Kowalevsky are indeed another cornerstone in the evolution of hemopoiesis. These dispersed and scattered islands of hemopoiesis are the most primitive form of the spleen which, in Pisces, becomes quite compact, attaining the definitive shape of the spleen.

The first settlement nodes of hemopoiesis occur in the wall of the gastrointestinal (GI) tract and particularly in the submucosa.[16] Here is where the first blood islands originate. A factor that may render this site suitable for the task of hemopoiesis is its proximity to absorptive surfaces where necessary nutrients can readily be obtained. Another factor may be the vascularity of this region. The common denominator of all hemopoietic tissues is the presence of an extensive vascular system providing a very slow rate of flow (sinusoidal microcirculation). The presence of this system may be essential for optimal hemopoietic function.

Settlement of Hemopoiesis

Even when migratory blood cell formation has evolved into a more "settled" form of hemopoiesis, intravascular division of blood cells does not cease to exist.[17-19] The two phenomena may coexist. The intravascular proliferation of blood cells may show a circadian rhythm[20] and may account entirely for the maintenance of hemocytes in certain insects.[21,22] Circulatory space continues to serve as a site of hemopoiesis in all piscines and even amphibians, although the magnitude of its contribution to hemopoiesis is gradually reduced. In fact, even in the settlement sites of hemopoiesis (hemopoietic organs), much of blood formation is intravascular and blood cells are formed within the lumen of large venous sinuses. As the cells mature, they become detached and move into the circulation. This pattern, seen even in birds,[23] is a continuation of the pattern first seen in segmented worms with the accumulation of cells in the blood islands of the lateral vascular channels.

Leukocytes are the first cell types that are produced extravascularly and this occurs in the least developed Pisces studied, the hagfish. In almost all piscines, amphibians, and avians, extravascular leukopoiesis and intravascular erythropoiesis may be considered the rule. Ultimately, erythropoiesis also moves to the extravascular site and the pattern of mammalian hemopoiesis is attained.

Sites of Hemopoiesis

In Pisces and Amphibia, hemopoiesis is settled in several discrete sites. The relative contribution of these sites to blood formation varies with the species and depends on the hematopoietic cell type. In addition to the circulatory space, these sites include the spleen and central lymphoid organ, intertubular stroma of the kidney, intestinal submucosa, and to a lesser degree the liver and the gonads. The spleen is the product of the compaction of "blood islands". In invertebrates, spleens of Kowalevsky are predominantly erythropoietic. The central lymphoid organ is in fact a region of the spleen that is primarily lymphocytopoietic. This organ, too, is the product of the compaction of the lymphogenous organs in invertebrates. It is the combination of these two functions (and organs) that, in vertebrates, gives rise to the dual structure of spleen with its red and white pulps.

Hemopoiesis in the intestinal submucosa may be considered a vestigial variation of splenic hemopoiesis. When blood islands undergo compaction to form the spleen, a few islands may escape and appear as separate hemopoietic foci in the intestinal submucosa. When kidney, gonads, or liver partake in the function of hemopoiesis, they are predominantly leukocytopoietic. Table 1 summarizes the sites of hemopoiesis, with the cell lineages produced in several major piscine and amphibian species.

A benchmark in the evolution of hemopoiesis occurs in amphibians where bone marrow is initiated, albeit transiently, into the function of hemopoiesis. The efficiency of bone marrow in hemopoiesis leads it to supplant other hemopoietic tissues in more advanced species. Thus, while in amphibians the bone marrow is only transiently hemopoietic, in reptiles it is a major site and, in some species, it is even a dominant site (Table 2). Reptiles in general present an evolutionary intermediate between amphibians and birds.

In birds hemopoiesis is restricted to the bone marrow, where erythropoiesis is still intravascular, but granulopoiesis is extravascular. Here the spleen retains only its lymphocytopoietic function. Thus, in birds lymphocytopoiesis and hemocytopoiesis begin to segregate, B-lymphocytes are produced in the bursa of Fabricius, and moreover the presence of lymph nodes in water birds (the only birds having lymph nodes) heralds the appearance of this organ in mammals.

With the evolution of Mammalia, the marrow retains the function of hemopoiesis in adults almost to the exclusion of other sites. In lower mammals, the spleen still has a limited hemopoietic function, but in man and other primates, extramedullary hemopoiesis is pathologic.

EMBRYONIC AND FETAL DEVELOPMENT

During embryonic and fetal development, hemopoiesis is also a migratory phenomenon.

Table 1
SITES OF HEMOPOIESIS IN PISCES AND AMPHIBIA

	Sites of hemopoiesis	Cell lineages produced[a]
Pisces		
Hagfish (cyclostome)	Spleen (diffuse)[b]	E,T,G,L
	Circulatory space (by division)	E,T
Lamprey (cyclostome)	Spleen (compact)[b]	E,T,G,L
	Intestinal submucosa	E,G
	Circulatory space	E,T
Lungfish (Dipnoi)	Spleen	E,T,L
	Intestinal submucosa	E,G,L
	Kidney	G,L
	Circulatory space	E
Ganoid fish	Spleen (discrete)	E,T,L
	Kidney (intertubular) stroma	G,L
	Intestinal submucosa	G,L
	Gonads (subcapsular area)	G
	Circulatory space	E,T
	Meninges and bone marrow	See text
Dogfish (Elasmobranch)	Spleen	E,T,G,L
	Gonads[b]	G
	Kidney	E,T,G,L
	Liver	E,T
	Pancreas (periportal space)	G
	Intestinal submucosa	G,T
	Circulatory space	E,T
Trout (Teleost)	Kidney[b]	E,T,G,L
	Spleen	E,T,L
	Pancreas (periportal space)	G
	Intestinal submucosa	T,G,L
	Circulatory space	T
Amphibia		
Salamander (urodeles)	Spleen	E,T,L
	Intestinal submucosa	G,L
	Liver (subcapsular)	G,L
Frog (Anurans)	Bone marrow	See text
	Spleen	E,L
	Kidney	G
	Intestinal submucosa	G
	Circulatory space	T

[a] E, erythrocyte; G, granulocyte; T, thrombocyte; L, lymphocyte.
[b] Clearly dominant in hemopoietic function.

Table 2
SITES OF HEMOPOIESIS IN REPTILES AND BIRDS

	Sites of hemopoiesis
Reptiles	
Horned toad	Spleen dominant
Turtle	Spleen equal to marrow
Lizard	Marrow dominant
Birds	Restricted to marrow

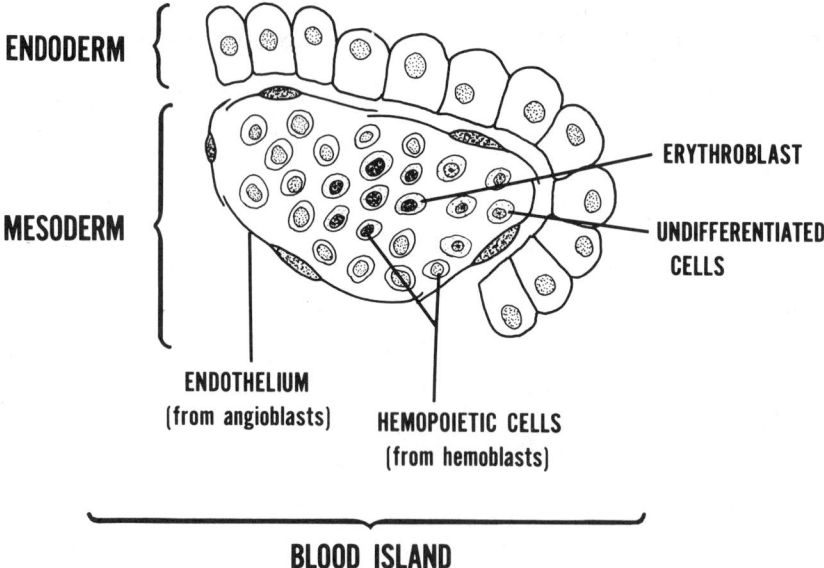

FIGURE 1. Diagrammatic representation of a blood island in the yolk sac. The island is adjacent to the endoderm which is required for the development of the island. The island itself consists of peripheral cells that are flattened to form the capillary endothelium (angioblasts) and central cells that are hemopoietic (hemoblasts). The latter cells are undifferentiated at the periphery, but they differentiate toward the center and become recognizable as erythroblasts. (From Tavassoli, M. and Yoffey, J., *Bone Marrow Structure and Function*, Alan R. Liss, New York, 1983. With permission.)

It actually begins in *extraembryonic* tissue, the yolk sac, then moves to *intraembryonic* sites, liver and spleen, and finally settles in the bone marrow.

In mammals, this migration continues in postnatal life by centripetal regression of hemopoiesis so that in adults of many species, including man, hemopoiesis is limited to the bones of the torso. The limb bones, except in their most proximal portions, are devoid of hemopoiesis.

This migration pattern is in some ways reminiscent of the evolutionary aspects of hemopoiesis, and it is often said that the ontogeny of hemopoiesis recapitulates its phylogeny.[24] This recapitulation is not exact. The two migration patterns differ in many respects, but they may be said to coincide in their essential features.

Embryogenesis of hemopoiesis is fundamentally comparable in all mammals. Thus, apart from the temporal frame, some generalization can be permitted. In all species, hemopoiesis begins in extraembryonic mesoblastic tissue and then migrates to intraembryonic tissues where it is again closely associated with the mesenchyme. It appears that all tissues of mesenchymal origin may be potentially hemopoietic under appropriate conditions, although these conditions are not always met and the potential is not always realized.

Yolk Sac Hemopoiesis

Hemopoiesis in the yolk sac occurs in distinct foci known as blood islands (Figure 1). In most species, only erythropoiesis is present as the environment of the yolk sac is restrictive in its inductive potential. The precursor cells responsible for the formation of these islands and capable of differentiating into erythroid cells are known as hemangioblasts. They migrate from the primitive streak region of the early blastoderm into the developing area opaca vasculosa.[25,26] They form a horseshoe-shaped region surrounding the posterior and posteriolateral region of area pellucida.[24,27] ^{3}H-thymidine-labeling studies have indicated that the cells destined to form blood islands invaginate through the primitive streak and then migrate laterally to produce the mesodermal layer.[28] These cells can differentiate in two directions (see Figure 1). The peripheral cells flatten to form capillary endothelial cells and are known as angioblasts whereas the central cells round up, develop intense basophilia, and become detached from peripheral cells. These cells are hemopoietic precursor cells and are true hemoblasts. These two lines of differentiation may occur together, but they are not necessarily associated: capillary endothelium can develop in the absence of hemopoiesis, and hemopoietic foci can be seen in the absence of endothelium. The current consensus maintains that the endothelium does not contribute to the differentiation of blood cells in the yolk sac[29,30] as was once thought to be the case.[31,32] The adjacent endoderm (see Figure 1), however, is required for the development of these islands.[33,34] In fact, the formation of blood islands in the mesoderm is the result of interaction of mesoderm and endoderm. The culture of experimentally separated mesoectoderm does not lead to the formation of many blood islands unless it is coincubated with the endoderm. This dependence of yolk sac hemopoiesis on endoderm and other morphological observations in human fetuses[35-39] has led some investigators to question the origin of blood cells from primitive mesenchyme, as ardently advocated by Maximow.[40,41]

Blood islands (see Figure 1) initially appear as thickening regions in the inner (mesodermal) layer of yolk sac, which is in contact with the extraembryonic endoderm. The undifferentiated hemangioblasts develop into an outer layer (in contact with the endoderm) of morphologically undifferentiated blast cells. The more centrally located cells soon become recognizable as erythroblasts (see Figure 1), which then undergo further maturation into nucleated red cells filled with embryonic hemoglobin and devoid of organelles. The nucleus, however, is not extruded, and the cells circulate in nucleated form.[42] Maturation may not be exactly synchronous, but apparently only a cohort of cells develops in the yolk sac,[43,44] and therefore yolk sac hemopoiesis is relatively homogeneous. *In vitro* studies indicate that yolk sac erythropoiesis cannot be stimulated by erythropoietin (EP).[45] This could mean either that yolk erythropoiesis is EP-independent or, less likely, that it is already maximally stimulated. With maturation, blood islands appear as large sinusoids filled with erythroid cells. These sinusoids then communicate with the circulation, and in fact, later stages of cell maturation take place within the circulation. In the mouse, whose gestation period is 21 d, yolk sac hemopoiesis begins at approximately the 8th day, proliferation proceeds from the eighth to the 10th day, and by the 9th day, cells begin to enter the circulation where they proliferate further. Mitosis may be observed through day 13 of gestation.[42] In man, with a gestation period of 38 weeks, yolk sac hemopoiesis is initiated between days 15 and 18 and continues for the first 6 weeks of gestation, then begins to decrease, and is undetectable by the 10th week (Figure 2).[46,47]

Primitive and Definitive Hemopoiesis

Primitive hemopoiesis differs from definitive hemopoiesis in that its products are restricted to the embryo and are not seen after birth. This difference is most easily distinguished in the case of erythroid tissue. Two characteristics make primitive erythropoiesis different from the definitive type: the end cell is nucleated, and the synthesized hemoglobin is of embryonic types. Moreover, the life span of the primitive erythroblasts appears to be shorter than that of definitive cells.[46,48,49] By contrast, red cells produced by definitive erythropoiesis undergo nuclear expulsion, thus forming end cells that are nonnucleated. They also synthesize hemoglobin types that are found in postnatal life (hemoglobins F, A, and A2), although in different proportions.[50] Post-yolk sac hemopoiesis is generally of definitive type. In man and most vertebrates, yolk sac hemopoiesis can embrace both primitive and definitive hemopoiesis. The mouse is an exception in which yolk sac hemopoiesis is highly restricted to primitive erythropoiesis.[42] The primitive type of hemopoiesis in the yolk sac does not depend on the

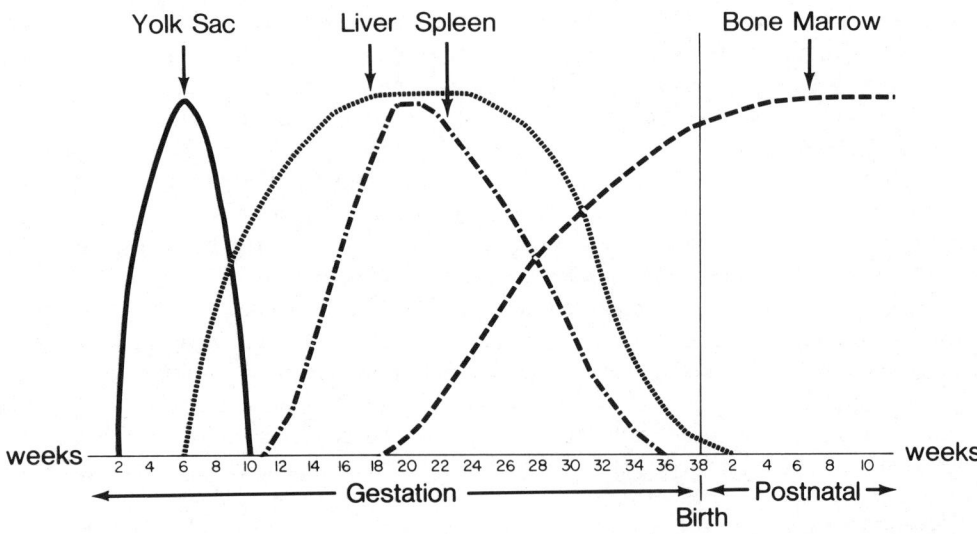

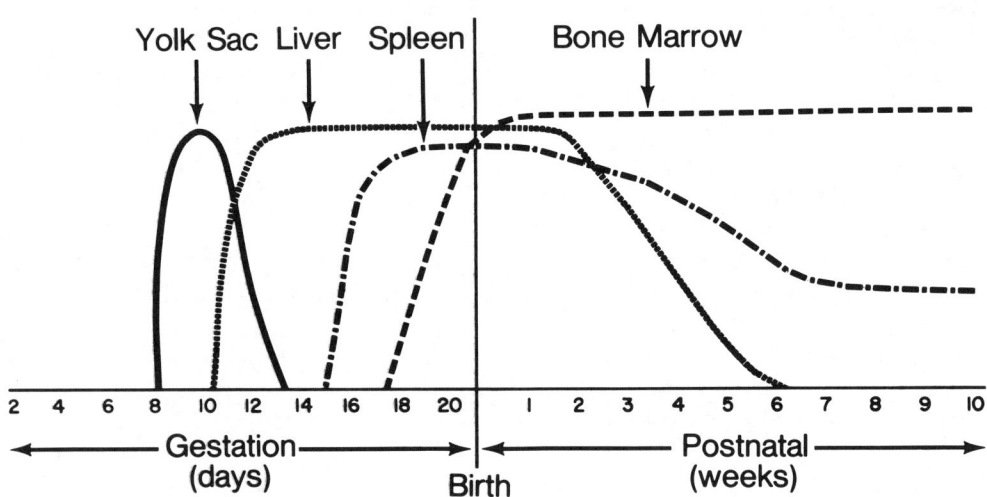

FIGURE 2. Sites of hemopoiesis during embryonic and postnatal life in man (top) and the mouse (bottom). (From Tavassoli, M. and Yoffey, J., *Bone Marrow Structure and Function*, Alan R. Liss, New York, 1983. With permission.)

presence of a normal intact embryo. This is evident from experiments using whole mouse embryos. In these cultures, the embryo can be induced to undergo degeneration while the yolk sac continues to grow and develops blood islands.[51] Thus, primitive hemopoiesis is strictly an extraembryonic phenomenon. By contrast, definitive hemopoiesis depends on the presence of a normal embryo. Recent evidence indicates that humoral factors, appearing in the embryo, act on hemopoietic stem cells to induce the onset of definitive hemopoiesis.[52]

The relationship between these two types of hemopoiesis has been the subject of some debate.[42] Are the primitive erythroid cells the precursors of definitive erythroid cells or are they both derived from a common precursor? Do the stem cells from the yolk sac migrate to seed the fetal liver or does the embryogenesis of stem cells proceed in the two tissues independently? There is no evidence that the primitive erythroid cells of the yolk sac actually seed the fetal liver and become involved in definitive erythropoiesis.[53] However, there is good evidence that both types of erythropoiesis may be the products of a common stem cell.[24] If the same stem cell can migrate between two compartments and differentiate to give rise to two different end products, one must assume that the two compartments provide different directives for the stem cell. In fact, there is some evidence to indicate that the environment of the yolk sac and liver control the differentiation of the stem cell.

There is good evidence to indicate that the yolk sac is the origin of all stem cell populations, both myeloid and lymphoid, that develop subsequently in the embryo and adult animal.[24] Thus, the mesenchymal cells, which differentiate into hemopoietic stem cells, migrate from one hemopoietic tissue to another, and the directive milieu of the hemopoietic tissue determines the subsequent path of development and consequently its end product. Accordingly, the origin of all stem cells is extraembryonic. There is, however, some evidence to indicate that stem cells can also arise intraembryonically.[54,55] By grafting cytogenetically labeled tissue anlagen in frogs, Turpen et al.[56,57] have recently demonstrated that the blood islands of the ventral mesoderm contribute to embryonic erythropoiesis which then declines. On the other hand, dorsal anterior mesoderm contributes to a population of stem cells that can give rise to different lineages of hemopoietic cells. Thus, in addition to the extraembryonic yolk sac stem cells, a population of embryonic cells may also be capable of colonizing hemopoietic tissues.

Liver Hemopoiesis

With the development of the liver, and when the formation of hepatic cords is in its definitive stages, hemopoiesis is transferred from the extraembryonic site to the intraembryonic liver. In the mouse, this occurs at day 10 of gestation, when the yolk sac erythropoiesis is at its peak. After hemopoiesis is settled in the new site, the yolk sac loses its hemopoietic cells. In man hemopoiesis is detectable in the liver at approximately 6 weeks of gestation,[48] expands exponentially for a few days (doubling time 8 hours), and then stabilizes with a doubling time of 2 d.[58] Liver remains hemopoietic during the entire fetal period and even during the first postnatal week,[59] although the magnitude of its hemopoietic activity is considerably reduced during the later part of this period (see Figure 2).

Hepatic hemopoiesis is initiated by the appearance of undifferentiated blast cells scattered in the liver cords.[60-62] This is followed by the appearance of erythroid cells and then megakaryocytes and granulocytic cells.[63] Erythropoiesis remains the dominant process with an erythroid-granulocytic ratio of 5:1. As discussed above, the origin of the precursors of these cells is probably extraembryonic, from the yolk sac, reaching the liver through a hematogenous route. Granulopoiesis and megakaryopoiesis indicate that the environment of the liver, unlike that of the yolk sac, is not restricted to the expression of erythropoiesis.

The extravascular site of hepatic erythropoiesis is reflected in the fact that its product is an anucleated red cell of 8 μm, somewhat larger than adult red cells (6 μm).[64] The endothelial barrier through which the nucleated cell must pass to enter the circulation can remove the nucleus.[65,67] The liver granulocyte-macrophage precursor cell (CFU-GM) is also larger than those in the marrow and homogeneous in volume,[68] indicating a single noncycling population.

Simultaneous with the appearance of hemopoiesis, stem cells become detectable in the liver. For a few days, this compartment enlarges to establish a pool. Although the size of this pool remains stable throughout the fetal life, the subsequent enlargement in the volume of hemopoietic tissue results in dilution and a relative fall in concentration of liver stem cells which remains fairly stable during the fetal life.[53,69-72] There is a slight increase in this concentration immediately before birth. With the loss of hemopoietic function in the liver, after birth, this concentration approaches zero.

There is evidence that stem cells in fetal liver differ somewhat from those of adult bone marrow in terms of cell density, size, and response to such hormones as erythropoietin.[73] These studies are, however, done in animal systems and little, if any, information exists with regard to humans.

Splenic Hemopoiesis

Embryologically, spleen originates from the thickening of mesenchymal tissue in the dorsal mesogastrium.[74,75] The development of spleen occurs in two distinct phases. In the first, the condensation of mesenchymal cells is interspersed with vascular spaces where the circulating blood comes into direct contact with mesenchymal cells. This corresponds to the primordial red pulp of spleen. The tissue is then "seeded" by hemopoietic cells to form hemopoietic areas. In the second phase, the white pulp develops through the formation of numerous circumscribed reticulum sheaths surrounding arterioles. These sheaths are then seeded by lymphocytes to form distinct demarcated nodules. The development of the white pulp and the lymphocyte seeding of the spleen are generally associated with a decline in its hemopoietic activity. The pattern of development is similar to the regenerative pattern of ectopic splenic implants during the postnatal life.[76] The latter recapitulates splenic embryogenesis. There is now good evidence based on cross-transplantation or experiments with parabiotic animals that in both systems the spleen is repopulated with hemopoietic cells or lymphocytes from the circulation.[24,77,78] In both the murine and human systems, the temporal appearance of hemopoiesis in spleen is subsequent to that in the liver. For instance, in the mouse embryo, liver hemopoiesis appears on day 10, whereas splenic hemopoiesis begins by day 15 (see Figure 2). A similar sequence is true for the loss of hemopoietic potential. Liver hemopoiesis regresses almost within 2 weeks after birth, whereas splenic hemopoiesis continues for a longer period and does not totally regress. In man, splenic hemopoiesis is detectable at approximately 12 weeks of gestation, reaches a peak by week 18, and declines thereafter. This decline coincides with the appearance of marrow hemopoiesis, but also with the development of the splenic white pulp and its lymphocyte seeding. It is not known which if any of these factors may contribute to the decline of splenic hemopoiesis. At any rate, splenic hemopoiesis ceases entirely by week 26 of gestation, long before the cessation of liver hemopoiesis. This is in contrast to the murine system in which splenic hemopoiesis does not cease entirely even after birth.

As in the case of liver, initiation of hemopoiesis is associated with the appearance of stem cells. However, the rate of expansion of stem cell compartments is not as rapid as in the liver (doubling time of 7 to 8 h in liver vs. 24 h in spleen), but more rapid than in the bone marrow.[73]

Bone Marrow Hemopoiesis

Bone marrow is the last site of hemopoiesis during the prenatal period. Development of marrow is generally thought to be the result of penetration of perichondrial mesenchymal cells, and their associated blood vessels, into the calcified zone of cartilage, which, in the tubular bones, is located in the central region of the shaft. This vascular mesenchyme then forms the reticular meshwork, which in hemopoietic marrow constitutes the frame upon which hemopoietic cells proliferate. The invasion from outside the calcified cartilage by the mesenchyme, although not uniformly observed,[38] is highly reminiscent of the development of bone marrow in ganoid fish described by Scharrer.[79] It is the invasion of vascular mesenchyme that leads to the resorption of the cartilage matrix and the establishment of the structural framework of the marrow. Nonetheless, organ culture studies indicate that there is a mutual interdependence between the vascular mesenchyme and the calcified cartilage or rudimentary bone.[73] The vascular mesenchyme is not hemopoietic before the invasion into the bone. On the other hand, the mesenchyme is necessary for the development of the rudimentary bone in organ culture. When the mesenchyme of the femoral rudiment is removed and the invasion of the bone and medullary cavity formation does not occur, the femoral rudiment degenerates.[80] However, when these two elements, rudimentary bone and the vascular mesenchyme, are cocultured, the femur develops a marrow cavity which can be induced to hemopoiesis by the addition of erythropoietin or thyroxin.[80]

In the mouse, marrow hemopoiesis appears by 17 to 18 d of gestation (see Figure 2). In man this is quite apparent at approximately 20 weeks.[48] Hemopoiesis is then initiated by the appearance of large numbers of undifferentiated basophilic cells within the dilated marrow sinuses. These are presumably hemopoietic stem cells which are populating the newly developed stromal meshwork. In the mouse embryo, marrow hemopoiesis is limited to granulopoiesis, and erythropoiesis does not appear until after birth. Granulopoiesis is entirely extravascular. Other mammals, such as rats, may share this pattern,[81] but in humans, embryonic marrow is erythropoietic as well. In the chick embryo, however, marrow undertakes both erythropoietic and granulopoietic function, erythropoiesis being intravascular, whereas granulopoiesis, as is the case with the mouse, is extravascular. The thin layer of marrow sinus endothelium separates these foci. This compartmentalization has been demonstrated after the injection of tritiated thymidine-labeled hemopoietic cells, which leads to the segregation of erythroid and granulocytic cells, respectively, to the intravascular and extravascular compartments.[24] This compartmentalized pattern (intravascular erythropoiesis, extravascular granulopoiesis) continues after hatching in birds. By contrast, in mammals, when the marrow becomes erythropoietic after birth, erythropoiesis switches to the extravascular compartment. The product of intravascular erythropoiesis in birds is nucleated red cells characteristic of the avian system. The product of extravascular erythropoiesis in mammals is anucleated red cells characteristic of the mammalian system. This relationship has been one of the bases of the formulation of the concept of a bone marrow-blood barrier.[67]

Evidence that the stromal meshwork is "seeded" by circulating hemopoietic stem cells is derived from sex chromosome studies in parabiosed chick embryos of opposite sex, demonstrating high reciprocal chimerism approaching 50% equilibrium level.[24]

The origin of the stem cells seeding the marrow in man is not known, but in chick embryo it is probably the fetal spleen and perhaps the yolk sac. In the mouse, fetal liver is thought to be the origin of the developing marrow stem cells. In fact, the rapid decline in the liver hemopoiesis is thought to be caused by mass migration of stem cells into the marrow.[53]

POSTNATAL DEVELOPMENT OF BONE MARROW

Not all of the bones, and consequently the bone marrow, are fully developed at the end of gestation. The development of marrow hemopoiesis, therefore, continues well into infancy and childhood. In man, the fully developed pattern is established only after puberty when the skeletal development is complete. In the interim, as the ossification of bones proceeds, the bone marrow cavity also develops and is seeded with hemopoietic stem cells and hemopoiesis begins.

In adults, on gross examination the bone marrow may appear red, indicating active hemopoiesis, or yellow, indicating hemopoietic inactivity. In the latter type of marrow, the yellow color is due to the presence of adipocytes that have obliterated active hemopoietic tissue. Actually adipocytes are seen in red marrow as well, but in fewer number so that they do not totally obliterate active hemopoiesis. The development of red and yellow marrow is the last step in the development of hemopoiesis.

At birth, all bones that contain marrow contain red marrow. Adipose tissue has not yet begun to develop. As adipose tissue development begins in the marrow, hemopoiesis undergoes regression in a centripetal direction so that in adults, the more peripherally located marrow consists entirely of adipose tissue (yellow marrow) and does not participate in hemopoiesis. This postnatal centripetal regression of active hemopoiesis is sometimes referred to as Neumann's law, after the investigator who first described it in 1882.[82] Yellow marrow can, under certain circumstances, become actively hemopoietic, particularly when the requirement of the body for hemopoiesis is enhanced.

There is some evidence that the centripetal regression of hemopoiesis may be the result of the development of a temperature gradient so that the temperature in the peripheral bones becomes suboptimal for hemopoiesis.[83] There is also some evidence that with the postnatal development of bones, the total marrow cavity is far more than the requirement of body for hemopoiesis and, therefore, part of the bone marrow cavity becomes hemopoietically inactive.[84,85] After the completion of skeletal development, the adult pattern of marrow is also completed with bones of the torso and the skull containing hemopoietically active red marrow, while bones of the limbs contain yellow marrow not participating in hemopoiesis.

REFERENCES

1. **Dorn, A.,** Ultrastructure of differentiating hemocytes in the embryo of *Oncopeltus fasciatus dallas* (Insecta, Heteroptera), *Cell Tissue Res.,* 187, 479, 1978.
2. **Cowden, R. R.,** Cytological and histochemical observations on connective tissue cells and cutaneous wound healing in the sea cucumber *Stichopus badionotus, J. Invest. Pathol.,* 10, 151, 1968.
3. **Cooper, E. L. and Stein, E. A.,** in Oligochaetis, *Invertebrate Blood Cell,* Ratcliffe, N. A. and Rowley, A. F., Eds., Vol. 1, Academic Press, New York, 1981, 75.
4. **Siminia, T.,** Haematopoiesis in the freshwater snail *Lymnaea stagnalis* studied by electron microscopy and autoradiography, *Cell Tissue Res.,* 150, 443, 1974.
5. **Siminia, T.,** Structure and function of blood and connective tissue cells of the fresh water pulmonate *Lymnaea stagnalis* studied by electron microscopy and enzyme histochemistry, *Z. Zellforsch. Microsk. Anat.,* 130, 497, 1972.
6. **Lie, K. J., Heyneman, D., Jeyarasasingam, U., et al.,** The life cycle of *Echinoparyphium ralphaudyi* sp. *n.* (Trematoda: Echinostomatidae), *J. Parasitol.,* 61, 59, 1975.
7. **Kinoti, G. K.,** Observations on the infection of bulinid snails with *Schistosoma mattheel.* II. The mechanism of resistance to infection, *Parasitology,* 62, 161, 1971.
8. **Jones, J. C.,** Hemocytopoiesis in insects, *Regulation of Hematopoiesis,* Vol. 1, Gordon, A. S., Ed., Appleton-Century-Crofts, New York, 1970, 7.
9. **Hoffman, J. A.,** The hemopoietic organs of the two orthopterans, *Locusta migratoria* and *Gryllus bimaculatus, Z. Zellforsch. Mikrosk. Anat.,* 106, 451, 1972.
10. **Hoffman, J. A.,** Blood-forming tissues in orthopteran insects: an analogue to vertebrate hemopoietic organs, *Experientia,* 29, 50, 1973.
11. **Nutting, W. L.,** A comparative anatomical study of the heart and accessory structures of the orthopteroid insects, *J. Morphol.,* 89, 501, 1951.
12. **Zachary, D. and Hoffman, J. A.,** The haemocytes of *Calliphora erythrocephala, Z. Zellforsch. Mikrosk. Anat.,* 141, 55, 1973.
13. **Brehelin, M.,** Presence d'un tissue hematopoietique chez le *Coleoptere melolontha, Experientia,* 29, 1539, 1973.
14. **Akai, H. and Sato, S.,** An ultrastructural study of the haemopoietic organ of the silk worm, *Bombyx mori, J. Invest. Physiol.,* 17, 1665, 1971.
15. **Monpeyssin, M. and Beaulaton, J.,** Hemocytopoiesis in the oak silkworm *Antheroae permyi* and some other Lepidoptera, *J. Ultrastruct. Res.,* 64, 35, 1978.
16. **Charmanier, M.,** Physiologie Des Invertebres — Effets de la regeneration intensive on de la presence d'une Sacculine sur la leucopoiese de *Pachygrapus marmoratus, C. R. Acad. Sci. Ser. D,* 276, 2553, 1973.
17. **Debaisieux, P.,** Histologie et histogenese chez *Chirocephalus diathanus prez. (Phyllopode anostrace), Cellule,* 54, 253, 1952.
18. **Debaisieux, P.,** Histologie et histogenese chez *Argulus foliaceus* L. (Crustace, branchiure), *Cellule,* 55, 245, 1953.
19. **Shapiro, M.,** Changes in the hemocyte population of the wax moth, *Galleria mellonella* during wound healing, *J. Insect Physiol.,* 16, 1725, 1968.
20. **Ravindranth, M. H.,** A comparative study of the morphology and behavior of granular haemocytes of arthropods, *Cytologia,* 42, 743, 1977.
21. **Jones, J. C. and Liu, D. P.,** A quantitative study of mitotic divisions of hemocytes of *Galleria mellonella* larvae, *J. Insect Physiol.,* 14, 1055, 1968.

22. **Jones, J. C. and Liu, D. P.**, Effect of ligaturing on total hemocyte counts of *Galleria mellonella* larvae, *J. Insect Physiol.*, 15, 1703, 1969.
23. **Campbell, F. R.**, Fine structure of the bone marrow of the chicken and pigeon, *J. Morphol.*, 123, 405, 1967.
24. **Metcalf, D. and Moore, M. A. S.**, *Haemopoietic Cells*, North-Holland, Amsterdam, 1971.
25. **Murray, P. D. F.**, The development in vitro of the blood of the early chick embryo, *Proc. R. Soc. (London) Ser. B*, 111, 497, 1932.
26. **Rudnick, D.**, Differentiation in culture of pieces of the early chick blastoderm, *Anat. Rec.*, 70, 351, 1938.
27. **Settle, G. W.**, Localization of the erythrocyte-forming areas in the early chick blastoderm cultivated in vitro, *Contrib. Embryol.*, 241, 223, 1954.
28. **Rosenquist, G. C.**, Radioautographic study of labeled grafts in the chick blastoderm. Development from primitive-streak stages to stage 12, *Contrib. Embryol.*, 262, 71, 1966.
29. **Edmonds, R. H.**, Areas of attachment between developing blood cells, *J. Ultrastruct. Res.*, 11, 577, 1964.
30. **Edmonds, E. H.**, Electron microscopy of erythropoiesis in the avian yolk sac, *Anat. Rec.*, 154, 785, 1966.
31. **Jordan, H. E.**, The microscopic structure of the yolk of the pig embryo with special reference to the origin of the erythrocytes, *J. Anat.*, 19, 227, 1916.
32. **Houser, J. W., Ackerman, G. A., and Knouff, R. A.**, Vasculogenesis and erythropoiesis in the living yolk sac of the chick embryo. A phase microscopic study, *Anat. Rec.*, 40, 29, 1961.
33. **Wilt, F. H.**, Erythropoiesis in the chick embryo: the role of endoderm, *Science*, 147, 1588, 1965.
34. **Miura, Y. and Wilt, F. H.**, The formations of blood islands in dissociated-reaggregated chick embryo yolk sac cells, *Exp. Cell Res.*, 59, 217, 1970.
35. **Thomas, D. B. and Yoffey, J. M.**, Human foetal haematopoiesis. I. The cellular composition of foetal blood, *Br. J. Haematol.*, 8, 290, 1962.
36. **Thomas, D. B. and Yoffey, J. M.**, Human foetal haematopoiesis. II. Hepatic haematopoiesis in the human foetus, *Br. J. Haematol.*, 10, 193, 1964.
37. **Yoffey, J. M., Thomas, D. B., Moffatt, D. E., Sutherland, I. H., and Rosse, C.**, *Biological Activity of Leukocytes*, Ciba Foundation Study Group #10, J & A. Churchill, London, 1961, 45.
38. **Yoffey, J. M. and Thomas, D. B.**, The development of bone marrow in the human foetus, *J. Anat.*, 98, 463, 1964.
39. **Yoffey, J. M.**, The stem cell problem in the fetus, *Isr. J. Med. Sci.*, 7, 825, 1971.
40. **Maximow, A.**, Experimentelle untersuchungen zur postfotalen histogenase des myeloiden gewebes, *Beitr. Z. Pathol. Anat.*, 41, 122, 1907.
41. **Maximow, A.**, Unter suchugen uber blut und bindegewebe. I. Die fruhesten entwicklungstadien der blut und bindegewezellen beim saugetier-embro, bis zum anfang der blutbildung in der leber, *Arch. Mikrosk. Anat.*, 73, 444, 1909.
42. **Marks, P. A. and Rifkind, R. A.**, Protein synthesis: its control in erythropoiesis, *Science*, 175, 955, 1972.
43. **Fantoni, A., Bank, A., and Marks, P. A.**, Globin composition and synthesis of hemoglobins in developing fetal mice erythroid cells, *Science*, 157, 1327, 1967.
44. **de la Chapelle, F. A. and Marks, P. A.**, Differentiation of mammalian somatic cells: DNA and hemoglobin synthesis in fetal mouse yolk sac erythroid cells, *Proc. Natl. Acad. Sci. U.S.A.*, 63, 812, 1969.
45. **Cole, R. J. and Paul, J.**, The effects of erythropoietin on haem synthesis in mouse yolk sac and cultured foetal liver cells, *J. Embryol. Exp. Morphol.*, 15, 245, 1966.
46. **Bloom, W. and Bartelmez, G. W.**, Hematopoiesis in young human embryos, *Am. J. Anat.*, 67, 21, 1940.
47. **Hesseldahl, H. and Falck, L. J.**, Hemopoiesis and blood vessels in human yolk sac. An electron microscopic study, *Acta Anat.*, 78, 274, 1971.
48. **Knoll, W. and Pingle, E.**, Der gang der erythropoese beim menschlichen embryo, *Acta Haematol.*, 2, 369, 1949.
49. **Maximow, A. A.**, Relation of blood cells to connective tissue and endothelium, *Physiol. Rev.*, 4, 533, 1924.
50. **Clarke, B. J., Nathan, D. G., Alter, B. P., Forget, B. G., Hillman, D. G., and Housemann, D.**, Hemoglobin synthesis in human BFU-E and CFU-E derived erythroid colonies, *Blood*, 54, 805, 1979.
51. **Chen, L. T. and Hsu, Y. C.**, Hemopoiesis of the cultured whole mouse embryo, *Exp. Hematol.*, 7, 231, 1979.
52. **Cudennec, C. A., Thiery, J. P., and LeDouarin, N. M.**, In vitro induction of adult erythropoiesis in early mouse yolk sac, *Proc. Natl. Acad. Sci. U.S.A.*, 78, 2412, 1981.
53. **Barker, J. E., Keenan, M. A., and Raphals, L.**, Development of the mouse hematopoietic system. II. Estimation of spleen and liver "stem" cell number, *J. Cell. Physiol.*, 74, 51, 1969.
54. **Dieterlen-Lievre, F.**, On the origin of haemopoietic stem cells in the avian embryo: an experimental approach, *J. Embryol. Exp. Morphol.*, 33, 607, 1975.
55. **Martin, C., Beaupain, D., and Dieterlen-Lievre, F.**, A study of the development of the hemopoietic system using quail-chick chimeras obtained by blastoderm recombination, *Dev. Biol.*, 75, 303, 1980.

56. **Turpen, J. B., Knudson, C. M., and Hoefen, P. S.**, The early ontogeny of hematopoietic cells studied by grafting cytogenetically labeled tissue anlagen: localization of a prospective stem cell compartment, *Dev. Biol.*, 85, 99, 1981.
57. **Turpen, J. B. and Knudson, C. M.**, Ontogeny of hematopoietic cells in *Rana pipiens*: precursor cell migration during embryogenesis, *Dev. Biol.*, 89, 138, 1982.
58. **Paul, J., Conkie, D., and Freshney, R. I.**, Erythropoietic cell population changes during the hepatic phase of erythropoiesis in the foetal mouse, *Cell Tissue Kinet.*, 2, 283, 1969.
59. **Borghese, E.**, The present state of research on W/W mice, *Acta Anat.*, 36, 185, 1959.
60. **Karrer, H. E.**, Electron microscope observations on chick embryo liver. Glycogen, bile canaliculi, inclusion bodies and hematopoiesis, *J. Ultrastruct. Res.*, 5, 116, 1961.
61. **Grasso, J. A., Swift, H., and Ackerman, G. A.**, Observations on the development of erythrocytes in mammalian fetal liver, *J. Cell Biol.*, 14, 235, 1962.
62. **Thomas, D. B. and Yoffey, J. M.**, Hepatic haematopoiesis in the foetus, *Br. J. Haematol.*, 10, 193, 1964.
63. **Mrlevic, D., Brown, J. S. and Howlott, B.**, The nucleus supragenualis, presumptive superior salivatory nucleus, *Acta Anat.*, 76, 35, 1970.
64. **Russell, E. S. and Bernstein, S. E.**, Blood and blood formation, in *Biology of the Laboratory Mouse*, Green, E., Ed., McGraw-Hill, New York, 1966, 351.
65. **Tavassoli, M. and Crosby, W. H.**, Fate of the nucleus of marrow erythroblast, *Science*, 173, 912, 1973.
66. **Tavassoli, M.**, Red cell delivery and the function of marrow-blood barrier, *Exp. Hematol.*, 6, 257, 1978.
67. **Tavassoli, M.**, The marrow-blood barrier, *Br. J. Haematol.*, 41, 297, 1979.
68. **Symann, M., Quesenberry, P. L., Fontebuoni, A., Howard, F., and Stohlman, F.**, Fetal hemopoiesis in diffusion chamber cultures. I. The pattern of pluripotent stem cell growth, *Cell Tissue Kinet.*, 9, 41, 1976.
69. **Silini, G., Andreozzi, U., and Pozzi, L.**, The role of the spleen in the repopulation of the haemopoietic system of heavily-irradiated mice, *Cell Tissue Kinet.*, 9, 341, 1976.
70. **Duplan, J. F.**, Efficacies therapeutiques comparees des injections de foie foetal or de moelle asseuse isogenique chez les souris irradiees, *C. R. Acad. Sci. Ser. D*, 267, 227, 1968.
71. **Moore, M. A. S. and Owen, J. J. T.**, Chromosome marker studies on the development of the haemopoietic system in the chick embryo, *Nature (London)*, 208, 956, 1965.
72. **Moore, M. A. S. and Metcalf, D.**, Ontogeny of the haemopoietic system: yolk sac origin of in vivo and in vitro colony forming cells in the developing mouse embryo, *Br. J. Haematol.*, 18, 279, 1970.
73. **Tavassoli, M. and Yoffey, J.**, *Bone Marrow Structure and Function*, Alan R. Liss, New York, 1983.
74. **Bloom, W.**, Embryogenesis in mammalian blood, in *Handbook of Hematology*, Vol. 2, Downey, H., Ed., Hoeber, New York, 1938, 865.
75. **Klemperer, P.**, The spleen, in *Handbook of Hematology*, Vol. 3, Downey, H., Ed., Hoeber, New York, 1938, 1587.
76. **Tavassoli, M., Ratzan, R. J., and Crosby, W. H.**, Studies on regeneration of heterotopic splenic transplants, *Blood*, 41, 701, 1973.
77. **Tavassoli, M., Ratzan, R. J., Maniatis, A., and Crosby, W. H.**, Regeneration of hemopoietic stroma in anemic mice of Sl/Sl^d and W/W^v genotype, *J. Reticuloendothel. Soc.*, 13, 518, 1973.
78. **Levy, S. B., Rubenstein, C. B., and Tavassoli, M.**, The spleen in Friend leukemia. II. Nonleukemic nature of spleen stroma, *J. Natl. Cancer Inst.*, 56, 1189, 1976.
79. **Scharrer, E.**, The histology of the meningeal myeloid tissue in the ganoids amia and lepisosteus, *Anat. Rec.*, 88, 291, 1944.
80. **Petrakis, N. L., Pons, S., and Lee, R. E.**, An experimental analysis of factors affecting the localization of embryonic bone marrow, *In Vitro*, 4, 3, 1969.
81. **Lucarelli, G., Ferrari, L., Porcellini, A., Carnevali, C., Rizzoli, V., and Stohlman, F.**, Fetal and neonatal hemopoietic tissue in the rat, *Exp. Hematol.*, 14, 7, 1967.
82. **Neumann, E.**, Das Gesetz der Verbreilung des gelben und roten Knochenmarkes, *Cent. Med. Wiss.*, 20, 321, 1882.
83. **Huggins, C. and Blocksom, B. H.**, Changes in outlaying bone marrow accompanying a local increase of temperature within physiological limits, *J. Exp. Med.*, 64, 253, 1936.
84. **Bigelow, C. L. and Tavassoli, M.**, Studies on conversion of yellow to red marrow using ectopic bone marrow implants, *Exp. Hematol.*, 12, 581, 1984.
85. **Tavassoli, M.**, Marrow adipose cells and hemopoiesis, an interpretative review, *Exp. Hematol.*, 12, 139, 1984.

AGING OF THE RED CELL: METABOLIC CHANGES DURING DEVELOPMENT AND SENESCENCE

Nancy A. Noble

INTRODUCTION

The purpose of this chapter is to review studies of metabolic changes in red cell metabolism both during development of the animal and during development and aging of the red cell in adult animals. Firstly, we will cover the main metabolic pathways of mature red cells and their functions. Then, comparative studies of fetal, newborn, and adult red cells will be reviewed. Finally, metabolic changes during erythropoiesis, reticulocyte maturation, and red cell senescence will be discussed. The emphasis will be primarily on cytosolic metabolites and proteins. Little will be said about red cell membranes and hemoglobin since they are covered in other sections of this book.

MAJOR METABOLIC PATHWAYS IN MATURE ERYTHROCYTES

Figure 1 shows the major metabolic pathways of mature red cells. These include the Embden-Meyerhof pathway, the pentose phosphate pathway, the Rapaport-Luebering shunt, and glutathione metabolism. In addition, the pathway by which the cell deals with toxic products of oxygen reduction is shown. An extensive list of enzymes present in mature red cell is given in Reference 1.

It should be noted that mature red cells contain no mitochondria and therefore no oxidative phosphorylation. They have no nuclei and cannot synthesize DNA or RNA. Therefore, they cannot divide or synthesize proteins. Glycolysis is the only source of energy which is produced in the form of ATP. Glucose enters human red cells by facilitated transport. However, normal red cells survive for about 120 d in the circulation and thus the metabolic pathways present are adequate to maintain the cell as a functional carrier of oxygen and carbon dioxide.

Embden-Meyerof Pathway

Under normal conditions, 89 to 97% of all glucose is metabolized by red cells via the Embden-Meyerof pathway.[2] The Embden-Meyerof pathway, or glycolysis, serves three vital functions. First, it provides energy in the form of ATP. ATP is necessary for the operation of the sodium-potassium pump. This pump maintains the deformability of the cell by actively pumping out sodium which leaks into the cell. The biconcave shape of the erythrocyte is thus maintained.[3,4] ATP is also necessary for glutathione synthesis[5] and incorporation of fatty acids into membrane phospholipids.[6] ATP is a substrate of the first reaction of glycolysis (see Figure 1). Since mature red cells have no other energy-producing pathways, an intact glycolytic system is essential to survival of the cell.

Second, glycolysis supplies NADH, the substrate for NADH-methemoglobin reductase (see Figure 1). This enzyme plays an important role in reduction of methemoglobin. Everyday, about 3% of hemoglobin is oxidized to methemoglobin.[7] When oxidized, hemoglobin cannot carry oxygen and thus is nonfunctional. This reductase is thought to account for about 67% of the reduction of methemoglobin while NADPH-methemoglobin reductase, which uses NADPH as a substrate, accounts for about 5%. Ascorbic acid (16%) and reduced glutathione (12%) also assist in methemoglobin reduction.[8]

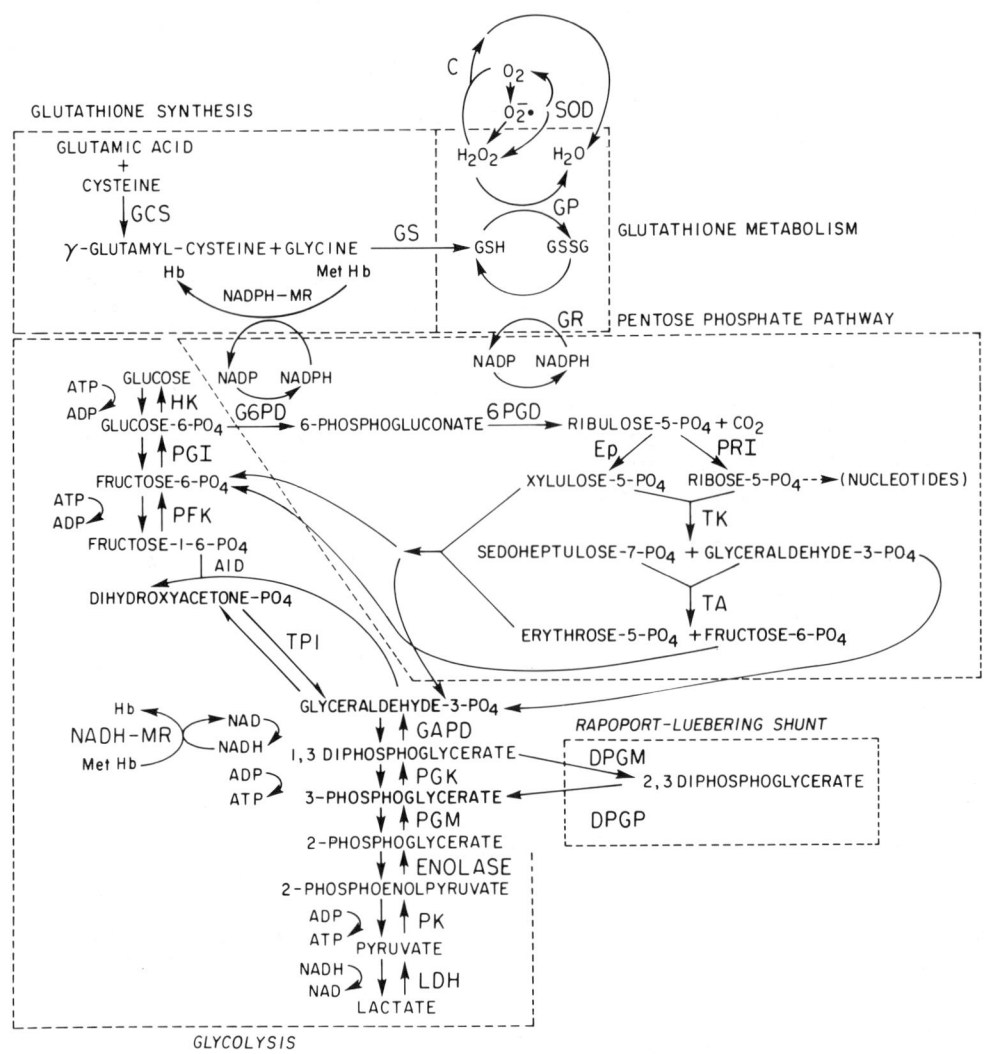

FIGURE 1. Major metabolic pathways in erythrocytes. *Enzymes*: HK, hexokinase; PGI, phosphoglucose isomerase; PFK, phosphofructokinase; Ald, aldolase; TPI, triosephosphate isomerase; GAPD, glyceraldehyde-phosphate dehydrogenase; PGK, phosphoglycerate kinase; PGM, phosphoglyceromutase; DPGM, diphosphoglycerate mutase; DPGP, diphosphoglycerate phosphatase; PK, pyruvate kinase; LDH, lactate dehydrogenase; G6PD, glucose-6-phosphate dehydrogenase; 6PGD, 6-phosphogluconate dehydrogenase; Ep, epimerase; PRI, phosphoribose isomerase; TK, transketolase; TA, transaldolase; GCS, glutamyl-cysteine synthetase, GS, glutathione synthetase; GP, glutathione peroxidase; SOD, superoxide dismutase; C, catalase; GR, glutathione reductase; NADPH-MR, NADPH-methemoglobin reductase; NADH-MR, NADH-methemoglobin reductase. *Cofactors*: ATP, adenosine triphosphate; ADP adenosine diphosphate; NAD-NADPH, nicotinamide adenine dinucleotide phosphate; NAD-NADH, nicotine adenine dinucleotide.

Rapaport-Luebering Shunt

The third function of glycolysis is to provide substrate for this shunt which produces 2,3-diphosphoglycerate or DPG (see Figure 1). When DPG is formed, there is no net energy production from glycolysis. However, this compound is produced in large quantities in red cells of most mammals and functions to reduce the affinity of hemoglobin for oxygen.[9,10] Deoxyhemoglobin A can bind DPG in a 1:1 molar ratio lowering the affinity of hemoglobin for oxygen and facilitating oxygen delivery to tissues. Without this compound, oxygen affinity would be very high and tissue oxygen tensions would have to be very low to effect adequate release of oxygen to tissues.

Under normal circumstances, there are three rate-limiting enzymes in the glycolytic pathway. These are hexokinase, phosphofructokinase, and pyruvate kinase.[11] These enzymes are very sensitive to the concentrations of cofactors and effectors and thus their *in vivo* activity can be considerably altered. During development, the environment of the red cell differs in many ways from the environment of adult animals. Red cell metabolism is therefore potentially altered by the external environment. In addition, fetal isozymes of red cell enzymes, with kinetic properties which differ from those of enzymes present in adult cells, can lead to alterations in metabolism. During red cell senescence, where the enzymes do not differ qualitatively and the surrounding environment is relatively constant, alterations in metabolic properties will generally result from secondary modifications of enzymes which render them unable to function adequately.

Pentose Phosphate Pathway

Normally, about 10% of glucose is metabolized through this, the major alternate pathway of glycolysis.[2] The major function of this pathway appears to be in the production of NADPH which is a necessary cofactor in several reactions, including glutathione reduction and methemoglobin reduction (see Figure 1).

ENZYMES IMPORTANT DURING OXIDANT STRESS

Reduced glutathione is a tripeptide of glycine, cysteine, and glutamic acid which is constantly turned over in the red cell.[12] Glutathione is kept in the reduced state (GSH) by the enzyme glutathione reductase (GR), which requires NADPH produced by the pentose-phosphate shunt. Glutathione is thought to play a major role in protection of cell components from oxidant damage.

Figure 1 indicates that molecular oxygen can undergo successive reduction to form superoxide radicals, hydrogen peroxide, hydroxyl radicals, and water. When hemoglobin is spontaneously oxidized, superoxide is generated. Superoxide is also the product of a number of enzymatic reactions and reactions of many oxidant drugs.[13,14] Once superoxide is formed, the remaining reductions can occur spontaneously. Thus, hydrogen peroxide and the very toxic hydroxyl radical can be formed. Three enzyme systems are present in mature erythrocytes to deal with these toxic oxygen products: superoxide dismutase (SOD), catalase (C), and glutathione peroxidase (GP) (see Figure 1). The importance of superoxide dismutase and catalase *in vivo* is not well established. At the present time, though, a number of deficiencies in glutathione peroxidase have been reported; none has been conclusively associated with hemolysis.[15]

From Figure 1 it can be seen that the important metabolic pathways of mature red cells are interrelated. Thus, any significant alteration in one pathway will likely affect the others. The possible alterations we will be discussing are those due to the unique environment of the fetus and those due to the physiological aging of the cell *in vivo*.

FETAL, NEWBORN, AND ADULT ERYTHROCYTE METABOLISM

In the transition from intrauterine to extrauterine life, human red blood cells undergo a number of changes. Characteristic is the substitution of adult hemoglobin A for fetal hemoglobin F, which progressively constitutes the major component of adult hemoglobin. *In utero*, hemoglobin F provides an advantage to placental oxygen transport because it interacts less with DPG. This decreased interaction results in fetal red cells with higher oxygen affinity which facilitates placental oxygen transfer.[16] Though it is unclear exactly what controls DPG levels in red cells, the decreased interaction of DPG with hemoglobin F may affect its production in fetal red cells by feedback mechanisms.

The unique metabolic characteristics of fetal red cells have been difficult to determine

Table 1
GENERAL CHARACTERISTICS OF FETAL AND NEWBORN ERYTHROCYTES[a] COMPARED TO ADULT CELLS

Age	Cell type		Cell size (mean corpuscular volume, fl)	Ref.
	Nucleated red cells (%)	Reticulocytes (%)		
12—16 weeks of gestation	3—6.5	18—40	180	16
24 weeks of gestation	0.5—1.0	5—10	123	16
Full-term infants	0.5	4.2—7.2	107—119	16,18
3 d of age	0	1—3	99	16
Adults	0	1—2	91	17

[a] Other general characteristics of fetal and newborn erythrocytes are the following: some unusual shapes are present;[19] red cell life span is shorter;[20-23] mechanical fragility is increased;[24] deformability decreases rapidly as cells age;[25] osmotic fragilities cover a wider range;[28] tendency to form Heinz bodies is increased;[30] methemoglobin is increased;[31,32] affinity of hemoglobin for oxygen[15] is increased.

with certainty for two reasons. First, because sampling of fetal blood was extremely difficult, most blood has been obtained from abortuses. Many metabolites change very rapidly, and results are called into question when the blood is not drawn before death and processed immediately. With technology for safe fetal blood sampling *in utero* immerging, more accurate data should be forthcoming. Samples from cord blood have been studied extensively, but, again, it is difficult to process samples rapidly enough for accurate metabolite determinations.

Second, the percentage of premature red cells is higher in fetal animals than in adult animals. In Table 1 are shown some values for the percentage of nucleated red cells and immature red cells, or reticulocytes, at several ages. Nucleated red cells are never seen in peripheral blood of normal adults. They are early forms that are limited to the bone marrow. Metabolically, they are very different from adult red cells. They contain a nucleus and can synthesize proteins. Though they cannot divide, they do have mitochondria and the enzymes necessary for oxidative phosphorylation.

Reticulocytes are also present in increased numbers in fetal blood. At 10 weeks of gestation, up to 80% of circulating cells are reticulocytes.[17] At 12 weeks, they can represent almost half of the red cells present (see Table 1).[17] In normal adults, only 1 to 2% of cells are reticulocytes.[18] Reticulocytes are cells newly released from the bone marrow which possess no nuclei and no mitochondria and have few intact metabolic pathways. They are able to synthesize proteins, but as they mature, they lose this ability also. The next section of this chapter will deal more thoroughly with the differences between reticulocytes and mature red cells.

Early studies of characteristics of fetal and newborn cells did not take into account the younger cell population and therefore it was not certain that differences between fetal and adult cells represented truly unique fetal characteristics or merely reflected the high percentage of younger cells. The use of blood from adult patients with high reticulocyte counts as controls has clarified the situation considerably.

In Table 1 are given some general characteristics of fetal and newborn red cells compared to those of adults. For most characteristics there is a continuum with red cells of the youngest fetuses being most different from adult cells. Also, there is considerable variability among normal full-term newborns in these characteristics.

Fetal red cells are larger than those of adults. Mean corpuscular volume is about 180 fl in 12- to 16-week fetuses and decreases to 99 fl at 3 d of age.[17,19] This compares with 91 fl for normal adults at sea level[18] (see Table 1).

Often at birth, unusual shapes are seen. The most common are echinocytes where projections are seen on the surface of the cell. Neither the significance or the cause for these changes is known.[20]

There is some question about whether red cells from healthy, full-term newborns have shortened survival in the circulation. It is, however, very clear that cells from premature infants have shortened survival. The most common method for determining red cell survival is by random labeling of an aliquot of cells with ^{51}Cr which binds to the non-alpha globin chain of hemoglobin.[21,22] Labeled cells are reinjected and the decay of label is followed. Data on newborn and fetal red cell survival are complicated by the finding that ^{51}Cr elutes from the gamma globin chains of fetal hemoglobin more easily than from the beta globin chain adult hemoglobin.[21] Therefore, at least part of the shortening of fetal red cell survival when chromium is used is due to the artifact of different chromium elution rates. The noneluting label diisopropyl flourophosphate (DF ^{32}P) has recently been used. Based on survival of radiolabeled cord blood erythrocytes, Bratteby and co-workers[23,24] concluded that cells produced during the last 60 d of fetal life lived between 45 and 60 d with the majority of cells dying before the mean life span. This increased rate of red cell destruction, combined with the need to increase red cell mass during growth, suggests that red cell production during the last 2 months of gestation is three to five times that found in adults.[25]

It appears that the membrane of fetal red cells differs considerably from that of adult cells. The cells are more sensitive to mechanical stress[26] and, as they age in the circulation, show decreased deformability.[27] It is important that red cells are able to both withstand the high shear forces in large vessels with rapid flows and be able to deform in order to pass through narrow capillaries and very small splenic slits. If the red cells cannot withstand mechanical stress and deform adequately, red cell damage may occur. Fetal cells are more sensitive to mechanical stress. Though mean cell deformability does not differ between newborns and adults, the most dense cells, or oldest cells in the circulation, show markedly decreased deformability compared to the oldest cells of adults. This is despite the fact that the oldest cells of the newborn are almost certainly younger than the oldest cells of adults. These data suggest that a rapid decrease in deformability during aging of neonatal red cells may play a role in splenic sequestration and destruction of these cells.

An interesting recent study looked at the possibility that increased free-DPG levels in fetal cells affect red cell deformability. Fetal hemoglobin binds DPG less readily than adult hemoglobin. Cells with high fetal hemoglobin levels therefore should have increased unbound or free DPG. Also, DPG had been reported to have a destabilizing effect on red cell membranes.[28] Adult and newborn red cell ghosts were loaded with different concentrations of DPG, and mechanical stability was measured in an ectacytometer. With increasing DPG concentration, mechanical stability decreased, but there were only small differences in adult and newborn cells at the same DPG level.[29] This suggests that at least part of the mechanical instability of fetal and newborn cells may be due to increased cytosolic DPG in these cells.

Osmotic fragility of red cells is determined by subjecting them to different salt concentrations and measuring the percent of cells that lyse at each concentration. The salt concentration at which 50% of the cells lyse does not differ between newborn and adult red cells, but there is a much broader range in the newborn; some cells are very sensitive to hypotonic lysis and some are very resistant. The reasons for this are unclear.[30]

There are many suggestions that fetal erythrocytes are more sensitive to oxidant stress than adults cells.[31] Fetal and newborn cells tend to form Heinz bodies, or precipitates of denatured hemoglobin at the cell membrane, more easily than adult cells.[32] In adults, Heinz bodies are formed during oxidant stress or if an unstable hemoglobin variant is present. Levels of methemoglobin, or hemoglobin where the heme iron is converted from ferrous to ferric, are higher in newborns. Part of the tendency to form Heinz bodies and to have increased methemoglobin levels seems to be due to the fact that hemoglobin F is more

Table 2
METABOLIC CHARACTERISTICS OF FETAL AND NEWBORN ERYTHROCYTES COMPARED TO ADULT CELLS

Metabolic characteristics
    Glucose consumption is increased, but is decreased when corrected for cell age.[37]
    DPG levels are normal or decreased.[40,41]
    Decline of ATP and DPG during incubation is accelerated.[42,43]
    Reduced glutathione is elevated.[45]
    Reduced glutathione is unstable.[46]
Enzyme alterations
    Activity increased when corrected for cell age:
        Phosphoglucose isomerase[45]
        Glyceraldehyde-3-phosphate-dehydrogenase[45]
        Phosphoglycerate kinase[47,48]
        Enolase[47,48]
    Activity decreased when corrected for cell age:
        Phosphofructokinase[53,54]
        NADPH-methemoglobin reductase[49,50]
        Catalase[51]
        Carbonic anhydrase[52]
    Altered isozyme patterns:
        Hexokinase[59]
        Phosphofructokinase[55,56]
        Phosphoglyceromutase[59]
        Lactic acid dehydrogenase[59]

sensitive to oxidation than hemoglobin A.[33,34] Also, these findings have been explained in part by the fact that newborn red cells have decreased activities of NADH-methemoglobin reductase, catalase, and low levels of tocopherol.[35] The significance of decreased plasma factors which protect against oxidant damage, such as tocopherol, is apparent from the fact that adult erythrocytes, transfused into newborn infants, appear more prone to form Heinz bodies.[36]

However, reduced glutathione levels in newborns are elevated for cell age[37] and the newborn pentose phosphate shunt appears to be normal and able to respond appropriately to oxidants such as methylene blue.[38]

A summary of metabolic characteristics of fetal and newborn erythrocytes compared to adult cells is given in Table 2. As with the characteristics given in Table 1, fetal and premature infant cells are more different from adult cells than those of healthy newborns.

The young cell age of newborn red cell populations is due to both the fact that growth requires a larger red cell mass and to the fact that fetal cells have shortened survival in the circulation. Adults are generally in a steady state. There is no growth and no development. As a fetus develops, it increases red cell mass and the cells produced survive longer and longer in the circulation. At about 2 months of age, red cell survival approaches the 120 d of adult cells, but red cell mass continues to increase. Therefore, it is difficult, if not impossible, to obtain adult red cell populations of similar age to those of fetal and newborn animals. Since numerous characteristics change with reticulocyte maturation and aging, it is important to try to control for cell age. The findings summarized in Table 2 are from studies in which the adult control cells had elevated reticulocyte counts. Though these controls are clearly not ideal, they do reveal a number of changes which occur during development.

Metabolism of glucose, the main metabolic substrate of red cells, is increased in newborn erythrocytes. The young age of the red cell population appears to explain this increase. Erythrocytes of premature infants, on the other hand, appear to consume less glucose than would be expected from young cell age.[39] Low inorganic phosphate levels may be partly responsible for this decrease in glucose consumption because phosphate activates phosphofructokinase (PFK), a major rate-limiting step of glycolysis.[40]

It is not clear whether ATP levels are higher for cell age in newborns, though there is evidence that they are very high in premature infants.[41] In healthy term infants, DPG levels are similar to those of adults, but during the first 10 months of life, DPG increases about 13%.[42] In contrast, in premature infants, levels of DPG are in the low normal range at birth and increase to above normal adult levels by 5 to 7 d after birth. In one study this rise averaged 56% of DPG levels at birth. From 2 to 4 weeks, DPG falls so that by 4 weeks of age levels are in the adult range.[43]

For reasons which are unclear, newborn erythrocytes show several abnormalities when incubated. They incorporate phosphorus more slowly than adult cells,[44] lose potassium at an increased rate,[30] and undergo marked morphological changes.[31,40] Also, upon incubation, DPG and ATP levels decline more rapidly than in adult cells incubated under similar conditions.[45] The decline in DPG levels appears to be due to increased breakdown of the compound by 2,3-DPG phosphatase.[46]

Four glycolytic enzymes appear to be inordinately high for cell age in newborn infants (see Table 2). It appears likely that these increases are due to increased expression of genes active in adult cells, though it is possible that the gene products are more stable in the environment of the fetus and newborn. Also, the metabolic consequences of these increased enzyme activities on newborn erythrocyte metabolism, if any, are unclear. All four of these enzymes are normally not rate limiting and therefore increased activity would not affect glycolytic rate, DPG or ATP levels.

A number of enzymes show decreased activity in newborn erythrocytes (see Table 2). Two of these enzymes, NADH-methemoglobin reductase[51,52] and catalase,[53] are involved in protection against oxidant stress.

The apparent deficiency of PFK in newborns has received much attention as a cause of decreased glycolytic rate of fetal and newborn erythrocytes.[55,56] PFK is a major rate-limiting step of glycolysis and is sensitive to many effectors. There is some evidence that newborn erythrocytes have increased expression of the liver-type homotetramer, L4, as well as decreased total activity of PFK.[57,58] An apparent instability of L4 in neonatal erythrocytes has been given significance as the cause for the low PFK activity *in vitro* as well as *in vivo*. However, the L4 isozyme is considerably more sensitive to inhibition by ATP and DPG than is the muscle-type homotetramer, M4, or the heterotetramers, M3L, M2L2, and ML3. A patient has recently been described with an unstable liver-type PFK subunit.[59] This man was totally asymptomatic and, while he had a 35 to 40% deficiency in red cell PFK activity, his ATP and DPG levels and red cell survival were normal. This and kinetic studies suggest that the liver-type PFK subunit may be so sensitive to product inhibition that it is nonfunctional in normal red cells. Therefore, overproduction of L4 or instability of this isozyme in newborn red cells should have no metabolic consequences. Rather, since the total PFK activity is decreased and the L4 isozyme is overrepresented, there must be a decrease in the activity of the metabolically active isozymes: M4, ML3, M2L2, and M3L. Therefore, though it appears that an *in vivo* decrease in PFK activity may explain the decreased glucose utilization in newborn cells, the exact mechanism by which this occurs has not been conclusively shown. It is also of interest that Travis and co-workers[60] have shown postnatal changes consistent with phosphofructokinase and hexokinase activation by a 50% rise in plasma inorganic phosphate during the first 4 d of life. During this period, few red cells are produced and these metabolic changes are responses of fetal erythrocytes to the extrauterine environment.

In addition to PFK, different isozyme patterns of hexokinase, phosphoglucomutase, lactic acid dehydrogenase, and enolase in newborn erythrocytes have been shown.[61] Though the isozyme distribution of hexokinase differs greatly for fetal and adult red cells, no differences have been found in the kinetic properties of these isozymes.[62] It is not yet clear whether the differences in isozyme pattern for enolase, lactic acid dehydrogenase, and phosphoglucomutase are unique fetal characteristics or the result of younger cell age.

In summary, fetal and newborn human erythrocytes show many differences from adult cells. Though none of the mechanisms underlying these differences are thoroughly understood, it is clear that fetal erythropoesis produces a unique "fetal" red cell, characterized by the presence of hemoglobin F. During the first few days of life, these cells do respond to changes in their extracellular environment, but remain "fetal" in character.[60] Gradually, over the first year of life, "adult" cells replace these fetal cells. As this occurs, the survival of the circulating erythrocytes is increased so that by the age of 1 year, the average age of the cells and their metabolic character approaches that of adults.

Finally, as an example of a developmental system which is quite well understood, changes in sheep erythrocytes during fetal, newborn, and adult life will be described. The gestation of sheep lasts about 150 d. Cells of fetal sheep of 125 to 135 d of gestation, newborn sheep cells from birth to 17 d of age, and adult sheep cells were studied.[63-65] Fetal sheep erythrocytes differ greatly from those of adults. As in humans, fetal sheep cells are larger and there are more reticulocytes. Reticulocyte count was elevated from the very low adult level of 0.4% to a level of 3.1% in fetal sheep.[63] Fetal sheep erythrocytes are very permeable to glucose; adult cells are essentially impermeable.[66] Like human cells, red cells of fetal sheep contain a high-oxygen-affinity fetal hemoglobin which does not interact with DPG.[66] Also, they contain elevated red cell enzyme activities:[63] 17 of 20 enzymes measured were dramatically increased.[63] The pattern of decline in these activities varied greatly. Seven enzymes showed an orderly decline in activity from fetal values through the postnatal period to adult levels. Activities of seven enzymes did not decrease between day 125 of gestation and birth, but declined after birth gradually to adult levels. Activities of pyruvate kinase were very high in fetal cells, but by birth had declined by half and were somewhat below adult activities. Two enzymes did not change during fetal, postnatal, and adult life. Activities of phosphoglucose isomerase and nucleoside phosphorylase were close to adult levels in the fetus, rose at birth, and fell again during the neonatal period. In contrast, activities of glutathione peroxidase were high in fetal sheep, were lowest at birth, and rose to adult levels during the first 3 weeks of life. Therefore, essentially every possible pattern of change was seen in these 20 enzymes.

However, when fetal values were compared with those of a serially bled ewe with a high reticulocyte count, only four enzymes were inordinately high in the fetus: DPGM, PFK, PK, GAPD.[64] It was concluded that these enzyme activities, like PGI, GAPD, PGK, and enolase, in human newborns (see Table 2) are unique characteristics of fetal erythrocytes.

In the lamb, during the first 5 d of life, there is a ten-fold increase in intraerythrocytic levels of DPG.[68-70] This is an important adaptation to extrauterine life because, in the lamb, this DPG acts to lower intracellular pH.[69,71] Hemoglobin-oxygen affinity is thereby lowered by the Bohr effect and tissue oxygen delivery is facilitated. It is assumed that the initial stimulus for this DPG increase is tissue hypoxia caused by the high hemoglobin-oxygen affinity of newborn sheep red cells. Neither fetal or adult sheep hemoglobin interacts with DPG as does human hemoglobin A.[67,70] Fetal sheep hemoglobin has intrinsically high oxygen affinity and adult hemoglobin has intrinsically low oxygen affinity. Adult sheep erythrocytes contain negligible DPG and very low levels of the DPG synthetic enzyme DPGM.[63,65]

The mechanism of this dramatic change in red cell metabolism was studied by bleeding lambs repeatedly after birth and studying the changes in glycolytic intermediates. The result suggested that the following sequence of events contributes to the postnatal rise in DPG levels in the lamb. Firstly, the concentration of plasma glucose more than doubles in the first 48 h of life, supplying substrate to the highly glucose permeable fetal red cell. Secondly, a transitory rise in blood pH is reflected intracellularly by activation of red cell PFK. However, because plasma phosphate levels are very low, there is a block to DPG synthesis because inorganic phosphate is a substrate of the GAPD reaction. Thirdly, as the plasma concentration of phosphate increases, GAPD is activated and DPG levels rise dramatically. Finally, the high activities of PFK, GAPD, and DPGM observed in fetal erythrocytes are probably necessary for such dramatic metabolic changes.[65]

This work provides as an example of characteristics of fetal erythrocytes which serve an important function in postnatal life. It exemplifies that both intracellular enzyme activities and extracellular pH and inorganic phosphate interact to produce an adaptive change immediately after birth. As fetal erythrocytes with high-oxygen-affinity fetal hemoglobin are replaced by adult cells containing low-affinity hemoglobin, the time passes when such a response is necessary. Because the adult erythrocyte is not permeable to glucose and does not contain the enzymes required, such a response is also no longer possible in adult animals.

METABOLIC CHANGES DURING ERYTHROPOIESIS, RETICULOCYTE MATURATION, AND SENESCENCE

For the purpose of discussion, we will divide the life span of the red cell into three stages. These are differentiation in the bone marrow, reticulocyte maturation and aging in the circulation, and removal from the circulation and catabolism in the reticuloendothelial system.

Differentiation in the Bone Marrow

Much progress has been made in the understanding of red cell development or erythropoiesis. For a detailed discussion, see Chapter in this book on the ontogeny of erythropoiesis. Briefly, progeny of the pluripotent stem cell, called colony-forming units-spleen or CFU-S, are committed to erythroid differentiation and form a primitive cell called Burst-forming unit-erythroid or BFU-E. They undergo rapid mitosis and through further amplification these BFU-E become colony-forming units-erythroid or CFU-E. The CFU-E is induced to form a pronormoblast. All subsequent processes may be considered maturational. The next stage, that of the basophilic normoblast, is characterized by disappearance of the nucleolus and a decline in RNA and DNA synthesis. Synthesis of hemoglobin A begins. During the next polychromatic normoblast stage, ribosomes, mitochondria, mRNA, and tRNA can no longer be replenished from synthetic processes in the nucleus. The amount of protein synthesis is therefore limited. Processes of degradation become prominent.[72] During the orthochromatic normoblast stage, the nucleus become increasingly pycnotic and is finally extruded forming a reticulocyte.[73] After release from the bone marrow, reticulocytes may be sequestered for several days in the spleen before entering the circulation.[74] They still contain ribosomes and mitochondria in the cytoplasm. Reticulocytes continue to synthesize protein, primarily hemoglobin, purines, and pyrimidines.[72] They contain only relics of lysosomes, though some of the lysosomal enzymes are present in the cytoplasm and on the cell membrane.[75] During maturation to erythrocytes, which takes place in 1 to 2 d in the circulation, ribosomes, mitochondria, and cytoplasmic vacuoles disappear. Reticulocytes are identified by the staining of ribosomes with brilliant cresyl blue.[76] When these ribosomes are lost, the cell is considered a mature erythrocyte.

Metabolism of Red Cell Precursors

With the exception of hemoglobin, little is known of developmental patterns of the cytosolic proteins which function in mature red cells. This is primarily because it has been difficult to obtain sufficient numbers of cells at specific stages of development to carry out biochemical analysis. Studies with K562 cells and Friend leukemia cell lines in culture have been done, but have the disadvantage that these cells never mature to enucleate erythroid cells.[77] Recently, a new method for study of enzymes during differentiation has been developed.[78-80] By pretreating mice with thiamphenicol, a drug which suppresses erythropoiesis, the percentage of spleen cells which are CFU-E can be greatly enhanced. Isolation and culture of these cells were carried out so that large numbers of cells at specific developmental stages from CFU-E to reticulocytes could be obtained. The results indicate that hemoglobin synthesis begins after one cell division. Activities of glycolytic enzymes hexokinase, phosphofructokinase, pyruvate kinase, and enolase are all highest in CFU-E and showed a rapid decline in activity after one cell division. Glucose-6-phosphate dehydrogenase and aldolase begin declining after several cell divisions, suggesting that these enzymes are still actively synthesized in CFU-E. 2,3-DPG mutase, the DPG synthetic enzyme, is almost absent in CFU-E, but progressively increases during differentiation. Unfortunately, the data were presented only as percentages of the highest activity obtained, so it is unclear how close to normal erythrocyte values these enzyme activities were 48 h after culture when they were classified as reticulocytes.

Two isozymes of pyruvate kinase were expressed in CFU-E, but during differentiation, one is lost and the other becomes prominent in reticulocytes and is the only isozyme of pyruvate kinase present in mature red cells. Because the total activity of pyruvate kinase declines during the time that this change in isozyme pattern is seen, it is unclear whether the red cell isozyme is synthesized and then partially degraded or whether the decrease is solely due to the degradation of the non red cell isozyme. The presence of two pyruvate kinase isozymes in red cell precursors has been reported by others.[77,81] Because this method provides cells at a specific stage of development, namely, CFU-E, it holds great promise for detailed studies of the mechanisms by which components of mature red cells are produced.

Reticulocyte Maturation and Red Cell Aging in the Circulation

Reticulocyte maturation and red cell aging in the circulation are considered together. For reasons that will become clear, it is difficult to distinguish between changes occurring during reticulocyte maturation and those occurring during senescence. A number of methods have been used to determine changes in red cell enzymes and metabolites with aging in the circulation. The first was the differential agglutination technique of Ashby where O-type red cells are transfused into a recipient of A blood type.[82,83] At intervals, a blood sample is drawn and recipient cells are agglutinated leaving donor cells for study. With time, donor cells age *in vivo* and changes associated with aging can be studied. Because it is very difficult to obtain samples of only donor cells, this approach is limited.

A second approach has been to relate the degree of reticulocytosis to the metabolic parameter. Using subjects with young red cell age due to hemolysis or bleeding has provided much data on enzymes which are altered in reticulocytes but not much data on those that occur after maturation during senescence.[84]

The majority of data have been derived from studies of red cells separated according to their density. Red cell density appears to increase during cell aging.[85] This technique is based on the finding that when a cohort of young cells is either radiolabeled in the bone marrow with ^{59}Fe or *in vitro* with ^{14}C, the label appears first at the top of a density gradient. With time, labeled cells become more dense. The density of rabbit red cells appears to increase fairly uniformly with age.[86] Mouse cells, on the other hand, are initially more dense than older cells, then become lighter, and finally show a pattern of red cell density largely independent of cell age.[87] In humans, very little work on density changes with age has been carried out. It does appear that very young cells are light, but then distribute themselves fairly evenly throughout a density gradient.[85]

Because so much data have been derived from density separated cells, the question of whether cell density does in fact decline in a uniform way is very important. It is clear that enzyme activities and metabolite levels are considerably higher in reticulocytes than in mature cells. The question is, "What is the pattern of decline of these variables as the cells mature and age?" Based upon the extensive work of Piomelli and colleagues,[88-90] it has been considered dogma for many years that many red cell enzyme activities decline in a logarithmic fashion as the cells age in the circulation. Their data are based on two assumptions. The first assumption is that density provides a perfect separation of cells by age. This is clearly not the case. Reticulocytes are present in decreasing frequency in layers of increasing density, but they are present in essentially all density-separated layers. The second assumption is that the decline of enzyme activities is a single, first-order decay from the beginning of the life of a cell in the circulation until it is removed at about 120 d of age. It is known from studies of reticulocytes and red cell precursors that dramatic changes in some enzyme activities occur during erythroid development.[80] That such large declines should suddenly cease as reticulocytes are released into the circulation and assume a uniform decline for the next 120 d would be somewhat surprising, especially because these newly released reticulocytes continue to synthesize and degrade proteins. An alternative to the assumption that enzyme activities show first-order decays with time is that reticulocytes are released with some enzyme activities being very high. During the 1 to 2 d of maturation in the circulation, these enzyme activities decay dramatically, leaving activities close to those of whole blood red cells. During the 120 d in the circulation as mature red cells, metabolic characteristics do not change appreciably. Newly released reticulocytes would be lighter on the average than mature cells, but some would be heavier than much older cells. Whole blood red cell populations could therefore be thought of as two populations: mature cells with a relatively constant metabolic picture and reticulocytes with rapidly changing metabolic picture. The decrease in metabolic variables in fractions of cells with increasing density would then be due to decreasing reticulocyte contamination in these fractions. The rate of change in density layers would be related to the magnitude of the change occurring as reticulocytes mature. For example, glucose-6-phosphate dehydrogenase is considered a very age-dependent enzyme; activity declines rapidly in density layers. The model mentioned here would predict that during reticulocyte maturation this enzyme activity decays rapidly and dramatically, but after maturation, very little change occurs.

Recently, several investigators have used new approaches to get at the question of the pattern of metabolic change with reticulocyte maturation and aging. Beutler investigated several age-related enzyme activities in patients with erythroblastopenia of childhood, where red cell production temporarily stops and circulating red cells become progressively aged.[81] If enzyme activities decline in a linear fashion throughout red cell life span, the cells of these individuals should have activities lower than normal. Surprisingly, activities of three enzymes considered to change greatly with cell age, hexokinase, pyruvate kinase, and glutamic-oxaloacetic transaminase, were all in the normal range.[81]

Morrison and co-workers have used the method of Gonzoni to look at *in vivo* aging of red cells.[87,92] Beginning with a large group of mice, erythropoiesis is suppressed by hypertransfusing half of the mice with the blood of the other half. At 2-week intervals, blood from half of the animals was given to the other half so that red cell populations of increasing age are present in the remaining animals. Interestingly, when these cells were separated by density, the mean density of cells did not decrease appreciably as the cells aged *in vivo*. The authors concluded that mean density does increase slightly, but only during the earliest stages of red cell life span.[87]

Also, we have induced hemolytic anemia in rats using the oxidant drug acetylphenylhydrazine and have followed the pattern of enzyme activity change as the animals recover from anemia. Enzyme activities, DPG and ATP levels were highest when reticulocyte counts were close to 100%. They then declined rapidly during recovery and showed a very small decline after reticulocyte count returned to normal levels.[93]

All of these suggest that, in fact, there is a biphasic decline of some metabolic variables during reticulocyte maturation and aging. This has been suggested by several investigators.[91,94-96] It has also been suggested that there may be a third phase of decline at the end of red cell lifespan.[94]

These new studies indicate that the pattern of change in red cell metabolism with reticulocyte maturation and aging is far from clear. This means that many changes which have been attributed to aging may in fact be part of the normal maturational process of reticulocytes. Methods need to be developed where pure reticulocyte populations can be separated and compared to populations of pure mature red cells. Until more is known, maturational and senescent changes cannot be easily distinguished.

Table 3
ENZYMES DECREASING WITH RED CELL AGE

Hexokinase[88,90,96,97]
Glucose-6-phosphate dehydrogenase[96,97]
6-Phosphoglucose dehydrogenase[96]
Phosphofructokinase[90]
Phosphoglucose isomerase[90]
Aldolase[90,96,97]
Glyceraldehyde-phosphate dehydrogenase[90]
Phosphoglyceromutase[90]
Phosphoglycerate kinase[90]
Triose-phosphate isomerase[90]
Enolase[90]
Pyruvate kinase[88,90,97]
Lactate dehydrogenase[90]
Ca^{2+}/Mg^{2+} ATPase[98,99]
Catalase[82]
Acetylcholinesterase[82,100]
Glutamic-oxaloacetic transaminase[101]

METABOLITES DECREASING WITH RED CELL AGE

ATP[96,100]
DPG[100-103]
Creatine[104]

METABOLIC RATES DECREASING WITH RED CELL AGE

Glucose consumption[90]
Ability for pentose phosphate shunt to respond to oxidant stress[90]

In Table 3 are given enzymes and metabolites which have been reported to decline during red cell aging. For consistency, only data from density-separated, human red cells are given, i.e., from normal, circulating red cell populations. It can be seen that a great number of enzyme activities decline with increasing cell density. The most striking of these are glutamic-oxalacetic transaminase, glucose-6-phosphate dehydrogenase, pyruvate kinase, and hexokinase. Other decreases are modest. Whether any of these activities decline enough that the enzymatic rate becomes limiting is unclear. Several investigators have suggested that hexokinase activity, which is relatively low in all red cells, decreases with aging to become a significant factor in energy metabolism.[95] Glucose consumption, DPG, and ATP levels do decrease in density-fractionated cells as does the ability of the pentose phosphate shunt to respond *in vitro* to the oxidant methylene blue.[90]

In Table 4 are shown characteristics of reticulocytes and those of aged red cells. As stated above, reticulocytes continue to synthesize and degrade proteins. They possess transferrin and adrenergic receptors which are clearly lost as part of the maturation process.[105,106] Permeability to Ca^{2+} is greatly increased in reticulocytes.[107] Sodium permeability is modestly increased.[107] During reticulocyte maturation, the membrane appears to be remodeled, membrane surface area is decreased, and phospholipid and cholesterol are lost.[108]

Though some membrane changes occur during maturation, others appear to proceed throughout *in vivo* aging, though at a slower rate. Several of the characteristics given in Table 4 are generally thought to be important in the senescence of red cells, in particular, decreased deformability and the appearance of surface antigens. Loss of erythrocyte deformability with age has been related to the decreased ATP levels with age. As mentioned, decreased ATP levels have been related to decreased hexokinase activity. ATP-depleted red

Table 4
CHARACTERISTICS OF RETICULOCYTES

Synthesize and degrade proteins
Possess transferrin and adrenergic receptors[105,106]
Ca^{2+} permeability and Na^+ permeability increases[107]
Volume increased[108]
Membrane phospholipid and cholesterol increased[108]

CHARACTERISTICS OF AGED RED CELLS

Membrane loss, decreased volume, and increased density[100,109]
Decreased deformability and filterability[113,115]
Decreased K^+ and Mg^{2+} and increased Na^+ [100,104]
Increased hemoglobin-oxygen affinity[102,103]
Decreased calmodulin activity[110,114]
Appearance of senescent cell antigen[116,117]

cells are very rigid and nondeformable.[111-113] Old cells with decreased ATP levels may become trapped in the splenic slits and small capillaries.

An antigen, referred to as senescent cell antigen, appears on the surface of senescent and damaged red cells.[116] Its appearance initiates immunological binding of physiological IgG autoantibodies. The presence of these antibodies appears to mark the cell for removal by macrophages. Since the red cell is incapable of synthesizing a new antigen, this antigen may merely be unmasked during aging. The antigen is immunologically related to Band 3. Recently it has been suggested that the appearance of this antigen is initiated by denatured hemoglobin binding to the red cell cytoskeleton. The conformational change brought about by this binding exposes the antigen, thus beginning the process of removal of the cell from the circulation.[117]

PROTEIN ALTERATIONS AND DEGRADATION DURING RETICULOCYTE MATURATION AND SENESCENCE

In this final section, mechanisms by which protein function is lost will be discussed. It appears that there are two ways in which protein function can be lost and both are probably initiated by spontaneous or accidental alterations in protein structure. This alteration can affect function by itself or it can serve as a marker for protein degradation or both.

In the reticulocyte and in mature red cells, abnormal proteins are susceptible to protein degradation primarily by an ATP-dependent proteolytic system.[118,119] A major question is, "How does the reticulocyte mature to contain a relatively constant compliment of active enzymes when a large number of the molecules are clearly degraded during maturation?" For example, glucose-6-phosphate dehydrogenase is the product of one locus, so presumably the vast majority of molecules synthesized are normal and identical. There should be no differences in stability among the native molecules. Yet, during maturation, a large percentage of this enzyme activity and protein content is lost through degradation. Most of what remains retains its activity through the next 120 d. How does this process stop at a programmed level? The answer to this question is not clear. Though it has not been conclusively demonstrated, it seems likely that molecules are always altered by chance. For some reason, reticulocyte proteins may be more sensitive to protein alteration than those of mature red cells. Reticulocytes may generate more oxidants for example. It is also possible that the proteolytic systems simply "wear out" and are not replenished because the capacity for protein synthesis has been lost. This appears unlikely since mild oxidation of mature red cells stimulates protein breakdown.[119] Therefore, mature red cells have the capacity to degrade abnormal proteins. The data suggest that under normal circumstances there are few abnormal proteins in erythrocytes. Even the reticulocyte has protein degradation rates lower than other mammalian tissues,[118] suggesting that mature red cells normally degrade proteins at a very low level.

There is considerable evidence that red cell proteins, including enzymes, do undergo post-translational modification during aging. *In vitro* lipid peroxidation results in red cell membrane changes very similar to those seen in aged cells.[120,121]

Turner et al.[122] looked at 16 red cell enzymes in density separated red cells and found that 5 had altered electrophoretic properties with increasing density. Similarly, glucose-6-phosphate dehydrogenase shows an additional band on electrofocusing with a lower pI during aging.[123]

Kahn and co-workers did an extensive study to determine the mechanism of loss of enzyme activity during aging.[57] They looked at six enzymes. Phosphoglucose isomerase, phosphofructokinase, and phosphoglycerate kinase all showed a decrease in the number of molecules with aging, suggesting that these proteins were degraded during aging. Decreased activity for the other three enzymes, glucose-6-phosphate dehydrogenase, 6-phosphogluconate dehydrogenase, and pyruvate kinase was due to protein loss as well as loss in enzyme activity per molecule. Pyruvate kinase activity declined 38% in the oldest compared to the youngest cells despite similar quantities of enzyme protein in all density layers. Paglia and Valentine[124] also found that PK was altered during aging. A new subunit of smaller molecular weight appears. They also found decreased electrophoretic mobility and decreased affinity for substrate phosphoenolpyruvate.[124]

In summary, there do appear to be considerable alterations in red cell proteins during aging. Whether any of these contribute to the destruction of the cell is unknown.

Covalent Modification of Red Cell Proteins During Aging

Finally, we will look at potential mechanisms of protein alteration which may lead to inactivation, degradation, or both. A number of spontaneous reactions occur which can covalently modify proteins. These include glycosylation, oxidation, racemization, and deamidation. None of these mechanisms have been shown to definitively lead to cell death under normal circumstances. We deal with them here because it appears likely that, with further study, a critical red cell protein will be shown to be altered in a way which clearly contributes to cell death.

Nonenzymatic Glycosylation

Glycosylation is a major nonenzymatic, post-translational modification reaction. Aldoses and ketoses have a carbonyl group which, when in chain form can undergo a neutrophilic attack by an amino group. Though many sugars can glycosylate proteins, because of its high plasma concentration, glucose is thought to be the main glycosylating sugar in erythrocytes.[126] The first reaction to form glycosylated proteins is rapid and reversible and the product is an aldimine. A very slow, multistep Amadori rearrangement can follow which leads to stable ketoamine.[126,127] Because this is a very slow reaction, the amount of glycosylated protein reflects the proteins exposure to glucose throughout its lifetime. Glycosylated proteins are commonly detected by electrophoresis, but since a number of glycosylations do not produce charge differences, complete analysis of bound sugars requires more sophisticated methods.[128]

In normal hemolysates, evidence of glycosylation is seen by the presence of a number hemoglobins which have greater anodic mobility than Hb A. These are commonly called "fast hemoglobins" and make up 6 to 8% of total hemoglobin.[129] Hb A_{1C}, the major anodal hemoglobin, where glucose is bound to the N terminus of beta globin,[130,131] accumulates throughout the lifespan of the red cell.[125] The lighter, younger cells on density gradients contain lower quantities of Hb A_{1C}.[132]

Because red cells do not depend on insulin for glucose transport, intracellular glucose reflects plasma glucose concentrations.[133] Therefore, in diabetics, red cell proteins can be exposed to very high concentrations of glucose (for historical review see Reference 134). Levels of Hb A1C reflect integrated plasma glucose concentrations over the weeks prior to measurement and have been an extremely useful index of long-term glucose control.[134,135] Hemoglobins altered by addition of sugars to the N terminus of the beta chains show decreased binding of the allosteric effector 2,3-DPG and are therefore less susceptible to the lowering of the oxygen affinity by DPG.[136] It is unlikely that increased oxygen affinity of these altered hemoglobins has a significant physiological effect though it should be kept in mind that many red cell proteins may be glycosylated with aging and the function of critical proteins may be affected.

Methylation of Spontaneously Altered Red Cell Proteins

Recently, an interesting area of red cell research has centered around spontaneous modifications of L-aspartyl and L-asparaginyl residues in red cell proteins. These reactions include racemization to D-asparaginyl and D-aspartyl residues as well as deamidation forming L-isoaspartyl residues where the peptide bond is made through the side chain carboxyl group.

Clarke and co-workers have shown that these altered residues accumulate significantly in the membrane fraction of aging red cells.[137] These spontaneous modifications occur in many long-lived proteins such as in the tooth and lens, but their rate of accumulation in red cell membranes appears to be at least 40 times faster than the rate reported for the lens and tooth.[138] This rate suggests that the average red cell proteins would have about 3.1% abnormal isomers after 120 d.[137]

The importance of these altered residues in red cell aging arises from the demonstration that enzymes are present in erythrocytes which specifically recognize and methylate these residues. A specific red cell enzyme catalyzes the synthesis of a methyl donor: S-adenosyl-L-methionine.[139] A group of protein carboxyl methyltransferases catalyzes the methylation of altered asparaginyl and aspartyl residues. Carboxyl methylated proteins have been found in the membrane[140] and the cytosol,[141] but the rate of methylation is very low. Methylated membrane proteins appear to increase with cell age though even in the oldest red cells it is very low.[142,143]

While initially it was thought that methylation might play a regulatory role, this now seems unlikely because these enzymes appear to act only on proteins which are altered by spontaneous, unpredictable events. Studies are now in progress to determine whether methylation serves to mark damaged proteins for proteolytic digestion or whether it serves as a first step in a process to partially repair altered residues by converting some of them to their native form.[142] A review of this area of investigation has recently appeared.[144]

Oxidation of Red Cell Proteins During Aging

The free radical generating reactions which occur in the red cell are mostly initiated through hemoglobin. Normally, about 3% of red cell hemoglobin is oxidized per day and as this occurs, superoxide radical is generated.[7] A number of redox drugs and xenobiotics also interact with hemoglobin and produce radicals often leading to the premature destruction of red cells.[145] Normal red cells, however, have many antioxidant systems including vitamin E, asorbic acid, reduced glutathione (GSH), and enzymes which break down oxygen radicals and hydrogen peroxide. With aging, the efficiency of a number of these systems may decrease and a key question is whether normal senescence mimics that produced by drugs or whether antioxidant systems are adequate throughout the red cell life span.

As discussed above, the appearance of the senescence cell antigen on the red cell surface appears to be initiated by denatured hemoglobin.[117] This denaturation is probably induced by oxidant damage to hemoglobin. Since denatured hemoglobin is a natural occurrence with cell aging, it appears that the protection mechanisms of the cell may fail as the cell ages. Also, the work on membrane lipid peroxidation suggests that commonly observed aging phenomena result from oxidant damage.[120,121]

In summary, there are clearly many changes in red cell properties with maturation and aging. For methodological reasons, it is difficult to dissociate maturational changes from changes due to aging. It appears most likely that metabolic characteristics of reticulocytes during maturation change rapidly until maturation, after which changes appear to be relatively slow.

Though no single process of aging is thoroughly understood, there are a number of secondary modifications of proteins which are likely to contribute to the ultimate demise of the cell. Knowledge of the senescent cell antigen provides one way, perhaps the major way, which red cells are removed from the circulation. As understanding of this and other mechanisms of modification of proteins increases, we will certainly gain a new perspective on the life and death of the red cell.

REFERENCES

1. **Friedman, H. and Rapaport, S. M.**, Enzymes of the red cell: a critical catalogue, in *Cellular and Molecular Biology of Erythrocytes*, Yoshikawa, H. and Rapaport, S. M., Eds., University Park Press, Baltimore, MD, 1974, 181.
2. **Murphy, J. R.**, Erythrocyte metabolism. II. Glucose metabolism and pathways, *J. Lab. Clin. Med.*, 55, 286, 1977.
3. **Nakao, M., Nakao, T., Yamazoe, S., and Yoshikawa, H.**, Adenosine triphosphate and shape of erythrocytes, *J. Biochem.*, 49, 487, 1961.
4. **Whittam, R.**, *Transport and Diffusion in Red Blood Cells*, Edward Arnold, London, 1964.
5. **Koj, A.**, Biosynthesis of glutathione in human blood, *Acta Biochim. Pol.*, 9, 11, 1962.
6. **Oliveira, M. M. and Vaughan,** Incorporation of fatty acids into membrane lipids of erythrocyte membranes, *J. Lipid Res.*, 5, 156, 1964.
7. **Winterbourn, C. C., McGrath, B. M., and Carrell, R. W.**, Reactions involving superoxide and normal unstable hemoglobins, *Biochem. J.*, 155, 493, 1976.
8. **Scott, E. M.**, Congenital methemoglobinemia due to DPNH-diaphorase deficiency, in *Hereditary disorders of erythrocyte metabolism*, Beutler, E., Ed., Grune & Stratton, New York, 1968, 103.
9. **Benesch, R. and Benesch, R. E.**, The effect of inorganic phosphates from the human erythrocyte on the allosteric properties of hemoglobin, *Biochem. Biophys. Res. Commun.*, 26, 162, 1967.
10. **Chanutin, A. and Curnish, R. R.**, Effect of inorganic phosphates on the oxygen equilibrium of human erythrocytes, *Arch. Biochem.*, 121, 96, 1967.
11. **Minakami, S. and Yoshikawa, H.**, Studies on erythrocyte glycolysis. II. Energy changes and rates limiting steps in erythrocyte glycolysis, *J. Biochem.*, 59, 159, 1966.
12. **Elder, H. A. and Mortensen, R. A.**, The incorporation of labelled glycine into erythrocyte glutathione, *J. Biol. Chem.*, 218, 261, 1956.
13. **Fridovich, I.**, Superoxide dismutases, *Annu. Rev. Biochem.*, 44, 147, 1975.
14. **Fridovich, I.**, Oxygen radicals, hydrogen peroxide and oxygen toxicity, in *Free Radicals in Biology*, Vol. 1, Pryor, W. A., Ed., New York, Academic Press, 1976, 239.
15. **Beutler, E.**, Red cell enzyme defects as nondiseases and as diseases, *Blood*, 54, 1, 1979.
16. **Bauer, C. H., Ludwig, M., Ludwig, I., and Bartels, H.**, Factors governing the oxygen affinity of human adult and foetal blood, *Respir. Physiol.*, 7, 271, 1969.
17. **Oski, F. A. and Naiman, J. L.**, Normal blood values in the newborn period, in *Hematological Problems in the Newborn*, W. B. Saunders, Philadelphia, 1982, 4.
18. **Wintrobe, M. M.**, Appendix A: Normal blood and bone marrow values in man, in *Clinical Hematology*, Lea & Febiger, Philadelphia, 1981, 1985.
19. **Zaizov, R. and Mattoth, Y.**, Red cell values on the first postnatal day as a function of gestational age, *Am. J. Hematol.*, 1, 276, 1976.
20. **Zipursky, A.**, Erythrocyte morphology in newborn infants: a new look, *Pediatr. Res.*, 11, 843, 1977.
21. **Suderman, H. J., White, F. D., and Israels, L. G.**, Elution of chromium-51 from labeled hemoglobins of human adults and cord blood, *Science*, 126, 650, 1957.

22. **Pearson, H. A. and Vertrees, K. M.,** Site of binding to chromium-51 by haemoglobin, *Nature (London),* 189, 1019, 1961.
23. **Bratteby, L. E.,** Studies of erythrokinetics in infancy, *Acta Paediatr. Scand.,* 57, 125, 1968.
24. **Bratteby, L. A., Garby, L., and Wadman, B.,** Studies of erythrokinetics in infancy. XII, *Acta Paediatr. Scand.,* 57, 305, 1968.
25. **Oski, F. A. and Naiman, J. L.,** Normal blood values in the newborn period, in *Hematological Problems in the Newborn,* W. B. Saunders, Philadelphia, 1982, 24.
26. **Goldbloom, R. B., Fisher, E., Reinhold, J., and Hsai, D. Y.,** Studies on mechanical fragility of erythrocytes. I. Normal values for infants and children, *Blood,* 8, 165, 1953.
27. **Linderkamp, O., Wu, P. Y. K., and Meiselman, H. J.,** Deformability of density separated red blood cells in normal newborn infants and adults, *Pediatr. Res.,* 16, 964, 1982.
28. **Sheetz, M. and Casaly, J.,** 2,3-Diphosphoglycerate and ATP dissociate erythrocyte membrane skeletons, *J. Biol. Chem.,* 255, 9955, 1980.
29. **Mentzer, W. C., Iarocci, T. A., and Mohandas, N.,** Modulation of RBC membrane mechanical fragility by 2,3 DPG in transient poikilocytosis, *Blood,* 64, 28A, 1984.
30. **Sjolin, S.,** The resistance of red cells in vitro: a study of osmotic properties, the mechanical resistance and the storage behavior of red cells of fetuses, children and adults, *Acta Paediatr.,* 43, 1, 1954.
31. **Gordon, H. H., Nitowski, H. M., and Cornblath, M.,** Studies of tocopherol deficiency in infants and children. I. Hemolysis of erythrocytes in hydrogen peroxide, *Am. J. Dis. Child.,* 90, 669, 1955.
32. **Tillman, W., Menke, J., and Schroter, W.,** The formation of Heinz bodies in ghosts of human erythrocytes of adults and newborn infants, *Klin. Wochenschr.,* 51, 201, 1973.
33. **Betke, K., Kleihauer, E., and Gartner, C.,** Von Verminderung, Methamoglobinreduktion, diaphoraseaktivitat und flavin in erythrozyten junger sauglinge, *Arch. Kinderh.,* 170, 66, 1964.
34. **Martin, H. and Huisman, T. H. J.,** Formation of ferrihaemoglobin of isolated human haemoglobin types by sodium nitrite, *Nature (London),* 200, 898, 1963.
35. **Nitowski, H. M., Cornblath, M., and Gordon, H. H.,** Studies of tocopherol deficiency in infants and children. II. Plasma tocopherol and erythrocyte hemolysis in hydrogen peroxide, *J. Dis. Child.,* 92, 164, 1956.
36. **Schroter, W. and Tillman, W.,** Heinz body susceptibility of red cells and exchange transfusion, *Acta Haematol.,* 49, 74, 1973.
37. **Konrad, P. N., Valentine, W. N., and Paglia, D. E.,** Enzymatic activities in the newborn: comparison with red cells of older normal subjects and those with comparable reticulocytosis, *Acta Haematol.,* 48, 193, 1972.
38. **Oski, F. A.,** Red cell metabolism in the premature infant. II. The pentose phosphate pathway, *Pediatrics,* 39, 689, 1965.
39. **Oski, F. A. and Smith, C.,** Red cell metabolism in the premature infant. III. Apparent inappropriate glucose consumption for cell age, *Pediatrics,* 41, 473, 1968.
40. **Bentley, H. P., Jr., Alford, C. A., Jr., and Diseker, M.,** Erythrocyte glucose consumption in the neonate, *J. Lab. Clin. Med.,* 76, 311, 1970.
41. **Oski, F. A. and Naiman, J. L.,** Red cell metabolism in the premature infant. I. Adenosine triphosphate levels, adenosine triphosphate stability and glucose consumption, *Pediatrics,* 36, 104, 1965.
42. **Delivoria-Papadopoulos, M., Roncevic, N. P., and Oski, F. A.,** Postnatal changes in oxygen transport in term, premature and sick infants: the role of 2,3-diphosphoglycerate and adult hemoglobin, *Pediatr. Res.,* 5, 235, 1971.
43. **Wimberly, P. D.,** Fetal Hemoglobin, 2,3-Diphosphoglycerate and Oxygen Transport in the Newborn, Premature Infant, Thesis, Copenhagen University, Copenhagen, 1982.
44. **Zipursky, A., LaRue, T., and Israels, L. G.,** The in vitro metabolism of erythrocytes from newborn infants, *Can. J. Biochem.,* 38, 727, 1960.
45. **Oski, F. A. and Komazawa, M.,** Metabolism of the erythrocytes of the newborn infant, *Semin. Hematol.,* 12, 49, 1975.
46. **Oski, F. A.,** The erythrocyte and its disorders, in *Hematology of Infancy and Childhood,* Nathan, D. G., and Oski, F. A., Eds., W. B. Saunders, Philadelphia, 1981, 26.
47. **Swierczewski, E., Gibelin, C., and Minkowski, A.,** Glutathion dans les globules rouges de la veine ombilicale: comparaison avec le glutathion dans les globules rouge de la femme enceinte, *Biol. Neonat.,* 3, 321, 1962.
48. **Zinkham, W. H.,** An in vitro abnormality of glutathione metabolism in erythrocytes from normal newborns: mechanism and clinical significance, *Pediatrics,* 23, 18, 1959.
49. **Komazawa, M. and Oski, F. A.,** Biochemical characteristics of "young" and "old" erythrocytes of the newborn infant, *J. Pediatr.,* 87, 102, 1975.
50. **Oski, F. A.,** Red cell metabolism in the newborn infant. V. Glycolytic intermediates and glycolytic enzymes, *Pediatrics,* 44, 84, 1969.

51. **Ross, J. D.,** Deficient activity of DPNH-dependent methemoglobin diaphorase in cord blood erythrocytes, *Blood,* 21, 51, 1963.
52. **Bartos, H. R. and Desforges, J. F.,** Erythrocyte DPNH-dependent diaphorase levels in infants, *Pediatrics,* 367, 991, 1961.
53. **Jones, P. E. H. and McCance, R. A.,** Enzymatic activities in the blood of infants and adults, *Biochem. J.,* 45, 646, 1949.
54. **Poblete, E., Thibeault, D. W., and Auld, P. A. M.,** Carbonic anhydrase in the premature, *Pediatrics,* 43, 4298, 1968.
55. **Gross, R. R., Schroeder, E. A. R., and Broulnstein, S. A.,** Energy metabolism in the erythrocytes of premature infants compared to full term newborn infants and adults, *Blood,* 21, 755, 1963.
56. **Travis, S. F., Kumar, S. P., Paez, P. C., and Delivoria-Papadopoulos, M.,** Red cell metabolic alterations in postnatal life in term infants: glycolytic enzymes and glucose-6-phosphate dehydrogenase, *Pediatr. Res.,* 14, 1349, 1980.
57. **Kahn, A., Boyer, C., Cottreau, D., Marie, J., and Boivin, P.,** Immunologic study of the age-related loss of activity of six enzymes in the red cells from newborn infants and adults — evidence for a fetal type of erythrocyte phosphofructokinase, *Pediatr. Res.,* 11, 271, 1977.
58. **Vora, S. and Piomelli, S.,** Fetal isozyme of phosphofructokinase in newborn erythrocytes, *Pediatr. Res.,* 11, 483, 1977.
59. **Vora, S., Davidson, M., Seaman, C., Miranda, A. F., Noble, N. A., Tanaka, K. R., Frenkel, E. P., and DiMaura, S.,** Hetcrogeneity of the molecular lesion in inherited phosphofructokinase deficiency, *J. Clin. Invest.,* 72, 1995, 1983.
60. **Travis, S. F., Kumar, S. P., and Delivoria-Papadopoulos, M.,** Red cell metabolic alterations in postnatal life in term infants: possible control mechanisms, *Pediatr. Res.,* 15, 133, 1981.
61. **Chen, S. H., Anderson, J. E., Giblett, E. R., and Stamatoyannopoulos, G.,** Isozyme patterns in erythrocytes from human fetuses, *Am. J. Hematol.,* 2, 23, 1977.
62. **Stocchi, V., Magnani, M., Canestrari, F., Dacha, M., and Forniani, G.,** Multiple forms of human red blood cell hexokinase: preparation, characterization and age dependence, *J. Biol. Chem.,* 257, 2357, 1982.
63. **Noble, N. A., Cabalum, T. C., Nathanielsz, P. W., and Tanaka, K. R.,** Erythrocyte enzymes in sheep: comparison of activity in fetal, newborn, maternal and nonpregnant ewe erythrocytes, *Biol. Neonat.,* 41, 61, 1982.
64. **Noble, N. A., Tanaka, K. R., and Nathanielsz, P. W.,** Erythrocyte enzymes in Ovis Aries: effect of cell age, *Comp. Biochem. Physiol. B,* 71, 1982.
65. **Noble, N. A., Jansen, C. A. M., Nathanielsz, P. W., and Tanaka, K. R.,** Mechanisms of red cell 2,3-diphosphoglycerate increase in neonatal lambs, *Blood,* 61, 920, 1983.
66. **Mooney, N. A. and Young, J. D.,** Nucleoside and glucose transport in erythrocytes from newborn lambs, *J. Physiol.,* 284, 229, 1978.
67. **Baumann, R., Bauer, C. H., and Rathschlag-Schraefer, A. M.,** Causes of the postnatal decrease in blood oxygen affinity in lambs, *Respir. Physiol.,* 15, 151, 1972.
68. **Kutas, F. and Stutzel, M.,** The organic-soluble phosphate contents of mammalian and avian erythrocytes at the beginning of postnatal life, *Experimentia,* 14, 214, 1958.
69. **Battaglia, F. G., McGaughy, H., Makowski, E. L., and Meschia, G.,** Postnatal change in oxygen affinity of sheep red cells: a dual role of diphosphoglyceric acid, *Am. J. Hematol.,* 219, 217, 1970.
70. **Bard, H., Fouron, J. C., Grothe, A. M., Soukini, M. A., and Cornet, A.,** The adaptation of the fetal red cells of newborn lambs to extrauterine life: the role of 2,3-diphosphoglycerate and adult hemoglobin, *Pediatr. Res.,* 10, 823, 1976.
71. **Bunn, H. F.,** Differences in the interaction of 2,3-diphosphoglycerate with certain mammalian hemoglobins, *Science,* 172, 1049, 1971.
72. **Granick, S. and Levere, R. D.,** Heme synthesis in erythroid cells, *Prog. Hematol.,* 4, 1, 1964.
73. **Awai, M., Okada, S., Takebayashi, J., Kubo, T., Inoue, M., and Seno, S.,** Studies on the mechanism of denucleation of the erythroblast, *Acta Haematol.,* 39, 193, 1968.
74. **Song, S. H. and Groom, A. C.,** Sequestration and possible maturation of reticulocytes in the normal spleen, *Can. J. Physiol. Pharmacol.,* 50, 400, 1972.
75. **Rapaport, S. M., Rosenthal, S., Schewe, T., Schultze, M., and Miller, M.,** The metabolism of the reticulocyte, in *Cellular and Molecular Biology of Erythrocytes,* Yoshikawa, H. and Rapaport, S. M., Eds., University Park Press, Baltimore, MD, 1974, 93.
76. **Heath, C. W. and Daland, G. A.,** The life of reticulocytes, *Arch. Intern. Med.,* 46, 533, 1930.
77. **Max-Audit, I., Testa, U., Kechemir, D., Titeux, M., Vainchenker, W., and Rosa, R.,** Pattern of pyruvate kinase isozymes in erythroleukemia cell lines and in normal human erythroblasts, *Blood,* 64, 930, 1984.
78. **Nijhof, W., Weirenga, P. K., and Goldlwasser, E.,** The regeneration of stem cells after a bone marrow depression induced by thiamphenicol, *Exp. Hematol.,* 10, 36, 1982.

79. **Nijhof, W. and Wierenga, P. K.**, Isolation and characterization of the erythroid progenitor cell: CFU-E., *J. Cell Biol.*, 96, 386, 1983.
80. **Nijhof, W., Wierenga, P. K., Staal, G. E. J., and Jansen, G.**, Changes in activities and isozyme patterns of glycolytic enzymes during erythroid differentiation in vitro, *Blood*, 64, 607, 1984.
81. **Takegawa, S. and Miwa, S.**, Change in pyruvate kinase (PK) isozymes in classical type PK deficiency and other PK deficiency cases during red cell maturation, *Am. J. Hematol.*, 16, 53, 1984.
82. **Allison, A. C. and Burn, G. P.**, Enzyme activity as a function of age in human erythrocytes, *Br. J. Haematol.*, 1, 291, 1955.
83. **Loehr, G. W. and Waller, H. D.**, Zellstoffwechsel und zellalterung, *Klin. Wochenschr.*, 37, 833, 1959.
84. **Chapman, R. G. and Schaumburg, L.**, Glycolysis and glycolytic enzyme activity of ageing red cells in man: changes in hexokinase, aldolase, glyceraldehyde-3-phosphate dehydrogenase, pyruvate kinase and glutamicoxaloacetic dehydrogenase, *Br. J. Haematol.*, 13, 655, 1967.
85. **Borun, E. R., Figueroa, W. G., and Perry, W. M.**, The distribution of Fe-59 tagged human erythrocytes as a function of cell age, *J. Clin. Invest.*, 36, 676, 1957.
86. **Piomelli, S., Lurinsky, G., and Wasserman, L. R.**, The mechanism of red cell aging. I. Relationship between cell age and specific gravity evaluated by ultracentrifugation in a discontinuous gradient, *J. Lab. Clin. Med.*, 69, 659, 1967.
87. **Morrison, M., Jackson, C. W., Mueller, T. J., Huang, T., Dockter, M. E., Walker, W. S., Singer, S. A., and Edwards, H. H.**, Does cell density correlate with cell age?, *Biomed. Biochim. Acta*, 42, 107, 1983.
88. **Corash, L. M., Piomelli, S., Chen, H. C., Seaman, C., and Gross, E.**, Separation of erythrocytes according to cell age on a simplified density gradient, *J. Lab. Clin. Med.*, 84, 147, 1974.
89. **Piomelli, S., Corash, L. M., Davenport, D. D., Miraglia, J., and Amorosi, E. L.**, In vivo lability of glucose-6-phosphate dehydrogenase in GdA- and Gd mediterranean deficiency, *J. Clin. Invest.*, 47, 940, 1968.
90. **Seaman, C., Wyss, S., and Piomelli, S.**, The decline in energetic metabolism and its relationship to cell death, *Am. J. Hematol.*, 3, 31, 1980.
91. **Beutler, E. and Hartman, G.**, Age-related red cell enzymes in children with transient erythroblastopenia of childhood and hemolytic anemia, *Pediatr. Res.*, 19, 44, 1985.
92. **Meuller, T. J., Jackson, C. W., Dockter, M. E., and Morrison, M.**, Use of an in vivo enrichment procedure to study membrane skeletal protein changes during red cell aging, in *Cellular and Molecular Aspects of Aging: the Red Cell as a Model*, Eaton, J. W., Konzen, D. K., and White, J. G., Eds., Alan R. Liss, New York, 1985, 277.
93. **Noble, N. A.**, The 2,3-DPG gene in rats: studies of mechanism and developmental stage of its action, in *Cellular and Molecular Aspects of Aging: the Red Cell as a Model*, Eaton, J. W., Konzen, D. K., and White, J. G., Eds., Alan R. Liss, New York, 1985, 389.
94. **Beutler, E.**, Biphasic loss of enzyme activity during in vivo aging, in *Cellular and Molecular Aspects of Aging: The Red Cell as a Model*, Eaton, J. W., Konzen, D., and White, J. G., Alan R. Liss, New York, 1985, 317.
95. **Beutler, E.**, Do red cell enzymes age? A new perspective, *Br. J. Haematol.*, 61, 377, 1985.
96. **Brok, F., Ramot, B., Zwang, E., and Danon, D.**, Enzyme activities in human red blood cells of different age groups, *Isr. J. Med.*, 2, 291, 1966.
97. **Salvo, G., Caprari, P., Samoggia, P., Mariani, G., and Salvati, A. M.**, Human erythrocyte separation according to age on a discontinuous Percoll density gradient, *Clin. Chim. Acta*, 122, 293, 1982.
98. **Luthra, M. G. and Kim, H. D.**, Ca^{2+} and Mg^{2+}-ATPase of density separated human red cell: effects of calcium and a soluble cytoplasmic activator (calmodulin), *Biochim. Biophys. Acta*, 600, 480, 1980.
99. **Hanahan, D. J. and Ekholm, J. E.**, The expression of optimum ATPase activities in human erythrocytes, *Arch. Biochem. Biophys.*, 187, 170, 1978.
100. **Cohen, N. S., Ekholm, J. E., Luthra, M. G., and Hanahan, D. J.**, Biochemical characterization of density separated human erythrocytes, *Biochim. Biophys. Acta*, 419, 229, 1976.
101. **Sass, M. D., Vorsanger, E., and Spear, D. W.**, Enzyme activity as an indicator of red cell age, *Clin. Chim. Acta*, 10, 21, 1964.
102. **Haidas, S., Labie, D., and Kaplan, J. C.**, II. 3-Diphosphoglycerate content and oxygen affinity as a function of red cell age in normal individuals, *Blood*, 38, 463, 1971.
103. **Edwards, M. J. and Rigas, D. A.**, Electrolyte labile increase in oxygen affinity during *in vivo* aging of hemoglobin, *J. Clin. Invest.*, 46, 1579, 1967.
104. **Vettore, L., DeMatteis, M. C., and Zampini, P.**, A new density gradient system for separation of human blood cells, *Am. J. Hematol.*, 8, 291, 1980.
105. **Jandl, J. H. and Katz, J. H.**, The plasma-to-cell cycle of transferrin, *J. Clin. Invest.*, 42, 314, 1963.
106. **Porzig, H., Baer, M., and Chanton, C.**, Properties of beta-adrenoceptor sites in metabolizing and non-metabolizing rat reticulocytes and in resealed reticulocyte ghosts, *Naunyn Schmiedebergs Arch. Pharmacol.*, 217, 286, 1981.

107. **Wiley, J. S. and Shaller, C. C.**, Selective loss of calcium permeability on maturation of reticulocytes, *J. Clin. Invest.*, 59, 1113, 1977.
108. **Shattil, S. J. and Cooper, R. A.**, Maturation of macroreticulocytes in vivo, *J. Lab. Clin. Med.*, 79, 215, 1972.
109. **Linderkamp, O. and Meiselman, H. J.**, Geometric and membrane mechanical properties of density separated human red cells, *Blood*, 59, 1121, 1982.
110. **Ekholm, J. E., Shukra, S. D., and Hanahan, D. J.**, Change in cytosolic activity of density (age) separated membrane Ca^{++}/Mg^{++} ATPase, *Biochem. Biophys. Res. Commun.*, 103, 407, 1981.
111. **LaCelle, P. L. and Arkin, B.**, Acquired rigidity: A possible determinant of normal red cell life span, *Blood*, 36, 837, 1970.
112. **Weed, R. I.**, Metabolic dependence of red cell deformability, *J. Clin. Invest.*, 48, 795, 1969.
113. **Weed, R. I.**, The importance of erythrocyte deformability, *Am. J. Med.*, 49, 147, 1970.
114. **Monzon, C. M., Penniston, J. T., Fairbanks, V. F., and Omer, E.**, Erythrocyte calmodulin correlates with red cell age, *Br. J. Hematol.*, 51, 261, 1982.
115. **Tillman, W., Levine, C., Prindull, G., and Shroter, W.**, Rheological properties of young and aged human erythrocytes, *Klin. Wochenschr.*, 58, 569, 1980.
116. **Kay, M. M. B.**, Senescent cell differentiation antigen, in *Cellular and Molecular Aspects of Aging: the Red Cell as a Model*, Eaton, J. W., Konsen, D. K., and White, J. G., Eds., Alan R. Liss, New York, 1985, 251.
117. **Low, P. S. and Waugh, S. M.**, The role of hemoglobin denaturation and band 3 clustering in red blood cell aging, *Science*, 227, 531, 1985.
118. **Boches, F. S. and Goldberg, A. L.**, Role for the adenosine triphosphate-dependent proteolytic pathway in reticulocyte maturation, *Science*, 215, 978, 1982.
119. **Goldberg, A. L. and Boches, F. S.**, Oxidized proteins in erythrocytes are rapidly degraded by the adenosine triphosphate-dependent proteolytic system, *Science*, 215, 1107, 1982.
120. **Pfeffer, S. R. and Swislocki, N. I.**, Role of peroxidation in erythrocyte aging, *Mech. Ageing Dev.*, 18, 355, 1982.
121. **Hochstein, P. and Jain, S. K.**, Association of lipid peroxidation and polymerization of membrane proteins with erythrocyte aging, *Fed. Proc.*, 40, 183, 1981.
122. **Turner, B. M., Fisher, R. A., and Harris, H.**, Post-translational alterations of human erythrocyte enzymes, *Isozymes*, 1, 781, 1975.
123. **Kahn, A., Biovin, P., Vibert, M., Cottreau, D., and Dreyfus, J. C.**, Post-translational modifications of glucose-6-phosphate dehydrogenase, *Biochimie*, 56, 1395, 1974.
124. **Paglia, D. E. and Valentine, W. N.**, Evidence for molecular alteration of pyruvate kinase as a consequence of erythrocyte aging, *J. Lab. Clin. Med.*, 76, 202, 1970.
125. **Bunn, H. F., Haney, D. N., Kamin, S., Gabbay, K. H., and Gallop, P. M.**, The biosynthesis of Hb A_{1C}: slow glycosylation of hemoglobin in vivo, *J. Clin. Invest.*, 57, 1652, 1976.
126. **Fluckiger, R. and Winterhalter, K. H.**, In vitro synthesis of hemoglobin ALC, *FEBS Lett.*, 71, 356, 1976.
127. **Koenig, R. J., Blobstein, S. H., and Cerami, A.**, Structure of the carbohydrate of hemoglobin A_{1C}, *J. Biol. Chem.*, 252, 356, 1976.
128. **Winterhalter, K. H.**, Determination of glycosylated hemoglobins, *Methods Enzymol.*, 76, 732, 1981.
129. **Trivelli, L. A., Ranney, H. M., and Lai, H. T.**, Hemoglobin components in patients with diabetes mellitus, *N. Engl. J. Med.*, 284, 3653, 1971.
130. **Bookchin, R. M. and Gallop, P. M.**, Structure of hemoglobin A_{1C}: nature of the N-terminal beta-chain blocking group, *Biochem. Biophys. Res. Commun.*, 32, 86, 1968.
131. **Bunn, H. F., Haney, D. N., Gabbay, K. H., and Gallop, P. M.**, Further identification of the nature and linkage of the carbohydrate in hemoglobin A_{1C}, *Biochem. Biophys. Res. Commun.*, 67, 103, 1975.
132. **Fitzgibbons, J. R., Koler, R. D., and Jones, R. T.**, Red cell age-related changes of hemoglobins Ala + b and A_{1C} in normal and diabetic subjects, *J. Clin. Invest.*, 58, 820, 1976.
133. **Higgins, P. J. and Bunn, H. F.**, Glycosylated hemoglobin in human and animal red cells: role of glucose permeability, *Diabetes*, 31, 743, 1982.
134. **Bunn, H. K., Gabbay, K., and Gallop, P.**, The glycosylation of hemoglobin: relevance to diabetes mellitus, *Science*, 200, 21, 1978.
135. **Nathan, D. M., Singer, D. E., Hurxthal, K., and Goodson, J. D.**, The clinical information value of the glycosylated hemoglobin assay, *N. Engl. J. Med.*, 310, 341, 1984.
136. **Bunn, H. F. and Briehl, R. W.**, The interaction of 2,3-diphosphoglycerate with various human hemoglobins, *J. Clin. Invest.*, 49, 1088, 1970.
137. **Clarke, S.**, The role of aspartic acid and asaragine residues in the aging of erythrocyte proteins, in *Cellular and Molecular Aspects of Aging: the Red Cell as a Model*, Eaton, J. W., Konzen, D. K., and White, J. G., Eds., Alan R. Liss, New York, 1985, 91.

138. **Masters, P. M.**, Senile cataracts and aging changes in human proteins, *J. Am. Gerontol. Soc.*, 31, 426, 1983.
139. **Oden, K. L. and Clarke, S.**, S-adenosyl-L-methionine synthetase from human erythrocytes: role in the regulation of cellular S-adenosylmethionine levels, *Biochemistry*, 22, 2978, 1983.
140. **Frietag, C. and Clarke, S.**, Reversible methylation of cytoskeleton and membrane proteins in intact human erythrocytes, *J. Biol. Chem.*, 56, 6102, 1982.
141. **O'Connor, C. M. and Clarke, S.**, Carboxyl methylation of cytosolic proteins in intact human erythrocytes, *J. Biol. Chem.*, 259, 2570, 1984.
142. **Barber, J. R. and Clarke, S.**, Inhibition of protein carboxyl methylation by s-adenosyl-L-homocysteine in intact erythrocytes, *J. Biol. Chem.*, 259, 7115, 1984.
143. **Galletti, P., Ingrosso, D., Nappi, A., Gragnaniello, V., Iolascon, A., and Pinto, L.**, Increased methyl esterification of membrane proteins in aged red blood cells: preferential esterification of ankyrin and band 4.1 cytoskeletal proteins, *Eur. J. Biochem.*, 135, 25, 1983.
144. **Clarke, S.**, Protein carboxyl methyltransferases: two distinct classes of enzymes, *Annu. Rev. Biochem.*, 54, 479, 1985.
145. **Kiese, M.**, *Methemoglobinemia: a Comprehensive Treatise*, CRC Press, Cleveland, Ohio, 1974.

DEVELOPMENTAL PATTERNS OF HUMAN HEMOGLOBIN SYNTHESIS

Sandra F. Schnall and Edward J. Benz, Jr.

INTRODUCTION

Several distinct hemoglobins are synthesized by human subjects in a developmentally regulated manner during embryonic, fetal, and adult life. The mechanisms controlling the switches from embryonic to fetal and then to adult hemoglobin production have become the focus of intense research efforts for two reasons. Firstly, human hemoglobin switching is an accessible and reasonably well-characterized system in which the developmental regulation of specific genes can be approached. Secondly, it is well established that humans exhibit heterogeneity for the amount of fetal hemoglobin produced in adult life[1] and that those individuals with sickle cell anemia and beta thalassemia who produce more hemoglobin F experience less morbidity. Thus, a better understanding of the mechanisms controlling hemoglobin switching can be expected to yield important new approaches for treatment of these disorders.

This brief survey of human hemoglobin switching introduces some of the cellular and molecular elements which have been demonstrated to be relevant to the process of developmental globin gene regulation. We shall focus almost exclusively on the switch from fetal (hemoglobin F) to adult (hemoglobin A) hemoglobin production, since this switch is the most thoroughly studied and the most relevant to therapy for hemoglobinopathies. We shall also consider briefly some inherited and acquired abnormalities of fetal hemoglobin production which suggest that therapeutic reactivation of hemoglobin F production is possible.

DEVELOPMENTAL BIOLOGY OF HEMOGLOBIN SWITCHING

The ontogenic development of the human red cell and its hemoglobin content exhibits several interrelated features including changes in the organ site of production, the morphology of the red cell, and the types of hemoglobin predominating in red cells at various developmental ages.[2] Human erythroid cells can first be detected in the primitive yolk sac as early as 2 to 3 weeks postconception (<10-mm crown rump length). Yolk sac-derived red cells appear to predominate in the circulation from the fourth to tenth weeks of gestation. These "primitive lineage" cells differ from adult red cells in that they consist of large basophilic, nucleated megaloblasts. These cells appear to arise from the mesenchymal layer.

The "definitive lineage" appears to be derived predominately from the fetal liver which is also a hematopoietic organ during gestation. Definitive cells are apparent as early as 6 to 8 weeks of gestation. These cells more closely resemble typical adult red cells, although they are clearly macrocytic when compared to adult cells. Like adult cells, they are enucleate, but they do exhibit several distinctive properties, such as predominance of the "i" rather than the "I" antigen and absence of carbonic anhydrase B.

Bone marrow-derived red cell production does not begin until the third trimester. The definitive red cell line predominates by the 10th to 12th week of gestation, so that only trace amounts of embryonic hemoglobin are evident beyond the first trimester.[3] The overall patterns of red cell and hemoglobin production during embryonic and fetal development are summarized in Figure 1.

Hemoglobin production during human development is remarkable for the diversity of hemoglobin subtypes produced. As in other species, each of these hemoglobins consist of a tetramer of two "alpha-like" and two "nonalpha" globin chains. The individual hemoglobins are identified by the composition of globin subunits they contain. The first hemo-

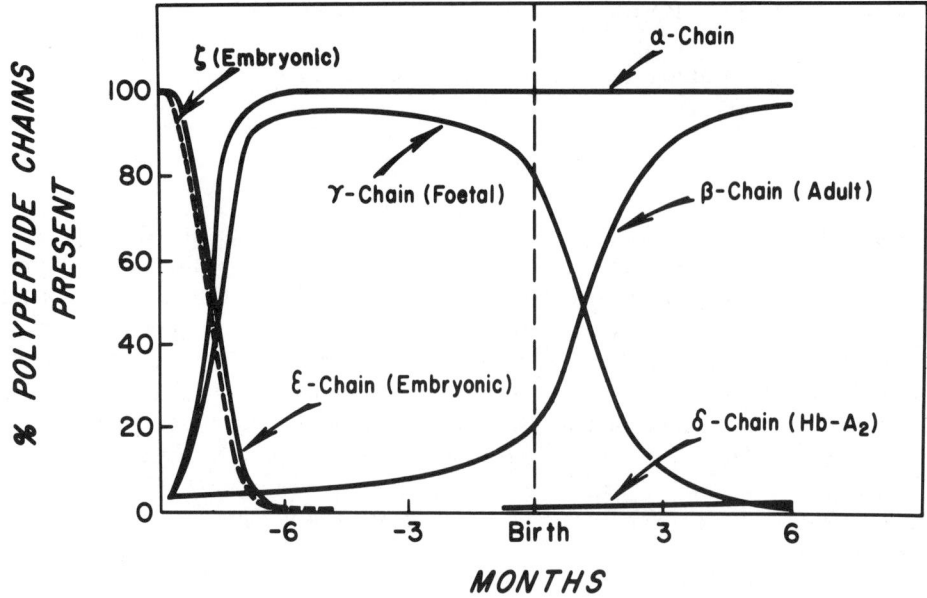

FIGURE 1. Pattern of hemoglobin chain production during development.

globins produced are three embryonic hemoglobins: Hb Gower I (zeta$_2$ epsilon$_2$), Hb Gower II (alpha$_2$ epsilon$_2$), and Hb Portland (zeta$_2$ gamma$_2$). The zeta and epsilon chains represent embryonic counterparts of the adult alpha and beta chains, respectively, each exhibiting about 60 to 65% nucleotide sequence homology with the adult genes. Gamma globin is the fetal counterpart of adult beta globin.

Recent evidence suggests that the primitive erythroblast line synthesizes predominately zeta and epsilon globin chains and that the presence of hemoglobin Portland in the primitive erythrocytes implies synthesis of at least small amounts of gamma chains as well.[4] These hemoglobins are predominant at about 5 weeks of gestation, but rapidly decline as a proportion of total hemoglobin production thereafter. Although this decline coincides with the ascendency of hepatic erythropoiesis, some embryonic hemoglobins are produced in early hepatic erythroblasts.[5] By the eighth or ninth week of gestation, embryonic hemoglobin virtually disappears.

Fetal Hemoglobin

Hemoglobin F is the predominant (>90%) hemoglobin produced throughout the remainder of fetal life until about 35 to 38 weeks of gestation. Fetal hemoglobin is itself heterogenous, consisting of two isoforms: the G_{gamma} and A_{gamma} forms. These two forms differ in the gamma chain component of the hemoglobin tetramer: G_{gamma} chains contain glycine at position 136, whereas A_{gamma} chains contain alanine at this position. The two types of gamma chains arise from two closely linked and highly homogolous gamma globin genes. The relative proportions of G_{gamma} and A_{gamma} hemoglobin F exhibit interesting variations during development. The G_{gamma}/A_{gamma} ratio in fetuses (from about 12 weeks of gestation until the third trimester) is 3:1, whereas in adults this ratio changes to 2:3, concomitant with a very dramatic decline in the total amount of fetal hemoglobin produced. The controls determining this change, and its purpose, if any, remain unknown.

Adult hemoglobins are synthesized in small amounts during gestation. No evidence for production of adult beta chains in the primitive cell line has been obtained, but beta chain synthesis has been detected as early as 10 to 12 weeks in the definitive line. This small

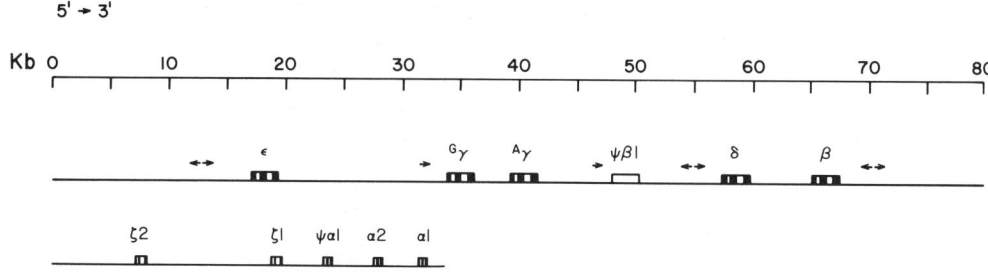

FIGURE 2. Structural characteristics of the β-globin and α-globin gene clusters

amount of adult hemoglobin synthesis in early fetuses forms the basis for antenatal diagnosis of beta chain hemoglobinopathies by measurements of globin chain synthesis in fetal blood samples.[6] This was the only technique available before the onset of recombinant DNA methods for direct examination of globin genes. Fetal blood sampling for this purpose is still done in occasional patients for disorders not readily diagnosed by gene mapping.

The "fetal to adult" hemoglobin switch begins almost invariably at about 35 weeks of gestation.[7] As indicated in Figure 1, a dramatic increase in the production of beta chains and, therefore, of hemoglobin A transpires at this point. A reciprocal decline in gamma chain and hemoglobin F production is also apparent, so that total hemoglobin production changes very little.[8] As indicated by the figure, this switch is essentially irreversible under normal physiologic conditions. It does not correlate well with the change from hepatic to marrow erythropoiesis. Indeed, production of both hemoglobins at both sites can be documented.[9]

The mechanisms controlling the complicated and exquisite switching system just outlined have proved to be exceedingly complex. Intense research has not yet totally clarified the underlying regulatory schemata. Nonetheless, two fundamental levels of control have been discerned. The first involves regulation at the level of the individual globin genes within each red cell precursor. The second involves determination of the potential for gamma and beta globin gene expression in different subpopulations of primitive erythroid stem cells. These cellular mechanisms determine which progenitors shall be "F-cells", having some capacity to produce hemoglobin F, and which shall be "A-cells", having no capacity to produce fetal hemoglobin.

MOLECULAR GENETIC BASIS FOR FETAL AND ADULT HEMOGLOBIN PRODUCTION

The human alpha-like genes (alpha and zeta) are closely linked to one another on chromosome 16; the alpha gene is duplicated. Only one functional zeta gene exists, although a "psuedo gene" rendered nonfunctional by accumulated mutations is also present in the cluster. The nonalpha genes (epsilon, G_{gamma}, A_{gamma}, delta, and beta) are located in a 60-kilobase cluster on chromosome 11. Two pseudo genes are also present in this cluster. The delta gene is expressed at low levels only in adult life, causing formation of small amounts (1 to 3%) of hemoglobin A2 ($alpha_2$ $delta_2$). It is interesting that both the alpha and nonalpha globin gene clusters exhibit a linkage pattern which correlates with their developmental patterns of expression. As shown in Figure 2, the embryonic genes appear "first" in the $5' \rightarrow 3'$ transcriptional orientation of the linkage maps and the adult genes appear last.

Evidence for intracellular regulation of globin genes during hemoglobin switching comes from several sources. Firstly, despite the remarkable changes in globin gene production occurring at different ages, the balance of alpha and nonalpha globin production remains

Table 1
HPFH VARIETIES

Type	HbF isoform elevated	HbF level (Heterozygote) (%)	Cellular distribution
Deletion			
HPFH I	$G_\gamma A_\gamma$	~30	Pancellular
HPFH II	$G_\gamma A_\gamma$	~30	Pancellular
Non deletion			
Greek	A_γ	10—20	Pancellular
Black	G_γ	15—20	Pancellular
Italian	A_γ	10—20	Pancellular
Hemoglobin Kenya[a]	G_γ	4—10	Pancellular
British, Swiss, etc.	Both variable	2—10	Heterocellular

Note: HPFH, hereditary persistence of fetal hemoglobin.

[a] Synthesis of Hb Kenya $[^G\gamma(\delta\beta)^+$ HPFH] is also persistent.

equal and constant. In other words, total nonalpha globin production is always equal to total alpha globin production. Secondly, several mutations exist which alter both the quantity and the cellular patterns of hemoglobin (HbF) production after birth. Thirdly, the perinatal switch from hemoglobin F to hemoglobin A is also accompanied by a switch from predominance of the G_{gamma} form to predominance of the A_{gamma} form.

Most adults synthesize only about 1% of hemoglobin F. As discussed later, the small amount of fetal hemoglobin is restricted to a distinct subpopulation of cells called "F"-cells.[10] Certain rare but well-documented individuals who produce larger amounts of hemoglobin F have been identified.[11] These patients have "hereditary persistence of fetal hemoglobin" (HPFH). Two categories have been defined: pancellular, characterized by uniformly high levels of hemoglobin F production in all of the patient's red cell populations, and heterocellular, characterized by an absolute increase in the number of F-cells. Patients with heterocellular HPFH exhibit a heterogenous cell-to-cell distribution of hemoglobin F and are thought to represent individuals with mutations in the genes which determine F-cell and A-cell number. In contrast, patients with pancellular HPFH exhibit mutations in the nonalpha globin gene cluster which seem to alter the molecular regulation of postnatal gamma and beta globin gene activity.

Two major types of pancellular HPFH have been defined: those arising from gene deletions in the nonalpha globin gene cluster and those not associated with deletions (nondeletion forms) (Table 1). The most important observation concerning all of these patients is that they are clinically well. These patients provide the best support for the hypothesis that abolition or reversal of switching will provide an effective therapy for hemoglobinopathies.

As shown by Figure 3, several types of gene deletions have been documented in the nonalpha gene cluster. Some deletions produce HPFH, whereas others produce a relatively mild form of thalassemia called delta beta thalassemia. In delta beta thalassemia, there is total absence of delta and beta globin synthesis, as well as moderate fetal hemoglobin production in a heterocellular distribution. The levels of hemoglobin F are significantly higher than those seen in severe "classical" forms of thalassemia, but not as high as seen in the classical forms of HPFH. This intermediate and heterogenous level of hemoglobin F production accounts for the milder degree of morbidity.

The DNA defects in these disorders correlate with the clinical features. HPFH is associated with larger deletions of DNA, whereas delta beta thalassemia patients exhibit smaller deletions which usually leave the gamma-delta intergenic region largely intact. This observation raised the exciting possibility that those regions of DNA deleted only in HPFH

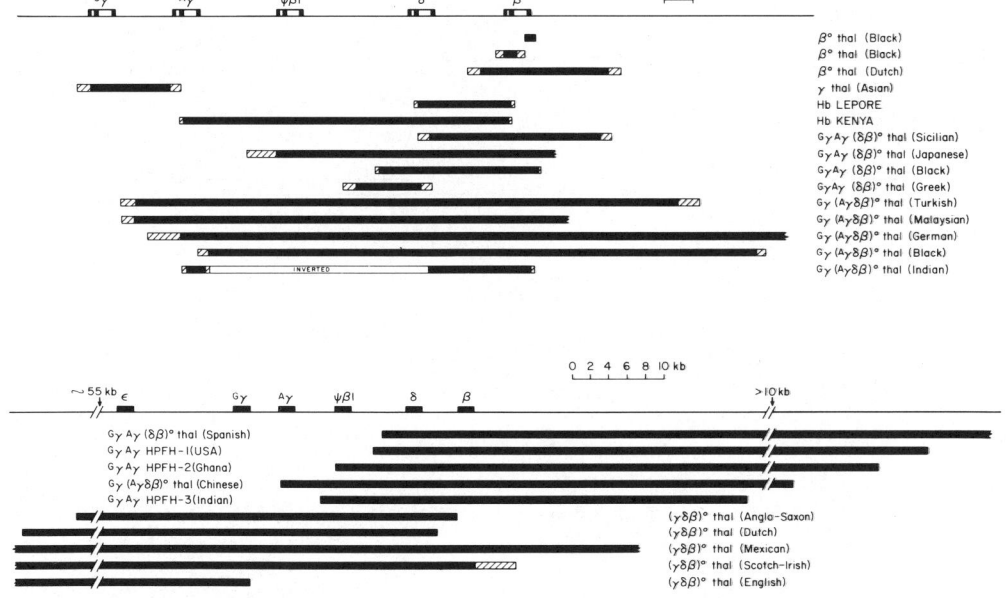

FIGURE 3. Gene deletions in the non-α-globin genecluster.

contained regulating sequences which promoted this perinatal switch. However, as indicated in Figure 3, no real conclusions can be drawn from the size of the deletions or the particular regions of DNA removed within the gene cluster.

The exact mechanism responsible for the deletion forms of pancellular HPFH remains unknown. However, innovative and promising hypotheses have emerged.[12] These focus on the content of the downstream DNA, just beyond the end point of the deletion. This DNA is brought into the proximity of the gamma globin genes by the deletion. One postulate suggests that removal of this large amount of DNA simply disrupts chromatin structure in such a way that the gamma globin genes can no longer remain repressed. Another suggests that the DNA at the end point of the deletion contains elements which promote activity of nearby genes. Activation of the gamma globin genes in adult life arises from placement of these stimulating elements in the vicinity of the gamma genes, as a result of the deletion. A number of studies also show that genes inserted into the active regions of chromatin become activated.[13] Thus, if the DNA at the deletion end points were transcriptionally active in adult erythroid cells, the gamma globin genes might become "passively" activated simply by being brought into the vicinity of these active regions.

Recent data[14] suggest strongly that pancellular HPFH active DNA is brought into the globin gene cluster as a result of at least one type of HPFH deletion. This DNA region, which is located only a few kilobases from the gamma genes in HPFH, is specifically hypomethylated in adult erythroid cells (hypomethylation of DNA correlates with transcriptional activity). This region also contains DNAse I hypersensitive sites which are erythroid specific, as well as a long "open reading frame" of DNA sequence. DNAse I hypersensitive sites are structural features of chromatin correlating with transcriptional activity, whereas open reading frames are sequences of DNA which can be translated into protein. Enhancer-like elements have also been found in this DNA. There is thus much circumstantial evidence that the configuration and activity of this DNA may influence the activity of the gamma globin genes.

Nondeletion forms of pancellular HPFH occur in black and Mediterranean populations.[15] The nonalpha globin gene cluster is largely intact in these individuals. Levels of fetal

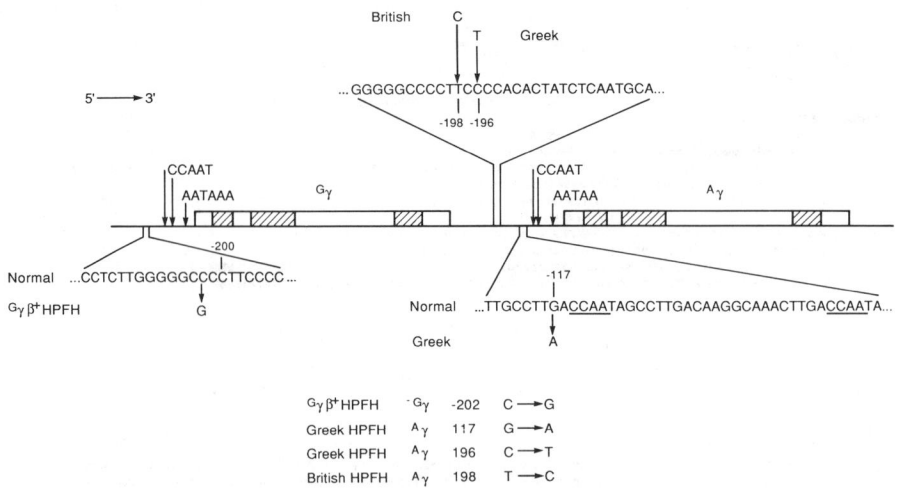

FIGURE 4. Nondeletion HPFH mutations.

hemoglobin production are, however, somewhat lower than in the deletion forms. Gamma globin genes isolated by molecular cloning from individuals with these forms of HPFH have revealed point mutations in the "upstream" or 5' flanking regions of the genes, in the regions known to contain promoters or tissue-specific regulatory signals. The location and nature of the mutations in several representative cases are shown in Figure 4.

The current hypothesis concerning the molecular basis of high fetal hemoglobin production in nondeletion pancellular HPFH is that these mutations increase the efficiency of the promoter regions in adult life, so that the affected gamma gene is transcribed at a higher rate. In each case, the type of fetal hemoglobin (A_{gamma} or G_{gamma}) overproduced by the patient correlates perfectly with the location of the mutation. For example, the single base change for a G_{gamma} form of HPFH is located 202 bases upstream; it creates a promoter which more closely resembles the highly active promoter seen attached to the thymidine kinase gene of herpes simplex virus.

In addition to pancellular and heterocellular forms of HPFH, polymorphisms determined by changes within the nonalpha globin gene cluster may exist in the "normal" members of at least some ethnic groups. Recent evidence suggests that patients with sickle cell anemia from Senegal exhibit a higher level of G_{gamma} hemoglobin F production than patients from Benin.[16,17] Moreover, globin gene mapping studies suggest that the beta^S gene in each of these populations arose independently during evolution on different "haplotypes" (see Chapter 9 by Steinberg). The amount of hemoglobin F produced correlates with the haplotype of the nonalpha globin gene cluster on which each sickle mutation is inherited. Mediterranean patients with thalassemia in whom the thalassemia gene arose on chromosomes with a haplotype resembling the Senegal form of sickle cell anemia also exhibit high levels of fetal hemoglobin production. Conversely Mediterranean thalassemic patients with haplotypes resembling the Benin chromosome exhibit lower levels of hemoglobin F. In all of these situations, the differences and the high levels of hemoglobin F production are apparent only in individuals homozygous for the hemoglobinopathy, suggesting that the potential of these haplotypes to stimulate hemoglobin F synthesis in adult life is apparent only under conditions of severe erythroid stress.

Another molecular mechanism which seems to modulate hemoglobin switching is based on the fact that human DNA is heavily methylated, largely by the addition of methyl groups to cytosines. Methylation of DNA, particularly in promoter regions of genes, correlates with transcriptional inactivity in some gene systems, whereas hypomethylation correlates with transcriptional activation. Both seem to be involved in globin gene regulation. In embryonic life, the epsilon gene is under methylated, whereas the gamma region is methylated. At 10 weeks of gestation, the gamma genes become under methylated at a time prestaging the switch to predominance of hemoglobin F. In the adult, the beta genes are hypomethylated, whereas the inactive gamma genes are methylated. Therefore, the methylation status of the globin genes appears to be at least one criterion determining their activity at different developmental ages.[18]

In summary, considerable progress has been made toward defining the structure of normal globin genes as well as characterizing the structural and functional consequences of mutations which alter the developmental regulation of hemoglobin switching. Unfortunately, the impressive progress made during the past few years has not yielded a unifying hypothesis about the molecular basis for developmental regulation of the globin genes. Further progress will require a far better understanding of the complex cellular mechanisms which also influence the types of hemoglobins produced during normal development.

CELLULAR BASIS FOR FETAL AND ADULT HEMOGLOBIN PRODUCTION

Trace amounts (approximately 1%) of hemoglobin F can be detected in normal human adults. This fetal hemoglobin is restricted to a small subpopulation of red cells called F-cells, which are detectable by immunologic tests which measure fetal hemoglobin present in individual red cells immobilized in semisolid media.

F-cells and A-cells (cells producing only adult hemoglobin) have been shown to arise from distinct subpopulations of primitive erythroid precursors. All human red cells arise from primitive committed progenitor cells which develop in the bone marrow from the pluripotent hematopoietic stem cell.[19,20] These cells, called burst forming units-erythroid (BFU-E) proliferate and undergo further differentiation. The capacity of the progeny of these stem cells to produce fetal hemoglobin appears to be determined primarily at this BFU-E stage. Later progenitors (colony forming units-erythroid, CFU-E, and recognizable erythroblasts) can only execute the program of fetal and adult hemoglobin distribution determined during the BFU-E stage.[21]

In general, studies of erythroid stem cells suggest that the potential of stem cell progenitors for hemoglobin F production from BFU-E and CFU-E diminishes as the stem cells differentiate and mature.[22] For example, when immature BFU-E from human peripheral blood are grown in tissue culture, the BFU-E skip the differentiation steps and proceed directly to the phase of maturation into erythroblasts; they thus produce very high levels of hemoglobin F. On the other hand, more mature BFU-E and CFU-E found primarily in the bone marrow produce predominately hemoglobin A, both in culture and *in vivo*. There thus appear to be regulatory mechanisms in stem cells whereby each cell may remain in the proliferating stem cell pool or be drawn out of that pool into the phase of maturation. Even in the proliferating stem cell pool, stem cells continue to differentiate with respect to globin gene expression, so that cells removed from the pool for maturation at late stages have less ability to make fetal hemoglobin than cells diverted toward the maturation pathway at earlier stages.[23,24]

It is not yet entirely clear how closely the complex cellular systems regulating low levels of HbF production in adults apply to the perinatal hemoglobin switch. Complex subpopulations of stem cell pools with varying hemoglobin F- and hemoglobin A- producing potentials develop around the time of the fetal hemoglobin switch. What is not clear is whether the signals which cause the appearance of adult cells are intrinsic or primarily extrinsic. Recent evidence from at least one laboratory suggests that a factor found in the plasma of fetal sheep (who have an analogous fetal hemoglobin switch) can stimulate the conversion of fetal BFU-E to a more adult program of globin gene expression.[25]

Table 2
ACQUIRED CONDITIONS ASSOCIATED WITH INCREASED HbF

Condition	Approximate level HbF (%)
JCML	Up to 70
Fanconi's anemia	2—85
Erythroleukemia	Up to 60
PNH	1—20
Aplastic anemia—acquired	1—20
Recovery from marrow engraftment/chemotherapy	1—20
Refractory anemia—preleukemia	1—20
Pregnancy — second trimester	1—10
Choriocarcinoma	1—10
Hydatidiform mole	1—10
Trophoblastic/testicular tumors	1—10
Cytotoxic agents ?DNA hypomethylation-ARA-C ?Stem cell disturbance-ARA-C, hydroxyurea, vinca alkaloids	Variable

Note: Abbreviations used: JCML, juvenile chronic myelogenous leukemia; PNH, paroxysmal nocturnal hemoglobinuria.

Direct understanding of the cell biology of fetal hemoglobin production has been hampered by the lack of a good *in vitro* cell model. Studies with stem cells obtained from patients have been elegantly performed and have yielded a rich harvest of information, but these assays do not permit one to conduct direct biochemical or molecular analysis of the differentiation events.

Several new avenues of research have developed which promise to deliver more definitive information. First, patients with mutations increasing the number of F-cells have been defined. These patients are said to have heterocellular HPFH. Attempts are already underway in several laboratories to clone the one or more genes which control this trait. Second, a human cell line called KMOE-2 has been developed which expresses predominately fetal hemoglobin when stimulated by incubation with hemin, but switches over to limited production of hemoglobin A when incubated with hemin and cytosine arabinoside.[26] This cell line should serve as a useful cellular host for analysis of genes which might affect the switching phenomenon. Finally, impressive progress is being made in a number of laboratories toward the goal of physically purifying erythroid stem cells.

REACTIVATION OF FETAL HEMOGLOBIN DURING POSTNATAL LIFE

Hemoglobin F is occasionally elevated in hereditary hemoglobinopathies (e.g., sickle cell disease and thalassemias) or certain acquired conditions (e.g., leukemias, aplastic anemias, and paroxysmal nocturnal hemoglobinuria) (Table 2). HbF also rises during pregnancy. In most of these situations, a rise in hemoglobin F is due primarily to an increase in the number of F-cells, rather than uniform increases in hemoglobin F production per red cell. Only in juvenile chronic myelogenous leukemia (JCML) (discussed later) is a uniform reversion of most red cell populations to true fetal erythropoiesis seen.

The rise in fetal hemoglobin seen in pregnancy has been regarded as evidence that humoral factors can modulate fetal and adult hemoglobin production.[27] During the first trimester of pregnancy (10 to 12 weeks), about 15 to 20% of women exhibit very modest increases in hemoglobin F, doubling the hemoglobin F percent to the 1 to 2% range. Another, seemingly independent, elevation in hemoglobin F occurs at about 20 to 25 weeks of pregnancy in 21% of females. In virtually all cases, however, fetal hemoglobin levels return to normal by the latter stages of the third trimester. Several studies have documented the fact that the fetal hemoglobin produced in these pregnant women is derived from maternal stem cells rather than from fetal red cells from transplacental bleeding or other mechanisms.[28,29] For example, the G_{gamma}/A_{gamma} ratio seen in these patients is the adult (2:3) rather than the fetal (3:1) ratio.

In most patients with hemoglobinopathies sufficient to produce a significant anemia, rises in hemoglobin F production are seen. These rises have been documented to be due to increases in F-cell number in the vast majority of these patients.[30] Some variability in fetal hemoglobin production can also be inherited on different nonalpha globin haplotypes, as noted earlier. These patients suffer from severe chronic erythroid stress due to hemolytic anemia and/or ineffective erythropoiesis. Consequently, they have massively expanded erythroid compartments in the bone marrow due to chronically high levels of both erythropoietin and burst-promoting activity. Presumably, these ''stress erythropoiesis'' conditions promote higher HbF levels.

A variety of acquired conditions associated with moderate levels of hemoglobin F are probably also due to the ''stress erythropoiesis'' phenomenon.[31] Thus, patients recovering from bone marrow transplantation, chemotherapy-induced hypoplasia, or aplastic anemia exhibit transient elevations in hemoglobin F which disappear as the hematocrit returns to normal. Some myelodysplastic disorders, such as paroxysmal nocturnal hemoglobinuria and preleukemias, also exhibit elevated hemoglobin F levels. The majority of these patients are thought to exhibit reversions to fetal phenotypes comparable to those seen in a variety of neoplastic conditions due to shortening of the cell cycle (Table 2).

A particularly interesting phenomenon of reversion to fetal hemoglobin production occurs in children with juvenile chronic myelogenous leukemia (JCML).[32] Some of these patients exhibit striking rises in the level of fetal hemoglobin. In contrast to the rise in F-cell number seen in other conditions, these patients exhibit a widespread increase in fetal hemoglobin production per cell with characteristics of true fetal erythropoiesis: absence of adult carbonic anhydrase and other isoenzymes, fetal G_{gamma} to A_{gamma} ratios, absence of hemoglobin A2, fetal surface antigen patterns, etc. Moreover, the quantitative extent of hemoglobin F and fetal red cell production correlates directly with the progress of the disease. Remission in JCML is associated with reversion to an adult erythropoiesis pattern, whereas progression of the disorder correlates with increased fetal erythropoiesis. One hypothesis concerning this condition is that the leukemic transformation represents reactivation of fetal stem cells.

Under most physiologic conditions, humans do not have the capacity to alter or rearrange their genetic program for HgF and HgA synthesis. Certain stimuli can redirect this program to a small degree. A variety of hormonal stimuli, such as thyroid hormone, growth hormone, and human chorionic gonadotrophic hormone, have also been said to alter fetal and adult hemoglobin production in culture systems.[33,34] The clinical relevance of these phenomena has as yet not been demonstrated.

THERAPEUTIC MANIPULATION OF FETAL HEMOGLOBIN PRODUCTION

One of the most exciting recent developments concerning fetal and adult hemoglobin production has been the publication of several reports suggesting that fetal hemoglobin levels can be manipulated in patients with beta chain hemoglobinopathies by the use of cytotoxic and chemotherapeutic agents (see below). At the present time, the effects obtained *in vivo* are insufficient to offset the unacceptable toxicity and long-term neoplastic risk. Nonetheless, the fact that hemoglobin F levels can be augmented in these patients offers great hope for the development of effective therapy.

As noted earlier, promoters near the human gamma globin genes are heavily methylated in their adult inactive state, but are hypomethylated in their fetal, active state. 5-Azacytidine, a cell cycle (S-phase) specific cytotoxic agent which inhibits methylation of DNA, has been shown to activate a variety of genes in tissue culture systems.[35,36] The activation of genes in these model systems correlates with the hypomethylating effect of the drug. 5-Azacytidine was thus given to baboons who had previously been shown to produce large amounts of HbF in response to severe anemias.[37] The drug alone reproduced this HbF reactivation without causing anemia. This suggested that hypomethylation alone could reactivate the gamma genes. When administered in controlled short-term trials to selected human patients, 5-azacytidine promoted significant increases in the amount of hemoglobin F.[38] In some patients with beta thalassemia, the drug generated enough hemoglobin F to eliminate transfusion requirements. However, the use of 5-azacytidine in therapeutic trials has generated considerable controversy about its mechanism of action.[39] In addition to its effect on methylation of genes, 5-azacytidine is also cytotoxic. Treatment with even modest doses of the drug has a cytoreductive effect on red cell progenitors, creating a state of stress erythropoiesis. Unfortunately, the drug is quite toxic, its beneficial effects are short lived, and the long-term risk of neoplastic transformation is theoretically substantial.

Other cytotoxic cell cycle specific agents, hydroxyurea and cytosine arabinoside (S-phase) and vinblastine (M-phase), have been shown to promote HgF production even at modest doses. A number of studies focusing on the mechanism of action of these agents have yielded a wealth of data which do not yet offer a clear picture as to the mechanisms underlying the stimulation of fetal hemoglobin production.[40,41] The magnitude of the effect on hemoglobin F levels is greater than one might expect from the modest marrow toxic effects. These drugs may thus have a stronger effect on the more rapidly cycling late erythroid progenitors (which have less capacity to express gamma globin genes) than on the more slowly cycling primitive progenitors. The direct hypomethylating effect of 5-azacytidine may augment the cellular effects seen with all the drugs. Even though these early studies represent an intellectual breakthrough and define a new approach to hemoglobinopathies, none of these agents are yet ready for widespread clinical trials.

SUMMARY

The developmental patterns of hemoglobin production during ontogeny offer a rich area for study of the mechanisms controlling developmentally regulated genes. Understanding of this system also offers the current best hope for effective "treatments" (cures) of sickle cell anemia, thalassemia, and other hemoglobinopathies. As indicated in this review, hemoglobin switching during human development proceeds through a number of well-regulated phases at the level of the individual globin genes and at the level of selective expansion and maturation of erythroid progenitor cell populations. The capacity of cells to produce fetal or adult hemoglobin appears to be determined not only by developmental age, but also by "differentiation age" of the progenitor cells. Although many questions remain to be answered, progress to date has been impressive and many other studies needed for further clarification of the question appear to be at hand. Further intense investigation of the system appears to be warranted, for it might provide major insights into the interaction of cellular and genetic mechanisms controlling development.

ACKNOWLEDGMENTS

The authors are supported by Grants HL 24385 and AM 28076 from the National Institutes of Health. Sandra F. Schnall is the recipient of a National Research Service Award Fellowship from The National Institutes of Health. Edward J. Benz, Jr. is the recipient of a Research Career Development Award from The National Institutes of Health and the Edward Paradiso Research Grant from The Cooley's Anemia Foundation.

REFERENCES

1. **Boyer, S. H., Belding, T. K., Margolet, L., Noyes, A. N., Burke, P. J., and Bell, W. R.**, Variations in the frequency of fetal hemoglobin-bearing erythrocytes (F-cells) in well adults, pregnant women and adult leukemias, *Johns Hopkins Med. J.*, 137, 105, 1975.
2. **Benz, E. J., Jr.**, Hemoglobin switching in animals, *Tex. Rep. Biol. Med.*, 40, 111, 1980.
3. **Fantoni, A., Farace, M. G., and Gambari, R.**, Embryonic hemoglobins in man and other mammals, *Blood*, 57(4), 623, 1981.
4. **Pesche, C., Mavilio, F., Care, A., Migliaccio, G., Migliacco, A. R., Salvo, G., Samoggia, P., Petti, S., Guerriero, R., Marinucci, M., Lazzaro, D., Russo, G., and Mastroberardino, G.**, Haemoglobin switching in human embryos: asynchrony of zeta → alpha and epsilon → gamma-globin switches in primitive and definitive erythropoietic lineage, *Nature (London)*, 313, 235, 1985.
5. **Pesche, C., Migliaccio, A. R., Migliaccio, G., Petrini, M., Calandrini, M., Russo, G., Mastroberardino, G., Presta, M., Gianni, A. M., Comi, P., Giglioni, B., and Ottolenghi, S.**, Embryonic → fetal Hb switch in humans: studies on erythroid bursts generated by embryonic progenitors from yolk sac and liver, *Proc. Natl. Acad. Sci. U. S. A.*, 81, 2416, 1984.
6. **Alter, B. P.**, Advances in the pre-natal diagnosis of hematologic diseases, *Blood*, 64(2), 329, 1984.
7. **Terrenato, L., Bertilaccio, C., Spinelli, P., and Colombo, B.**, The switch from haemoglobin F to A: the time course of qualitative and quantitative variations of haemoglobins after birth, *Br. J. Hematol.*, 47, 31, 1981.
8. **Colombo, B., Kim, B., Atencio, R. P., Molina, C., and Terrenato, L.**, The pattern of fetal haemoglobin disappearance after birth, *Br. J. Hematol.*, 32, 79, 1976.
9. **Wood, W. G. and Weatherall, D. J.**, Haemoglobin synthesis during human foetal development, *Nature (London)*, 244, 162, 1973.
10. **Boyer, S., Belding, T. K., Margolet, L., and Noyes, A. N.**, Fetal hemoglobin restriction to a few erythrocytes (F cells) in normal human adults, *Science*, 188, 361, 1975.
11. **Wood, W. G., Stamatoyannopoulos, G., Lim, G., and Nute, P. E.**, F-cells in the adult: normal values and levels in individuals with hereditary and acquired elevations of HgF, *Blood*, 46(5), 671, 1975.
12. **Weatherall, D. J., Wood, W. G., Jones, R. W., and Clegg, J. B.**, The developmental genetics of human hemoglobin, in *Experimental Approaches for the Study of Hemoglobin Switching*, Stamatoyannopoulos, G. and Neinhus, A., Alan R. Liss, New York, 1985, 3.
13. **Tuan, D., Feingold, E., Newman, M., Weissman, S. M., and Forget, B. G.**, Different 3' end points of deletions causing delta beta-thalassemia and hereditary persistence of fetal hemoglobin: implications for the control of gamma-globin gene expression in man, *Proc. Natl. Acad. Sci. U. S. A.*, 80, 6937, 1983.
14. **Feingold, E. A., Collins, F. S., Metherall, J. E., Stoeckert, C. J., Weissman, S. M., and Forget, B. G.**, Molecular analysis of deletion and non-deletion hereditary persistence of fetal hemoglobin and identification of a new mutation causing beta-thalassemia, *Ann. N.Y. Acad. Sci.*, 445, 159, 1985.
15. **Collins, F. S., Stoeckert, C. J., Serjeant, G. R., Forget, B. G., and Weissman, S. M.**, G/gamma beta+ hereditary persistence of fetal hemoglobin: cosmid cloning and identification of a specific mutation 5' to the G/gamma gene, *Proc. Natl. Acad. Sci. U. S. A.*, 81, 4894, 1984.
16. **Harano, T., Reese, A. L., Ryan, R., Abraham, B. L., and Huisman, T. H. J.**, Five haplotypes in black beta-thalassemia heterozygotes: three are associated with high and two with low G/gamma values in fetal haemoglobin, *Br. J. Hematol.*, 59, 333, 1985.
17. **Labie, D., Pagnier, J., Lapoumeroulie, C., Rouabhi, F., Dunda-Belkhodja, O., Charclin, P., Beldjord, C., Wajcman, H., Fabry, M. E., and Nagel, R. L.**, Common haplotype dependency of high G/gamma-globin gene expression and high HbF levels in beta-thalassemia and sickle cell anemia patients, *Proc. Natl. Acad. Sci U. S. A.*, 82, 2111, 1985.

18. **Mavilio, F., Giampaolo, A., Care, A., Miglialccio, G., Calandrini, M., Russo, G., Pagliardi, G. L., Mastroberardino, G., Marinucci, M., and Peschle, C.,** Molecular mechanisms of human hemoglobin switching: selective undermethylation and expression of globin genes in embryonic, fetal and adult erythroblasts, *Proc. Natl. Acad. Sci. U. S. A.,* 80, 6907, 1983.
19. **Nienhuis, A. W. and Benz, E. J., Jr.,** Regulation of hemoglobin synthesis during the development of the red cell. I, *N. Engl. J. Med.,* 297(24), 1318, 1977.
20. **Bunch, C., Wood, W. G., and Weatherall, D. J.,** Cellular origins of the fetal-hemoglobin-containing cells of normal adults, *Lancet,* 1, 1163, 1979.
21. **Stamatoyannopoulos, G., Papayannopoulou, T., Brice, M., Kurachi, S., Nakamoto, B., Lim, G., and Farquhar, M.,** Cell biology of hemoglobin switching. I. The switch from fetal to adult hemoglobin formation during ontogeny, in *Hemoglobins in Development and Differentiation,* Alan R. Liss, New York, 1981, 287.
22. **Papayannopoulou, T., Kalmantis, T., and Stamatoyannopoulos, G.,** Cellular regulation of hemoglobin switching: evidence for inverse relationship between fetal hemoglobin synthesis and degree of maturity of human erythroid cells, *Proc. Natl. Acad. Sci. U. S. A.,* 76(12), 6420, 1979.
23. **Stamatoyannopoulos, G., Papayannopoulou, T., Nakamoto, B., and Kurachi, S.,** Hemoglobin switching activity, in *Globin Gene Expression and Hematopoietic Differentiation,* Stamatoyannopoulos, G. and Neinhus, A., Alan R. Liss, New York, 1983, 347.
24. **Wood, W. G., Bunch, C., and Kelly, S.,** The cellular basis of haemoglobin switching: further studies of fetal-to-adult haemopoietic cell transplantation, in *Experimental Approaches for the Study of Hemoglobin Switching,* Stamatoyannopoulos, G. and Neinhus, A., Alan R. Liss, New York, 1985, 369.
25. **Stamatoyannopoulos, G., Nakamoto, B., Kurachi, S., and Papayannopoulou, T.,** Direct evidence for interaction between human erythyroid progenitor cells and a hemoglobin switching activity present in fetal sheep serum, *Proc. Natl. Acad. Sci. U. S. A.,* 80, 5650, 1983.
26. **Kaku, M., Yagawa, K., Nakamura, K., and Okano, H.,** Synthesis of adult-type hemoglobin in human erythemia cell line, *Blood,* 64(1), 314, 1984.
27. **Popat, N., Wood, W. G., Weatherall, D. J., and Turnbull, A. C.,** Pattern of maternal F-cell production during pregnancy, *Lancet,* 2, 377, 1977.
28. **Chui, D. H. K., Wong, S. C., Enkin, M. W., Patterson, M., and Ives, R. A.,** Proportion of fetal hemoglobin synthesis decreases during erythroid cell maturation, *Proc. Natl. Acad. Sci. U. S. A.,* 77, 2757, 1980.
29. **Pembrey, M. E., Weatherall, D. J., and Clegg, J. B.,** Maternal synthesis of haemoglobin F in pregnancy, *Lancet,* 1, 1350, 1973.
30. **Weatherall, D. J., Clegg, J. B., and Wood, W. G.,** A model for the persistence or reactivation of fetal haemoglobin production, *Lancet,* 2, 660, 1973.
31. **Stamatoyannapoulos, G., Veith, R., Galanello, R., and Papayannopoulou, T.,** HbF production in stressed erythropoiesis: observations and kinetic models, *Ann. N.Y. Acad. Sci.,* 445, 188, 1985.
32. **Weatherall, D. J., Pembray, M. E., and Pritchard, J.,** Fetal haemoglobin, *Clin. Haematol.,* 3, 467(3), 1974.
33. **Hoffman, R., Dainiak, N., Coupal, E., Maffai, L., Ritchey, K., and Forget, B. G.,** Hormonal influences on globin chain synthesis of fetal cord blood BFU-E *in vitro,* in *Cellular and Molecular Regulation of Hemoglobin Switching,* Grune & Stratton, New York, 1979, 389.
34. **Zanjani, E. D., Gormus, B. J., Bhakthavathsalan, A., Engler, T. M., McHale, A. P., and Mann, L. I.,** Effect of thyroid hormone on erythropoiesis and the switch from fetal to adult hemoglobin synthesis in sheep, in *Cellular and Molecular Regulation of Hemoglobin Switching,* Grune & Stratton, New York, 1979, 169.
35. **Dover, G. J., Charache, S., Nora, R., and Boyer, S. H.,** Progress toward increasing fetal hemoglobin production in man: experience with 5-azacytidine and hydroxyurea, *Ann. N.Y. Acad. Sci.,* 445, 218, 1985.
36. **Nienhuis, A. W., Ley, T. J., Humphries, R. K., Young, N. S., and Dover, G.,** Pharmacological manipulation of fetal hemoglobin synthesis in patients with severe beta-thalassemia, *Ann. N.Y. Acad. Sci.,* 445, 198, 1985.
37. **DeSimone, J., Heller, P., Hall, L., and Zwiers, D.,** 5-Azacytidine stimulates fetal hemoglobin synthesis in anemic baboons, *Proc. Natl. Acad. Sci. U. S. A.,* 2, 79, 4428,
38. **Charache, S., Dover, G., Smith, K., Talbot, C. C., Moyer, M., and Boyer, S.,** Treatment of sickle cell anemia with 5-azacytidine results in increased fetal hemoglobin production and is associated with nonrandom hypomethylation of DNA around the gamma-delta, beta globin gene complex, *Proc. Natl. Acad. Sci. U. S. A.,* 80, 4842, 1983.
39. **Ley, T. J., DeSimone, J., Anagnou, N. P., Keller, G. H., et al.,** 5-Azacytidine selectively increases gamma-globin synthesis in a patient with beta-thalassemia, *N. Engl. J. Med.,* 307(24), 1469, 1982.

40. **Galanello, R., Verth, R., Papayannopalou, T., and Stamatoyannopoulos, G.,** Pharmacologic stimulation of HbF in patients with sickle cell anemia, in *Experimental Approaches for the Study of Hemoglobin Switching,* Stamatoyannopoulos, G. and Neinhus, A., Alan R. Liss, New York, 1985, 433.
41. **Nathan, D. G.,** Regulation of fetal hemoglobin synthesis by cell cycle specific drugs, in *Experimental Approaches for the Study of Hemoglobin Switching,* Alan R. Liss, New York, 1985.

GENETIC ABNORMALITIES OF HEMOGLOBIN

Martin H. Steinberg

INTRODUCTION

A convenient way to classify genetically determined abnormalities of hemoglobin is to divide these disorders into ones associated with abnormalities in the primary structure of globin and ones caused by reduced synthesis of structurally normal globin. The former have been termed hemoglobinopathies[1,2] and the latter have been termed thalassemias.[3-5] Thalassemic hemoglobinopathies may also occur and imply reduced synthesis of a structurally abnormal globin.[6,7] The hemoglobinopathies and thalassemias comprise the most commonly inherited disorders of man.

The hemoglobinopathies are most often due to a single base change altering a normal codon and causing a single amino acid substitution. However, variant hemoglobins have been described that result from deletion or addition of one or more amino acids. These arise as a result of chromosomal crossing-over with deletions or addition of coding DNA, reading frame-shifts, and termination codon mutations.[1,2]

Thalassemias are caused by mutations in promoters of gene transcription, defects in processing of nuclear mRNA due to a variety of mRNA splicing mutations, nonsense mutations, polyadenylation signal mutation, extensive rearrangements within globin gene clusters, and deletion of one or more globin genes.[1-7] The organization of globin genes and the pathways leading from globin genes to the assembly of the hemoglobin tetramer are discussed in the previous chapter.

GENETICS

As α- and non-α-genes are on different autosomal chromosomes, they segregate independently.[8,9] There are two identical functional α-globin genes per haploid cell[10] and a single ζ-chain[11] gene that is normally expressed only in the yolk sac cells of the early embryo. The theta (θ)-globin gene lies 3' to the α_1-genes. θ-Globin protein has yet to be isolated and the function of this new "α-like" element remains mysterious. The single ϵ-gene[12] is also expressed in the embryo, while the duplicated $^G\gamma$- and $^A\gamma$-genes[13] reach the zenith of their function during fetal life. The adult δ-[14] and β-globin genes[15] are expressed at low levels during the first trimester of pregnancy, increase during the remainder of intrauterine development, and peak during the first year of life.[1,2,16] These developmentally regulated genes and their hemoglobin products are shown in Figure 1. The ζ- and ϵ-embryonic globin chains are found in megaloblastic nucleated primitive erythroid progenitor cells derived from the yolk sac. The $\zeta \rightarrow \alpha$ and $\epsilon \rightarrow \gamma$ switches occur asynchronously in megaloblasts and while erythroid development proceeds through the hepatic stage to definitive normocytic bone marrow erythropoiesis. The $\zeta \rightarrow \alpha$ switch switch precedes the $\epsilon \rightarrow \gamma$ one. The synthesis is of β-globin is restricted to erythroblasts of liver and marrow, while the ζ-globin gene is not normally expressed in these cells.[1,2]

The classical genetic designations of dominance or recessivity are difficult to apply to disorders of hemoglobin and their utility is largely supplanted by our very detailed understanding of normal and abnormal globin genes, the pathophysiology of variant hemoglobins, and the relative ease of detecting the final products of globin genes. The phenotypic expression of globin disorders is usually not clinically apparent unless the affected individual is a homozygote or a mixed heterozygote. An exception to this is found in cases of unstable hemoglobins or hemoglobins with high oxygen affinity. The inheritance of hemoglobino-

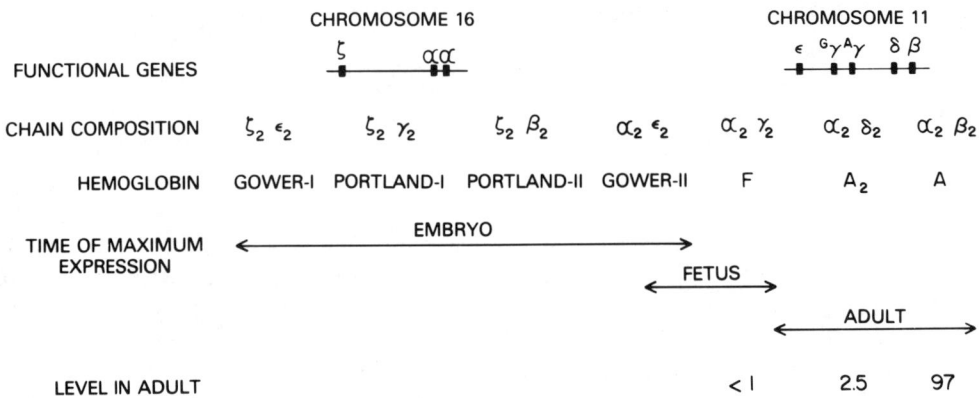

FIGURE 1. Developmentally regulated hemoglobins. The α-gene cluster is present on chromosome 16 and the β-gene cluster is present on chromosome 11. Each has genes that are maximally expressed in the embryo during fetal development and postnatally. There are four primarily embryonic, one fetal, and two adult hemoglobins.

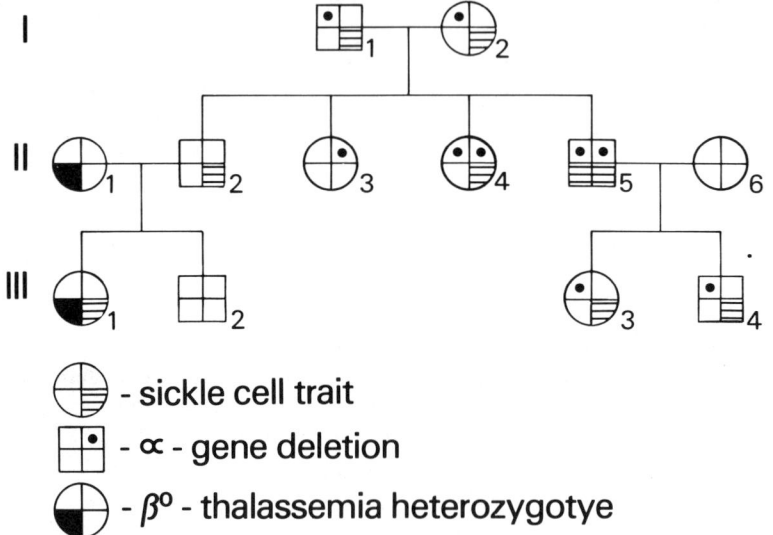

FIGURE 2. Pedigree of family with HbS, β-thalassemia, and α-thalassemia. Both individuals in the first generation (I-1, I-2) have sickle cell trait (HbAS) and are heterozygous for a single deleted α-globin gene (-α/αα). In the idealized four offspring (II-2, II-3, II-4, II-5), half will have HbAS (II-2, II-4), a quarter will have sickle cell anemia (II-5), HbSS, and a quarter will have normal hemoglobin (II-3), HbAA. In addition, some will have four α-genes (II-2, αα/αα), some will have a deleted gene (II-3, -α/αα), and some will have two deleted genes (II-4, II-5; -α/-α). II-1, the spouse of II-2, has heterozygous β°-thalassemia (β-thal). A quarter of the offspring of this mating will be mixed heterozygotes for β°-thal and HbAS (III-1). The spouse of II-5 is normal (II-6). All offspring of this mating will have HbAS and will be heterozygous for a single α-gene deletion (III-3, III-4).

pathies and thalassemias, while obeying Mendelian laws, should be evaluated on the basis of globin genotype and the qualitative characteristics and amount of the globin that is present. A hypothetical pedigree of a family that has genes for hemoglobin S(HbS), β°-, and α-thalassemia is shown in Figure 2 and illustrates the inheritance of a β-globin variant with β- and α-thalassemias.

Table 1
MEDICALLY IMPORTANT ABNORMAL HEMOGLOBINS

Class	Molecular pathology	Example	Ref.
Reduced solubility	Gel formation upon deoxygenation	HbS	1,2
Unstable	a. Substitutions in heme pocket	Hb Köln	70
	b. α-Helix disruption by prolyl residues	Hb Genova	71
	c. Substitutions in the interior of the molecule altering tertiary structure or α,β, contacts	Hb Philly	68
	d. Deleted amino acid residues	Hb Gun Hill	72
Increased O_2 affinity	Stabilized oxy-structure, destabilized deoxy-structure, reduced 2,3-DPG binding	Hb Chesapeake	67
Met-hemoglobin (Hb M)	Heme binding residue substitutions causing ferriheme formation	HbM-Boston	73

HEMOGLOBINOPATHIES

There have been nearly 400 variants of hemoglobin described.[1] The majority of variants involve the β-globin chain, a somewhat smaller number of variants involve the α-globin chain, and much fewer variants involve the δ- and γ-globin chain. By far the bulk of hemoglobin variants have been detected because of a change in electrophoretic mobility rather than the production of clinical symptoms or hematological abnormalities. The vast majority of hemoglobinopathies are medically innocuous. Yet some produce clinically important diseases and the sickle cell disorders are major causes of mortality and morbidity in many countries. The different classes of medically important abnormal hemoglobins are shown in Table 1. The process of evolution has conserved the structure of selected portions of globin. These regions of the molecule play important roles in binding heme, associations between unlike globin chains, oxygen transport, and hemoglobin solubility. Neutral mutations involving the exterior surfaces of globin are unlikely to be associated with symptomatic hemoglobinopathies.

Sickle Cell Disease

The mutation responsible for sickle hemoglobin occurs in the sixth codon of the β-globin chain,[17] where the GAG specifying glutamic acid[15] is changed to a GTG that codes for valine.[18,19] Sickle hemoglobin (HbS; $\alpha_2^A \beta_2^S$; glu → val) arose multicentrically[20] within equatorial Africa and became established in populations where *falciparium* malaria was endemic as a balanced polymorphism. The heterozygote for HbS has a selective advantage under the pressure of malarial infestation.[21]

Molecular Aspects

The sickle mutation permits deoxy HbS to form a gel or liquid crystal.[22] Upon reoxygenation HbS once again goes into solution. While the precise physical chemistry of HbS gelation has not been worked out, only a single β^S-chain per tetramer is required for the intermolecular contacts to occur that allow deoxy HbS polymerization.[23] HbS polymer is generally believed to consist of elongated fibers composed of 14 intertwining strands of HbS tetramer.[24]

The gelation of deoxygenated HbS solutions or HbS within erythrocytes is not instantaneous. A delay occurs between deoxygenation and polymerization that is dependent upon the concentration and solubility of HbS.[25-28] Polymer formation appears to be a nucleation-dependent process with the delay time corresponding to the time required for critical nuclei to form.[29]

The solubility of HbS is dependent upon a number of factors that include pH, ion concentration, temperature,[29,30-32] and the composition of other hemoglobins that may be present. The latter is of greatest clinical importance as other hemoglobins, most notably hemoglobin F (HbF), are excluded from the HbS polymer or are impaired in their copolymerization with HbS.[33-36] For a number of reasons, when hemoglobins like hemoglobin A (HbA) and HbF are present with HbS, the pathologic consequences of this variant are blunted. There are examples in which a second mutation in a β^S-gene enhances the propensity for polymerization.

Cellular Aspects

While studies of HbS in dilute solutions and nonphysiologic buffers have provided important insights into the structure of HbS polymer and gelation, the polymerization of HbS within the erythrocyte determines the pathology of sickle cell disease. Nuclear magnetic resonance spectroscopy has shown measurable levels of intracellular HbS polymer in well-oxygenated sickle erythrocytes.[37,38] There appears to be a general relationship between the level of HbS polymer, as calculated from the cellular hemoglobin concentration, and the severity of different sickle cell syndromes.[39] For example, sickle cell trait (HbAS) carriers are hematologically normal, with no vaso-occlusive symptoms, and have no intracellular polymer. HbS-β^+-thalassemia patients with moderately severe disease have some polymer, and individuals with sickle cell anemia have the most severe disease and the highest levels of polymer.

The erythrocytes in sickle cell anemia (HbSS) are heterogeneous in a number of respects.[40-43] There are large fractions of dense cells and light cells. The former represent irreversibly sickled cells and other dense erythrocytes while the latter are reticulocytes, young red cells that have just escaped the bone marrow. Between these extremes are a spectrum of cells of different densities.

Sickle erythrocytes have a number of well-defined abnormalities. They may have low K^+ and H_2O content,[41] increased Ca^{2+} levels,[44] damaged membrane skeletal proteins,[45,46] high levels of activated oxygen radicals,[46] twice normal levels of lipid peroxidation products,[48] and abnormal lipid topography,[49] increased hemichrome content,[47,50] and altered surface charge topography.[51] All abnormalities are not present to the same extent in all cells. It has been proposed that combinations of these defects allow the sickle erythrocyte to adhere to vascular endothelial cells and that transitory adherence may afford the necessary time for progressive HbS polymerization and subsequent microvascular obstruction.[51-53] There is a good correlation between the adherence of sickle cells to monolayers of human endothelial cells and the vaso-occlusive complications of sickle cell disease.[52,54] While adherence to endothelial cell cultures does not relate to the hemolytic severity of disease,[52] recent studies have shown that sickle erythrocytes also adhere to and are engulfed by macrophages.[55] Sickle red cells have increased levels of surface immunoglobulin, and adherence to macrophages is related to the intensity of hemolysis.

The pathophysiology of sickle cell disease may be viewed as follows, although, as with all such schema, extrapolation of *in vitro* data to complex physiologic systems undoubtedly represents an oversimplification. Deoxygenation of HbS induces polymer formation. Polymerization, sickling and unsickling of cells, and the interaction of HbS polymer with cellular constituents that include membrane skeletal proteins and ion channels lead to membrane damage. Some cells adhere to endothelia and this influences the vaso-occlusive severity of disease, while interactions with reticuloendothelial cells may determine the extent of hemolysis and anemia.

Table 2
SICKLE CELL DISEASE

	PCV	MCV	HbA$_2$	HbF	Severity
Sickle cell trait (HbAS)	N	N	N	N	0
Sickle cell anemia (HbSS)	↓↓↓	N	N	↑	++++
HbSC disease (HbSC)	↓↓	N	N	N/↑	+++
HbS-β$^+$-thalassemia	↓	↓	↑	N/↑	++
HbS-β$^\circ$-thalassemia	↓↓	↓	↑	↑↑	++++
HbSS-α-thalassemia	↓↓	↓	↑	↑	++++
HbS-HPFH	N	N	↓	↑↑↑	0

Note: Abbreviations used: N, normal; ↑-↑↑↑, minor to marked increases (or decreases); 0 to ++++, no clinical disease to severe vaso-occlusive disease.

Clinical Aspects

Some of the more common and medically important sickle cell disorders are shown in Table 2. Sickle cell trait (HbAS) is found in about 8% of black Americans, but is found in over 50% of selected African populations.[1,2] These carriers of sickle hemoglobin are well and only rarely have problems that can be related to the abnormal hemoglobin.[56] Sickle cell anemia (HbSS), present in 1 of 600 to 800 American blacks, is the most severe of the sickling hemoglobinopathies and is typified by moderate to marked anemia and a multitude of vaso-occlusive events.[12] These include, but are not limited to, repetitive bouts of pain, destruction of bone, ulceration of skin, pulmonary damage, and progressive loss of functioning splenic tissue. The mixed heterozygous sickle disorders, HbSC disease and HbS-β$^+$-thalassemia, are milder than HbSS because of the reduction of HbS concentration by hemoglobin C (HbC) and HbA, respectively, HbS-hereditary persistence of fetal hemoglobin (HPFH) is not accompanied by hematological or clinical abnormalities of any consequence.[1] In this disorder, all erythrocytes contain about 25% HbF. In addition to a significant reduction in HbS concentration in cells of these individuals, HbF does not participate in polymer formation.[33]

HbC and HbE

These are the second and third most common variant hemoglobins, although just which one predominates in the world is not clear. HbC ($\alpha_2\beta_2^C$; β^6 glu → lys) arose in Africa where certain groups have a heterozygote frequency of about 40%. About 2 to 3% of black Americans have HbC trait. HbE ($\alpha_2\beta_2^E$;β^{26}glu → lys) arose in the Orient where some populations have a heterozygote frequency of nearly 50%.[1,2]

Heterozygotes for HbC and HbE are not symptomatic and homozygotes have minimal or no anemia. Mixed heterozygotes with HbSC disease and HbE-β$^\circ$-thalassemia have clinically significant disease.

Both variants have interesting properties. HbC appears to promote the leak of K$^+$ and H$_2$O from the erythrocyte,[57] increasing the cellular hemoglobin concentration. This is evident from the increased density of cells in HbCC[58] and HbSC disease[59] and helps explain the pathophysiology of the latter disorder.

HbE is synthesized at a reduced rate when compared to HbA in individuals with HbE trait. While a portion of the reduced synthesis of HbE may be explained by the reduced affinity of the positively charged β^E chain for α-globin,[60] the synthesis of β^E is intrinsically retarded.[61,62] Thus, the β^E-allele is functionally thalassemic (see section entitled Thalassemic Hemoglobinopathies).

Other Hemoglobinopathies

Compared to the disorders associated with HbS, C, and E, other clinically important hemoglobinopathies are very rare (see Table 1). Two groups are of special interest: the unstable hemoglobins and hemoglobins with increased oxygen affinity.

Unstable hemoglobins can arise from a number of different classes of mutation.[63] The degree of molecular instability conferred by the structural change can be very mild, so that virtually no clinical effects are present, and the defect is best demonstrated by thermal denaturation of hemoglobin in low-ionic-strength buffers.[64] On the other hand, molecular instability can be so dramatic that the variant hemoglobin is indetectable by almost all techniques[65,66] and is rapidly and completely destroyed by the erythroid proteolytic enzyme systems. Most unstable hemoglobins are associated with mild curtailment of red cell survival. In some instances, drugs that have "oxidant" properties can induce unstable hemoglobins to precipitate intracellularly and shorten cell life. The most marked of the unstable hemoglobins are accompanied by moderate degrees of hemolytic anemia, while the rare hyperunstable variants have the clinical phenotype of thalassemia (see the section entitled Thalassemic Hemoglobinopathies).

The relationship between hemoglobin saturation with oxygen and the partial pressure of oxygen is described by the sigmoidal-shaped oxyhemoglobin dissociation curve.[1] The point on the curve where hemoglobin is half saturated with O_2 is called the P50. While numerous factors may affect the P50, the shape and position of the curve are determined by the hemoglobin molecule. The normal curve requires a heterotetramer with two pairs of unlike chains. Homotetramers like β_4 or γ_4, called hemoglobin H (HbH) and Hb Barts, have hyperbolic oxygen dissociation curves and remain fully saturated with O_2 under physiologic conditions. They are therefore useless as respiratory molecules. HbH and Hb Barts are seen in the α-thalassemias where there is a severe deficit or total absence of α-globin (see the section entitled Gene Deletion). Mutant hemoglobins may have abnormal curves. A number of variants reduce the P50 indicating an increased avidity for O_2.[67] Since oxygen delivery to tissues is impaired because of high hemoglobin-oxygen affinity, hypoxia results. The normal homeostatic mechanisms lead to increased production of the regulatory hormone erythropoietin and increased quantities of red cells are made to compensate for the impairment of oxygen transport. The clinical result is polycythemia, an increase in the red cell mass.

Less commonly, variant hemoglobins can have reduced oxygen affinity.[64,69] Tissue oxygen transport is enhanced, erythropoietin levels fall, and the red cell mass is reduced.

THALASSEMIA

This very large and heterogeneous group of disorders is typified by unbalanced globin synthesis.[1-5] Normally, nearly equal quantities of α- and non-α-globin chains are synthesized in developing red cells. This balanced globin synthesis ensures that no chain is present in excess and that each α-chain has a non-α-mate. Dimers of α- and β-chains or α- and γ-chains then combine to form HbA or HbF. Uncombined globin chains are generally unstable and susceptible to proteolysis, or they may form unstable dysfunctional homotetramers (see section entitled other Hemoglobinopathies). In thalassemia, the biosynthesis of one or more normal globin chains is impaired. Synthesis of the uninvolved chains proceeds normally leading to their excess. These redundant chains provoke cellular injury and premature death by virtue of their intracellular precipitation into insoluble aggregates and promotion of oxidant-induced damage to the cell contents and membrane. The damaged cells either die within the bone marrow or are removed from the circulation by a hypertrophied spleen and other reticuloendothelial cells. Thalassemias can suppress α- and β-globin chain synthesis and are termed α- and β-thalassemias, respectively. Impaired synthesis of γ- and δ-chains may occur but has far less medical importance. A general classification of the thalassemias is shown in Table 3.

Table 3
A CLINICAL CLASSIFICATION OF THE THALASSEMIAS

I. β-Thalassemia — absent or reduced β-globin synthesis
   A. β^+-Thalassemia (β-globin synthesis present, but reduced)
   B. β^o-Thalassemia (β-globin synthesis absent)
   C. δ,β-Thalassemia (β- and δ-globin synthesis absent with increased levels of HbF)
   D. Lepore hemoglobin (β- and δ-globin synthesis absent with reduced synthesis of a δβ-fusion globin)
   E. HPFH (deletion) (β- and δ-globin synthesis absent with full compensation by increased γ-globin synthesis)

II. α-Thalassemia[a] — absent or reduced α-globin synthesis
   A. Silent carrier, heterozygous α-thalassemia-2 (-α/αα)
   B. α-Thalassemia trait (-α/-α; --/αα)
   C. HbH disease (-α/--)
   D. Hydrops fetalis (--/--)

Note: All entities can be present as the heterozygote, homozygote, or mixed heterozygote.

[a] α-Globin genes: normal (αα/αα); one missing gene (-α/αα); two missing genes (-α/-α; --/αα); three missing genes (-α/--); and four missing genes (--/--).

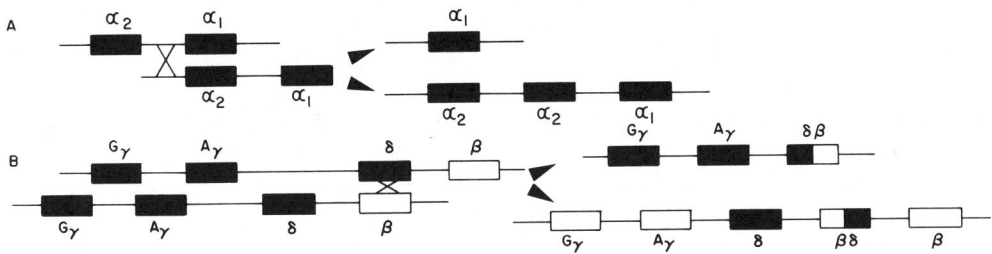

FIGURE 3. Crossing-over as the cause of gene deletion and duplication. (A) Unequal crossing-over between α-globin genes generating loss of the α_2-gene from the thalassemia-producing chromosome and its addition to the triplicated locus chromosome; (B) unequal crossing-over between partially homologous δ- and β-globin genes generating the thalassemic Lepore (δβ) gene and the anti-Lepore (βδ) gene. The Lepore chromosome has no functional β-gene and the δβ-gene is poorly expressed because of weak promotors and other factors; thus the thalassemic phenotype results from this gene fusion.

Molecular Causes of Thalassemia

Some knowledge of gene transcription, mRNA processing, and translation is vital to understanding the molecular causes of the thalassemias. These are discussed in the preceding chapter. Rather than catalog the multitudinous genetic lesions in the α- and β-thalassemias, examples of general classes of defects will be described. Each type of abnormality has the potential of impairing either α- or β-globin synthesis, although to date all lesions have not been described in each type of thalassemia. In addition, it is highly likely that these types of mutations will be expanded in number and will be generalized to other genetic disorders.

Gene Deletion

The simplest cause of thalassemia to understand is gene deletion. Deletions are most common in the α-globin gene cluster as a result of the considerable homology of both α-globin genes and their 5′ noncoding regions.[74-78] Misalignment of the α-gene-containing chromosomes during meiosis and crossing-over can lead to the loss of one or both α-genes from a chromosome (Figure 3). The reciprocal product of gene deletion is a chromosome with an extra gene.[79] Extensive, but less complete, homology between genes in the β-gene cluster also leads to instances of deletion of the δ- and β-globin genes as well as the generation

Table 4
THALASSEMIC HEMOGLOBINOPATHIES

Type	Examples	Features	Ref.
Fusion genes	Lepore hemoglobins δβ-Chains (three described)	β-Thalassemia trait	80
Termination codon mutants	Hb Constant Spring α^{142} UAA → CAA (glutamine) (four α-variants, one β-variant)	α-Thalassemia-2	118
Hyper-unstable	Hb Indianapolis (112^{arg})	Severe β-thalassemia	65
	Hb Quong Sze (125^{pro})	α-Thalassemia-1	66
Alternative mRNA splicing	HbE (β26^{lys})	Mild β-thalassemia	61
Reduced mRNA	Hb Vicksburg	β^+-Thalassemia	123
(? Thalassemia mutation in cis)	Hb Northshore	Mild β^+-thalassemia	124

of fusion genes[80-82] (Table 4; Figure 2).[83-91] Loss of a gene leads to a β°- or α°-type of thalassemia where no gene product is detectable.

Nonsense Mutation

The translation of mRNA continues until a termination codon is encountered. Base substitution introducing a termination codon into the coding portion of the gene prematurely stops translation. Nucleotide deletion changing the codon reading frame may also lead to an in-phase terminator.[92-94] The result is failure to synthesize a normal-sized globin. There appears to be rapid degradation of whatever globin polypeptide is synthesized and a defect in the transcription of the mutant gene.[95,96] *In vitro* studies have shown that a suppressor tRNA can lead to normal mRNA accumulation and globin synthesis in the presence of a nonsense mutation.[97] Several nonsense mutations have been described, and as they totally inhibit normal globin synthesis, they are associated with the β°-phenotype.

mRNA Processing Mutations

To serve as an effective template for translation to a polypeptide, nuclear pre-mRNA must be processed.[98-100] This involves capping the 5' terminus with distinctive methylated nucleotides that appear to enhance the efficiency of translation and polyadenylation of the 3' terminus that promotes stability of the message. A mutation involving the poly (A) addition sequence has been associated with α-thalassemia.[101] However, splicing out of intervening sequences and ligation of the three exons of globin mRNA provide the greatest chance for thalassemia-inducing mutation. Highly conserved dinucleotides, GT and AG, embedded in conserved consensus sequences flank the 5' donor and 3' acceptor sites of all introns.[100] Mutation of one of the conserved dinucleotides can totally abolish normal splicing and cause an α°- or β°-phenotype.[102-105] Other splice junction mutations may impair normal splicing and cause a cryptic, hitherto unutilized 3' splice site to be activated.[105-108] In these instances, the β^+-phenotype is usually produced since some normal splicing occurs and normal mRNA is made. Other mutations may by themselves activate cryptic splice sites by causing sequence changes that make them more closely resemble the physiologic splice sequence.[104,110] Again, as some mRNA is correctly spliced and translated, the β^+-phenotype occurs.

Promoter Mutations

Sets of conserved sequences 5' to the site of initiation of mRNA translation are known as promoters. These ubiquitous features of eukaryotic genes lie about 30, 80 and 100 to 200 base pairs 5' to coding DNA and are known as the TATA, CCAAT, and CACCC boxes, respectively. In addition there may be less well-defined more distal promoters and enhancers 5' and 3' to the gene. Promoters play important roles in the fidelity and frequency of transcriptional events by serving as areas of binding or penetration of the polymerases that are involved in gene transcription. Enhancers appear to be critical for tissue-specific gene expression. Nucleotide substitutions in the TATA and CCAAT boxes have been associated with reduced transcription and the β+-thalassemia phenotype.[111-114]

Recent studies suggest that promoter mutations can upregulate as well as suppress gene expression. The nongene deletion hereditary persistence of fetal hemoglobin syndromes is characterized by continued expression of the γ-globin genes during adult life. A G → A transition in the distal CCAAT box of the ^{A}γ-globin gene has been associated with 10- to 20-fold increases in HbF,[115,116] while mutations involving more upstream sequences have also been found to increase ^{G}γ- and ^{A}γ-gene expression.[117]

Termination Codon Mutation

Mutation in the normal termination codon permits continued read-through of mRNA until the next in-phase termination codon is encountered. A series of such mutants involving the α-globin gene has been described.[6,118] All have a different amino acid residue at position 142 (normally the α-chain has 141 residues) followed by 30 additional identical residues. There is a deficit of mRNA for one well-studied variant of this type and an α+-phenotype.[118,119]

Miscellaneous Mutations

In one instance no mutation was found in a gene associated with β-thalassemia. This suggests that DNA changes some distance from the gene can influence its expression.[120] In a nondeletion α-thalassemia, the normal initiation codon of the α_2-locus, ATG was replaced by ACG. This abolished any output of globin from this gene.[121]

Restriction Fragment Length Polymorphism

There are a number of polymorphic sites in and around the β-like globin gene cluster that lead to DNA length changes when cleaved with restriction endonucleases. Specific thalassemia mutations tend to be associated with particular β-globin gene cluster haplotypes as defined by restriction endonuclease digestion. While a perfect correlation between haplotype and thalassemia mutation does not exist even within ethnic groups, study of these DNA polymorphisms has provided important information on the origins and spread of mutations in this gene cluster.[113]

Clinical Aspects

Thalassemias are most prevalent in areas of the world where *falciparium* malaria is or was endemic.[1,3] Emerging evidence indicates the carriers of a thalassemia gene, like those with HbAS, have a reproductive advantage under the selective pressure of malaria. The heterozygote with β- or α-thalassemia has either minimal or no anemia and other than having small, poorly hemoglobinized cells, may be indistinguishable from normal. Homozygotes or mixed heterozygotes for β+- or β°-thalassemias have a severe disease with a greatly curtailed life span unless treatment is instituted.[1,3] The most efficacious form of treatment includes blood transfusion to keep the hemoglobin level near normal and the regular administration of iron chelating agents to help rid the body of excessive iron stores that cause serious damage to a number of organs.[1,3,122]

Of the α-thalassemias (see Table 4), the loss of one or two genes is innocuous; three missing genes are associated with a mild to moderate hemolytic anemia; and deletion of all α-genes causes intrauterine death.[1-3,78]

THALASSEMIC HEMOGLOBINOPATHIES

This relatively new class of disorders is defined as structural abnormalities of globin (hemoglobinopathies) that are synthesized at suboptimal levels (thalassemia).[6,7] A general classification is shown in Table 4. This is evidently a heterogeneous group due to a variety of molecular causes. The fusion genes and termination codon mutants were discussed earlier, but are also included in this classification.

HbE

This variant (see section entitled HbC and HbE) is especially instructive for the insights it provided into mRNA splicing. The β^E-mutation (GAG-AAG; glu $\rightarrow$ lys) occurring within the codon for amino acid residue 26 produces the sequence GGTAAGG.[110] This closely resembles the customary consensus 5' splice sequence C/A AGGT A/G AGT. In addition to normally spliced mRNA, an mRNA that utilizes this signal and deletes a portion of the first exon is also made.[110] Thus, the β^E-mutation retards the normal removal of the first intervening sequence, causes aberrant splicing to take place, and reduces β^E-mRNA accumulation. Other examples of this class of mutation have been described.[125]

Hyper-unstable Variants

These are so unstable that a globin protein cannot be detected by the usually used methods. Paradoxically, these variants cause more cellular injury than expected in the simple heterozygote when their phenotype is compared to heterozygous β^0-thalassemia or to α-thalassemia-2.[65,66] This may be a result of proteolytic destruction of the unstable variant as well as the excessive uncombined normal globin chains that usually are present in thalassemia.[126-128]

CONCLUSION

Hemoglobinopathies continue to be discovered, albeit at a reduced rate, but even now occasional new variants are of exceptional interest for the insights they provide into protein synthesis, subunit assembly and interaction, and globin gene function. Further study of thalassemia causing mutations should continue to provide information on the role of distal promoters, intervening sequences, and "remote" DNA sequences on the expression of globin and other eukaryotic genes.

REFERENCES

1. **Bunn, H. F. and Forget, B. G.**, *Hemoglobin: Molecular, Genetic and Clinical Aspects*, W. B. Saunders, Philadelphia, 1985.
2. **Honig, G. R. and Adams, J. G.**, III, *Human Hemoglobin Genetics*, Springer-Verlag, New York, 1985.
3. **Weatherall, D. J. and Clegg, J. B.**, *The Thalassaemia Syndromes*, 3rd ed., Blackwell Scientific, Oxford, 1981.
4. **Benz, E. H. and Forget, B. G.**, The thalassemia syndromes: models for the molecular analysis of human disease, *Annu. Rev. Med.*, 33, 363, 1982.
5. **Weatherall, D. J. and Clegg, J. B.**, Thalassemia revisited, *Cell*, 29, 7, 1982.
6. **Steinberg, M. H. and Adams, J. G.**, Thalassemic hemoglobinopathies, *Am. J. Pathol.*, 113, 396, 1983.
7. **Nienhuis, A. W., Anagnou, N. P., and Ley, T. J.**, Advances in thalassemia research, *Blood*, 63, 738, 1984.
8. **Deisseroth, A., Nienhuis, A., Turner, P., Velez, R., Anderson, W. F., Ruddle, F., Lawrence, J., Creagan, R., and Kucherlapati, R.**, Localization of the human α-globin structural gene to chromosome 16 in somatic cell hybrids by molecular hybridization, *Cell*, 12, 205, 1977.
9. **Deisseroth, A., Nienhuis, A., Lawrence, J., Giles, R., Turner, P., and Ruddle, F. H.**, Chromosomal localization of human β-globin gene on human chromosome 11 in somatic cell hybrids, *Proc. Natl. Acad. Sci. U. S. A.*, 75, 1456, 1978.

10. **Orkin, S. H.**, The duplicated human α-globin genes lie close together in cellular DNA, *Proc. Natl. Acad. Sci. U. S. A.,* 75, 5950, 1978.
11. **Proudfoot, N. J., Gill, A., and Maniatis, T.**, The structure of the human zeta-globin gene and a closely linked nearly identical pseudo-gene, *Cell,* 3, 553, 1982.
12. **Baralle, F. E., Shoulders, C. C., and Proudfoot, N. J.**, The primary structure of the human ε-globin gene, *Cell,* 21, 621, 1980.
13. **Slightom, J. L., Blechl, A. E., and Smithies, O.**, Human fetal $^G\gamma$ and $^A\gamma$ globin genes: complete nucleotide sequences suggest that DNA can be exchanged between these duplicated genes, *Cell,* 21, 627, 1980.
14. **Spritz, R. A., de Riel, J. K., Forget, B. G., and Weissman, S. M.**, Complete nucleotide sequence of the human δ-globin gene, *Cell,* 21, 639, 1980.
15. **Lawn, R. M., Efstratiadis, A., O'Connell, C., and Maniatis, T.**, The nucleotide sequence of the human β-globin gene, *Cell,* 21, 647, 1980.
16. **Efstratiadis, A., Posakony, J. W., Maniatis, T., Lawn, R. M., O'Connell, C., Spritz, R. A., de Riel, J. K., Forget, B. G., Weissman, S. M., Slightom, J. L., Blechl, A. E., Smithies, O., Baralle, R. E., Shoulders, C. C., and Proudfoot, N. J.**, The structure and evolution of the human β-globin gene family, *Cell,* 21, 653, 1980.
17. **Ingram, V. M.**, Abnormal human haemoglobins. III. The chemical difference between normal and sickle haemoglobins, *Biochem. Biophys. Acta,* 36, 402, 1959.
18. **Chang, J. C. and Kan, Y. W.**, A sensitive new prenatal test for sickle cell anemia, *N. Engl. J. Med.,* 307, 30, 1982.
19. **Conner, B. J., Keys, A. A., Morin, C., Itakura, K., Teplitz, R. L., and Wallace, R. B.**, Detection of sickle cell β^S-globin allele by hybridization with synthetic oligonucleotides, *Proc. Natl. Acad. Sci. U. S. A.,* 80, 278, 1983.
20. **Kan, Y. W. and Dozy, A. M.**, Evolution of the hemoglobin S and C genes in world populations, *Science,* 209, 388, 1980.
21. **Allison, A. C.**, Malaria in carriers of the sickle-cell trait and in newborn children, *Exp. Parasitol.,* 6, 418, 1957.
22. **Perutz, M. F. and Mitchinson, J. M.**, State of haemoglobin in sickle cell anaemia, *Nature (London),* 166, 677, 1950.
23. **Wishner, B. C., Hanson, J. C., Ringle, W. M., and Love, W. E.**, Crystal structure of sickle-cell deoxyhemoglobin, in Proc. Symp. on Molecular and Cellular Aspects of Sickle Cell Disease, Hercules, J. I., Cottam, G. L., Waterman, M. R., and Schechter, A. N., Eds., DHEW Publ. No. (NIH) 76-1007, U.S. Department of Health, Education, and Welfare, Bethesda, MD, 1975.
24. **Dykes, G. W., Crepeau, R. H., and Edelstein, S. J.**, Three-dimensional reconstruction of the 14-filament fibers of hemoglobin S, *J. Mol. Biol.,* 130, 451, 1979.
25. **Moffat, K. and Gibson, Q. H.**, The rates of polymerization and depolymerization of sickle cell hemoglobin, *Biochem. Biophys. Res. Commun.,* 61, 237, 1974.
26. **Malfa, R. and Steinhardt, J.**, A temperature-dependent latent-period in the aggregation of sickle-cell deoxyhemoglobin, *Biochem. Biophys. Res. Commun.,* 59, 887, 1974.
27. **Zarkowsky, H. S. and Hochmuth, R. M.**, Sickling times of individual erythrocytes at zero PO_2, *J. Clin. Invest.,* 56, 1023, 1975.
28. **Adachi, K. and Asakura, T.**, Demonstration of a delay time during aggregation of diluted solutions of deoxyhemoglobin S and hemoglobin C Harlem in concentrated phosphate buffer, *J. Biol. Chem.,* 253, 6641, 1978.
29. **Hofrichter, J., Ross, P. D., and Eaton, W. A.**, A physical description of hemoglobin S gelation, in Proc. Symp. on Molecular and Cellular Aspects of Sickle Cell Disease, Hercules, J. I., Cotton, G. L., Waterman, M. R., and Schechter, A. N., Eds., DHEW Publ. No. (NIH) 76-1007, U.S. Department of Health, Education, and Welfare, Bethesda, MD, 1975.
30. **Behe, M. J. and Englander, S. W.**, Mixed gelation theory: kinetics equilibrium and gel incorporation in sickle hemoglobin mixtures, *J. Mol. Biol.,* 133, 137, 1979.
31. **Magdoff-Fairchild, B., Poillon, W. N., Li, T. I., and Bertles, J. F.**, Thermodynamic studies of polymerization of deoxygenated sickle cell hemoglobin, *Proc. Natl. Acad. Sci. U. S. A.,* 73, 990, 1976.
32. **Itano, H. A.**, Solubilities of naturally occurring mixtures of human hemoglobins, *Arch. Biochem. Biophys.,* 47, 148, 1953.
33. **Nagel, R. L., Bookchin, R. M., Johnson, J., Labie, D., Wajcman, H., Isaac-Sodeye, W. A., Honig G. R., Schiliro, G., Crookston, J. H., and Matsutomo, K.**, Structural basis of inhibitory effects of hemoglobin F and hemoglobin A_2 on the polymerization of hemoglobin S, *Proc. Natl. Acad. Sci. U. S. A.,* 76, 670, 1979.
34. **Benesch, R. E., Edalju, R., Benesch, R., and Kwong, S.**, Solubilization of hemoglobin S by other hemoglobins, *Proc. Natl. Acad. Sci. U. S. A.,* 77, 5130, 1980.
35. **Moffat, K.**, Gelation of sickle cell hemoglobin: effects of hybrid tetramer formation in hemoglobin mixtures, *Science,* 185, 274, 1974.

36. **Cheetham, R. C., Huehns, E. R., and Rosemeyer, M. A.,** Participation of haemoglobins A, F, A_2 and C in polymerization of haemoglobin S, *J. Mol. Biol.*, 129, 45, 1979.
37. **Noguchi, C. T., Torchia, D. A., and Schechter, A. N.,** Determination of deoxyhemoglobin S polymer in sickle erythrocytes upon deoxygenation, *Proc. Natl. Acad. Sci. U. S. A.*, 77, 5487, 1980.
38. **Noguchi, C. T. and Schechter, A. N.,** The intracellular polymerization of sickle hemoglobin and its relevance to sickle cell disease, *Blood*, 58, 1057, 1981.
39. **Brittenham, G. M., Schechter, A. N., and Noguchi, C. T.,** Hemoglobin S polymerization: primary determinant of the hemolytic and clinical severity of sickling syndromes, *Blood*, 65, 183, 1985.
40. **Bertles, J. F. and Milner, P. F. A.,** Irreversibly sickled erythrocytes: a consequence of the heterogeneous distribution of hemoglobin types in sickle-cell anemia, *J. Clin. Invest.*, 47, 1731, 1968.
41. **Clark, M. R., Unger, R. C., and Shohet, S. B.,** Monovalent cation composition and ATP and ligand content of irreversibly sickled cells, *Blood*, 51, 1169, 1978.
42. **Fabry, M. E. and Nagel, R. L.,** The effect of deoxygenation on red cell density: significance for the pathophysiology of sickle cell anemia, *Blood*, 60, 1370, 1982.
43. **Kaul, D. K., Fabry, M. E., Windisch, P., Baez, S., and Nagel, R. L.,** Erythrocytes in sickle cell anemia are heterogeneous in their rheological and hemodynamic characteristics, *J. Clin. Invest.*, 72, 22, 1983.
44. **Eaton, J. W., Skelton, T. D., Swofford, H. S., Kolpin, C. E., and Jacob, H. S.,** Elevated erythrocyte calcium in sickle cell disease, *Nature (London)*, 246, 105, 1973.
45. **Lux, S. E.,** Spectrin-actin membrane skeletons of normal and abnormal blood cells, *Semin. Hematol.*, 16, 21, 1979.
46. **Platt, O. S., Falcone, J. F., and Lux, S. E.,** Molecular defect in the sickle erythrocyte skeleton: abnormal spectrin binding to sickle inside-out vesicles, *J. Clin. Invest.*, 75, 266, 1985.
47. **Hebbel, R. P., Eaton, J. W., Balasingam, M., and Steinberg, M. H.,** Spontaneous oxygen radical generation by sickle erythrocytes, *J. Clin. Invest.*, 70, 1253, 1982.
48. **Das, S. K. and Nair, R. C.,** Superoxide dismutase, glutathione peroxidase, catalase, and lipid peroxidation of normal and sickled erythrocytes, *Br. J. Haematol.*, 44, 87, 1980.
49. **Lubin, B. and Chiu, D.,** Abnormalities in membrane phospholipid organization in sickle erythrocytes, *J. Clin. Invest.*, 67, 1643, 1981.
50. **Asakura, T., Minakata, K., Adachi, K., Russell, O., and Schwartz, E.,** Denatured hemoglobin in sickle erythrocytes, *J. Clin. Invest.*, 59, 633, 1977.
51. **Hebbel, R. P., Yamada, O., Moldow, C. F., Jacob, H. S., White, J. G., and Eaton, J. W.,** Abnormal adherence of sickle erythrocytes to cultured vascular endothelium: a possible mechanism for microvascular occlusion in sickle cell disease, *J. Clin. Invest.*, 65, 154, 1980.
52. **Hebbel, R. P., Boogaerts, M. A. B., Eaton, J. W., and Steinberg, M. H.,** Erythrocyte adherence to endothelium in sickle cell anemia. A possible determinant of disease severity, *N. Engl. J. Med.*, 302, 992, 1980.
53. **Steinberg, M. H. and Hebbel, R. P.,** The clinical diversity of sickle cell anemia: genetic and cellular modulation of disease severity, *Am. J. Hematol.*, 14, 405, 1983.
54. **Hebbel, R. P., Moldow, C. F., and Steinberg, M. H.,** Modulation of erythrocyte-endothelial interactions and the vaso-occlusive severity of sickling disorders, *Blood*, 58, 947, 1981.
55. **Hebbel, R. P. and Miller, W. J.,** Phagocytosis of sickle erythrocytes: immunologic and oxidative determinants of hemolytic anemia, *Blood*, 64, 733, 1984.
56. **Heller, P., Best, W. R., Nelson, R. B., and Becktel, J.,** Clinical implications of sickle cell trait and glucose-6-phosphate dehydrogenase deficiency in hospitalized black male patients, *N. Engl. J. Med.*, 300, 1001, 1979.
57. **Brugnara, C., Kopin, A., Bunn, H. F., and Tosteson, D. C.,** Regulation of cation content and cell volume in hemoglobin erythrocytes from patients with homozygous hemoglobin C disease *J. Clin. Invest.*, 75, 1608, 1985.
58. **Fabry, M. E., Kaul, D. K., Raventos, C., Baez, S., Reider, R., and Nagel, R.,** Some aspects of the pathophysiology of homozygous HbCC erythrocytes, *J. Clin. Invest.*, 67, 1284, 1981.
59. **Fabry, M. E., Kaul, D. K., Raventos-Suarez, C., Chang, H., and Nagel, R.,** SC erythrocytes have an abnormally high intracellular hemoglobin concentration, *J. Clin. Invest.*, 70, 1315, 1982.
60. **Bunn, H. F. and McDonald, M. J.,** Electrostatic interactions in the assembly of haemoglobin, *Nature (London)*, 306, 498, 1983.
61. **Traeger, J., Wood, W. G., Clegg, J. B., Weatherall, D. J., and Wasi, P.,** Defective synthesis of HbE is due to reduced levels of β^E mRNA, *Nature (London)*, 288, 497, 1980.
62. **Benz, E. J., Jr., Berman, B. W., Tonkonow, B. L., Coupal, E., Coates, T., Boxer, L. A., Altman, A., and Adams, J. G.,** Molecular analysis of the β-thalassemia phenotype associated with inheritance of hemoglobin E ($\alpha_2\beta_2^{26}$ glu→ lys), *J. Clin. Invest.*, 68, 118, 1981.
63. **Reider, R. F.,** Human hemoglobin stability and instability: molecular mechanisms and some clinical correlations, *Semin. Hematol.*, 11, 423, 1974.

64. **Steinberg, M. H., Adams, J. G., Thigpen, J. T., Morrison, F. S., and Dreiling, B. J.**, Hemoglobin Hope ($\alpha_2\beta_2^{136}$ gly→asp)-S disease: clinical and biochemical studies, *J. Lab. Clin. Med.*, 84, 632, 1974.
65. **Adams, J. G., Boxer, L. A., Baehner, R. L., Forget, B. G., Tsistrakis, G. A., and Steinberg, M. H.**, Hb Indianapolis (β112(G14) arginine): an unstable β-chain variant producing the phenotype of severe β-thalassaemia, *J. Clin. Invest.*, 63, 931, 1979.
66. **Goosens, M., Lee, K. Y., Liebhaber, S. A., and Kan, Y. W.**, Globin structural mutant α^{125}leu→pro is a novel cause of α-thalassaemia, *Nature (London)*, 296, 864, 1982.
67. **Charache, S., Weatherall, D. J., and Clegg, J. B.**, Polycythemia associated with a hemoglobinopathy, *J. Clin. Invest.*, 45, 813, 1966.
68. **Reider, R. F., Oski, F. A., and Clegg, J. B.**, Hemoglobin Philly (β^{35} tyrosine→phenylalanine): studies in the molecular pathology of hemoglobin, *J. Clin. Invest.*, 48, 1627, 1969.
69. **Imamura, T., Fujita, S., Ohta, Y., Hanada, M., and Yanase, T.**, Hemoglobin Yoshizuka (G10(108) β asparagine→aspartic acid): a new variant with a reduced oxygen affinity from a Japanese family, *J. Clin. Invest.*, 48, 2341, 1969.
70. **Carrell, R. W., Lehmann, H., and Hutchinson, H. E.**, Hemoglobin Koln (β-98 valine→methionine): an unstable protein causing inclusion-body anemia, *Nature (London)*, 210, 915, 1966.
71. **Cohen-Solal, M. and Labie, D.**, A new case of hemoglobin Genova $\alpha_2\beta_2^{28}$(B10)leu→pro. Further studies on the mechanism of instability and defective synthesis, *Biochim. Biophys. Acta*, 295, 67, 1973.
72. **Bradley, T. B., Wohl, R. C., and Reider, R. F.**, Hemoglobin Gun Hill: deletion of five amino acid residues and impaired hemoglobin binding, *Science*, 157, 1581, 1967.
73. **Gerald, P. S. and Efron, M. L.**, Chemical studies of several varieties of HbM, *Proc. Natl. Acad. Sci. U. S. A.*, 47, 1758, 1961.
74. **Higgs, D. R., Old, J. M., Clegg, J. B., Pressley, L., Hunt, D. M., Weatherall, D. J., and Serjeant, G. R.**, Negro α-thalassaemia is caused by deletion of a single α-globin gene, *Lancet*, II, 272, 1979.
75. **Embury, S. H., Lebo, R. V., Dozy, A. M., and Kan, Y. W.**, Organization of the α-globin genes in the Chinese α-thalassaemia syndromes, *J. Clin. Invest.*, 63, 1307, 1979.
76. **Pressley, L., Higgs, D. R., Clegg, J. B., and Weatherall, D. J.**, Gene deletions in α-thalassaemia prove that the 5' ζ locus is functional, *Proc. Natl. Acad. Sci. U. S. A.*, 77, 3586, 1980.
77. **Liebhaber, S. A., Goosens, M., and Kan, Y. W.**, Homology and concerted evolution at the α_1 and α_2 loci of human α-globin, *Nature (London)*, 280, 26, 1981.
78. **Higgs, D. R. and Weatherall, D. J.**, Alpha-thalassemia, *Curr. Top. Hematol.*, 4, 37, 1983.
79. **Goosens, M., Dozy, A. M., Embury, S. H., Zachariades, Z., Hadjiminas, M. G., Stamatoyannopoulos, G., and Kan, Y. W.**, Triplicated α-globin loci in humans, *Proc. Natl. Acad. Sci. U. S. A.*, 77, 518, 1980.
80. **Baglioni, C.**, The fusion of two polypeptide chains in hemoglobin Lepore and its interpretation as a genetic deletion, *Proc. Natl. Acad. Sci. U. S. A.*, 48, 1880, 1962.
81. **Flavell, R. A., Kooter, J. M., DeBoer, E., Little, P. F. R., and Williamson, R.**, Analysis of the human $\delta\beta$ globin gene loci in normal and hemoglobin Lepore DNA: direct determination of gene linkage and intergenic distance, *Cell*, 15, 25, 1978.
82. **Adams, J. G., Morrision, W. T., and Steinberg, M. H.**, Hb Parchman: a double crossover within a single human gene, *Science*, 218, 291, 1982.
83. **Kan, Y. W., Holland, J. P., Dozy, A. M., Charache, S., and Kazazian, H. H.**, Deletion of α-globin structure gene in hereditary persistence of foetal haemoglobin, *Nature (London)*, 258, 162, 1975.
84. **Forget, B. G., Hillman, D. G., Lazarus, H., Baralle, E. F., Benz, E. J., Jr., Caskey, C. T., Huisman, T. H. J., Schroeder, W. A., and Housman, D.**, Absence of messenger RNA and gene DNA for β-globin chains in hereditary persistence of fetal hemoglobin, *Cell*, 7, 323, 1976.
85. **Ottolenghi, S., Comi, P., Giglioni, B., Tolstoshev, P., Lanyon, W. G., Mitchell, G. J., Williamson, Russo, G., Musumeci, S., Schiliro, G., Tsistrakis, G. A., Charache, S., Wood, W. G., Clegg, J. B., and Weatherall, D. J.**, $\delta\beta$-Thalassemia is due to a gene deletion, *Cell*, 9, 71, 1976.
86. **Ramirez, F., O'Donnell, J. V., Marks, P. A., Bank, A., Musumeci, S., Schiliro, G., Pizzarelli, G., Russo, G., Luppis, B., and Gambino, R.**, Abnormal or absent βmRNA in β°-Ferrara and gene deletion in $\delta\beta$ thalassaemia, *Nature (London)*, 263, 471, 1976.
87. **Mears, J. G., Ramirez, F., Liebowitz, D., Nakamura, F., Bloom, A. D., Konotey-Ahulu, F., and Bank, A.**, Changes in restricted human cellular DNA fragments containing globin gene sequences in thalassemia and related disorders, *Proc. Natl. Acad. Sci. U. S. A.*, 75, 1222, 1978.
88. **Orkin, S. H., Old, J. M., Weatherall, D. J., and Nathan, D. G.**, Partial deletion of β-globin gene DNA in certain patients with β°-thalassaemia, *Proc. Natl. Acad. Sci. U. S. A.*, 76, 2400, 1979.
89. **Orkin, S. H., Alter, B. P., and Altay, C.**, Deletion of the Aγ-globin gene in Gγ-$\delta\beta$-thalassemia, *J. Clin. Invest.*, 64, 866, 1979.
90. **Fritsch, E. F., Lawn, R. M., and Maniatis, T.**, Characterizations of deletions which affect the expression of fetal globin genes in man, *Nature (London)*, 279, 598, 1979.

91. **Orkin, S. H., Goff, S. C., and Nathan, D. G.,** Heterogeneity of DNA deletion of γδβ-thalassemia, *J. Clin. Invest.*, 67, 878, 1981.
92. **Chang, J. C. and Kan, Y. W.,** β°-thalassemia, a nonsense mutation in man, *Proc. Natl. Acad. Sci. U. S. A.*, 76, 2886, 1979.
93. **Trecartin, R. F., Liebhaber, S. A., Chang, J. C., Lee, K. Y., and Kan, Y. W.,** β°-Thalassemia in Sardinia is caused by a nonsense mutation, *J. Clin. Invest.*, 68, 1012, 1981.
94. **Orkin, S. H. and Goff, S. C.,** Nonsense and frameshift mutations in β°-thalassemia detected in cloned β-globin genes, *J. Biol. Chem.*, 256, 9782, 1981.
95. **Humphries, R. K., Ley, T. J., Anagnou, N. P., Baur, A. W., and Nienhuis, A. W.,** β°-39 Thalassemia gene: a premature termination codon causes β-mRNA deficiency without affecting cytoplasmic β-mRNA stability, *Blood*, 64, 23, 1984.
96. **Takashita, K., Forget, B. G., Scarpa, A., and Benz, E. J., Jr.,** Intranuclear defect in β-globin mRNA accumulation due to a premature translation termination codon, *Blood*, 64, 13, 1984.
97. **Chang, J. C., Temple, G. F., Trecartin, R. F., and Kan, Y. W.,** Suppression of the nonsense mutation in homozygous β°-thalassaemia, *Nature (London)*, 281, 602, 1979.
98. **Breathnach, R., Benoist, C., O'Hare, K., Ganon, F., and Chambon, P.,** Ovalbumin gene: evidence for a leader sequence in mRNA and DNA sequences at the exon-intron boundaries, *Proc. Natl. Acad. Sci. U. S. A.*, 75, 4853, 1978.
99. **Abelson, J.,** RNA processing and the intervening sequence problem, *Annu. Rev. Biochem.*, 48, 1035, 1979.
100. **Breathnach, R. and Chambon, P.,** Organization and expression of eukaryotic split genes coding for proteins, *Annu. Rev. Biochem.*, 50, 349, 1981.
101. **Higgs, D. R., Goodbourn, S. E. Y., Lamb, J., Clegg, J. B., Weatherall, D. J., and Proudfoot, N. J.,** α-Thalassemia caused by a polyadenylation signal mutation, *Nature (London)*, 306, 398, 1983.
102. **Baird, M., Driscoll, C., Schreiner, H., Sciarrata, G., Sansone, G., Niazi, G., Ramirez, F., and Bank, A.,** A nucleotide change at a splice junction in the human β-globin gene is associated with β°-thalassemia, *Proc. Natl. Acad. Sci. U. S. A.*, 78, 4218, 1981.
103. **Treisman, R. A., Proudfoot, N. J., Shander, M., and Maniatis, T.,** A single base change at a splice site in a β°-thalassemic gene causes abnormal RNA splicing, *Cell*, 29, 903, 1982.
104. **Ley, T. J., Anagnou, N. P., Pepe, G., and Nienhuis, A. W.,** RNA processing errors in patients with β-thalassemia, *Proc. Natl. Acad. Sci. U. S. A.*, 79, 4775, 1982.
105. **Felber, B. K., Orkin, S. H., and Hamer, D. H.,** Abnormal RNA splicing causes one form of α-thalassemia, *Cell*, 29, 895, 1982.
106. **Westaway, D. and Williamson, R.,** A intron nucleotide sequence variant in a cloned β⁺-thalassemia globin gene, *Nucleic Acids Res.*, 9, 1777, 1981.
107. **Orkin, S., Goff, S. C., and Hechtman, R. L.,** Mutation in an intervening sequence splice junction in man, *Proc. Natl. Acad. Sci. U. S. A.*, 78, 5041, 1981.
108. **Spritz, R. A., Jagadeeswaran, P., Choudary, P. V., Biro, P. A., Elder, J. T., de Riel, J. K., Manley, J. L., Gefter, M. L., Forget, B. G., and Weissman, S. M.,** Base substitution in an intervening sequence of a β⁺-thalassemia human globin gene, *Proc. Natl. Acad. Sci. U. S. A.*, 78, 2455, 1981.
109. **Humphries, R. K., Ley, T. J., Goldsmith, M. E., Cline, A., Kantor, J. A., and Nienhuis, A. W.,** Silent nucleotide substitution in β⁺-thalassemia gene activates a cryptic splice site in β globin RNA coding sequence, *Blood*, 60(Suppl. 1), 54a, 1982.
110. **Orkin, S. H., Kazazian, H. H., Antonarakis, S. E., Ostrer, H., Goff, S. C., and Sexton, J. P.,** Abnormal RNA processing due to the exon mutation of the β^E-globin gene, *Nature (London)*, 300, 768, 1982.
111. **Poncz, M., Ballantine, M., Solowiejczyk, D., Barak, I., Schwartz, E., and Surey, S.,** β-Thalassemia in a Kurdish Jew. Single base change in the T-A-T-A box, *J. Biol. Chem.*, 257, 5994, 1982.
112. **Antonarakis, S. E., Orkin, S. H., Cheng, T., Scott, A. F., Sexton, J. P., Trusko, S., Charache, S., and Kazazian, H. H.,** β-Thalassemia in American blacks: novel mutations in the TATA box and IVS-2 acceptor splice sites, *Proc. Natl. Acad. Sci. U. S. A.*, 81, 1154, 1984.
113. **Orkin, S. H., Kazazian, H. H., Antonarakis, S. E., Goff, S. C., Boehm, C. D., Sexton, J. P., Waber, P. G., and Giardina, P. J. V.,** Linkage of β-thalassaemia mutations and β-globin gene polymorphisms with DNA polymorphisms in the human β-globin gene cluster, *Nature (London)*, 296, 627, 1982.
114. **Orkin, S. H., Antonarakis, S. E., and Kazazian, H. H., Jr.,** Base substitution in position-88 in a β-thalassemia globin gene, *J. Biol. Chem.*, 259, 8679, 1984.
115. **Collins, F. S., Methrall, J. E., Yamakawa, M., Pan, J., Weissman, S. M., and Forget, B. G.,** A point mutation in the A$_\gamma$-globin gene promotor in Greek hereditary persistence of fetal haemoglobin, *Nature (London)*, 313, 325, 1985.
116. **Gelinas, R., Endlich, B., Pfeiffer, C., Yagi, M., and Stamatoyannopoulos, G.,** G to A substitution in the distal CCAAT box of the A$_\gamma$-globin gene in Greek hereditary persistence of fetal haemoglobin, *Nature (London)*, 313, 323, 1985.

117. **Collins, F. S., Stoeckert, C. J., Jr., Serjeant, G. R., Forget, B. G., and Weissman, S. M.,** $G_\gamma\beta^+$ Hereditary persistence of fetal hemoglobin: cosmid cloning and identification of a specific mutation 5' to the G_γ gene, *Proc. Natl. Acad. Sci. U. S. A.*, 81, 4894, 1984.
118. **Clegg, J. B., Weatherall, D. J., and Milner, P. F.,** Haemoglobin Constant Spring — a chain termination mutant, *Nature (London)*, 234, 337, 1971.
119. **Liebhaber, S. A. and Kan, Y. W.,** Differentiation of the mRNA transcripts originating from the α_1- and α_2-globin loci in normals and α-thalassemias, *J. Clin. Invest.*, 68, 439, 1981.
120. **Semenza, G., Delgrosso, K., Poncz, M., Malladi, P., Schwartz, E., and Surrey, S.,** The silent carrier allele: β-thalassemia without a mutation in the β-globin gene or its immediate flanking regions, *Cell*, 39, 123, 1984.
121. **Pirastu, M., Sagieo, G., Chang, J. C., Cao, A., and Kan, Y. W.,** Initiation codon mutation as a cause of α-thalassemia, *J. Biol. Chem.*, 259, 12, 315, 1984.
122. **Ley, T. J., Griffith, P., and Nienhuis, A. W.,** Transfusion haemosiderosis and chelation therapy, *Clin. Haematol.*, 11, 437, 1982.
123. **Adams, J. G., Steinberg, M. H., Newman, M. V., Morrison, W. T., Benz, E. J., and Iyer, R.,** Beta-thalassemia present in cis to a new β-chain structural variant: Hb Vicksburg (β^{75}(E19)leu→o), *Proc. Natl. Acad. Sci. U. S. A.*, 78, 469, 1981.
124. **Smith, C. M., Hedlund, B., Cich, J. A., Turkey, D. P., Olson, M., Steinberg, M. H., and Adams, J. G.,** Hemoglobin North Shore: a variant hemoglobin associated with the phenotype of β-thalassemia, *Blood*, 61, 378, 1983.
125. **Fessas, Ph., Loukopoulos, D., Loutradi-Anagnostou, A., and Komis, G.,** "Silent" β-thalassemia caused by a "silent" β-chain mutant: the pathogenesis of a syndrome of thalassaemia intermedia, *Br. J. Haematol.*, 51, 577, 1982.
126. **Steinberg, M. H., Adams, J. G., Morrison, W. T., Pullen, D. J., Abney, R., Ibrahim, A., and Rieder, R. F.,** Hemoglobin Mississippi (β^{44} sers → cys): studies of the thalassemic phenotype in a mixed heterozygote with β^+-thalassemia, *J. Clin. Invest.*, 79, 746, 1987.
127. **Vissers, M. C. M., Winterbourn, C. C., and Carrell, R. W.,** Rapid proteolysis of unstable globins in human bone marrow, *Br. J. Haematol.*, 53, 417, 1983.
128. **Steinberg, M. H.,** Role of hemoglobin instability in premature red cell destruction, in *Cellular and Molecular Aspects of Aging: The Red Cell as a Model*, Eaton, J. W., Konzen, D. K., and White, J. G., Eds., Alan R. Liss, New York, 1985, 149.

# DEVELOPMENTAL PATTERN OF RED CELL MEMBRANE

Lee-Nien Lillian Chan

INTRODUCTION

The cellular plasma membrane has gained increasing attention since it has become evident that it performs many important functions. The cell membrane not only regulates the passage of materials in and out of cells, it also mediates the transfer of informational signals between the cell and its extracellular environment. Protein, lipids, and carbohydrates are the three basic components of cell membranes. Each cell type, however, has its own repertoire of different membrane components which are arranged in special architectural organization, resulting in a membrane that has unique structural, physiological, as well as immunological properties. The differentiation of plasma membranes during development is a subject of considerable interest and a large number of studies have been made. A summary of what is currently known about human red blood cell membrane differentiation and development is given in this chapter.

HUMAN ERYTHROCYTE MEMBRANE

The structure and organization of the human erythrocyte plasma membrane have been extensively examined, probably because of the ease with which pure cell populations and membrane preparations can be obtained. recent reviews on this subject include articles by Cohen,[1] Marchesi,[2] Tanner,[3] Schrier,[4] and Schwartz et al.[5] The human erythrocyte plasma membrane contains 52% protein, 40% lipid, and 8% carbohydrate (Table 1). Most of the carbohydrates (about 93%) are associated with proteins, and the remainder of them are associated with lipids.

Proteins

The proteins of the human red cell membrane have been operationally separated into two categories. The *extrinsic* or *peripheral* components are those which are easily extracted from the membrane ghosts, whereas the *intrinsic* or *integral* membrane proteins are those which are more tightly associated with the membrane lipid bilayer and can only be extracted by treatment with detergents.[6] Most of the former components constitute the membrane skeleton which is a multiprotein network lining the cytoplasmic side of the lipid bilayer.[1-4] The major elements of this network are spectrin (bands 1 and 2), actin (band 5), ankyrin (band 2.1), and band 4.1. The predominant integral membrane proteins are band 3 (the anion transporter) and glycophorin or PAS-1 (the major membrane sialoglycoprotein). Most of the integral membrane proteins extend through the lipid bilayer and are therefore transmembranous. It is now evident that the red cell membrane skeleton interacts with the membrane bilayer *via* linkages between spectrin and band 3, mediated by ankyrin.

Lipids

Some salient features of the lipids of human red cell membrane include[5] (1) there is asymmetric distribution of phospholipids between the inner and outer leaflets of the lipid bilayer: about two thirds of phosphatidylcholine and 80 to 85% of sphingomyelin are in the outer leaflet, whereas 80% of phosphatidylethanolamine and 100% of phosphatidylserine are present in the inner (cytoplasmic) leaflet, and (2) the cholesterol content is about equimolar with phospholipids. The distribution of cholesterol in the bilayer, as well as relative to phospholipids, is as yet unknown.

Table 1
COMPARISON OF MEMBRANE PROPERTIES BETWEEN HUMAN NEONATAL AND ADULT ERYTHROCYTES

		Neonatal to adult	
I.	Proteins		
	Membrane protein composition	Similar	
	Spectrin extractability	<<	
	Resistance of triton shell to urea	>>	
	Immunoreactivity of myosin	>>	
II.	Lipids		
	Cholesterol	>>	
	Phospholipid	>>	
	Cholesterol/phospholipid ratio	Similar	
	Sphingomyelin	>>	
	Phosphatidylinositol	>>	
	Saturated fatty acids	>>	
	Phosphatidylcholine	<<	
	Unsaturated fatty acids	<<	
	Viscosity	Similar	
III.	Blood groups and antigens		
	A,B, Lutheran, S, Bg antigens	<<	
	Lewis antigen	Absent	Present
	I/i antigen	i	I
IV.	Receptors		
	Insulin receptors	>>	

Note: References for this table are cited in the text.

Carbohydrates

The carbohydrates are primarily linked to integral membrane proteins and are usually present on the outer surface of the cell. The red cell membrane has three different distinct types of carbohydrate chains:[7]

1. Asparagine-linked lactosaminoglycans with a core portion and large branched side chains rich in galactose and *N*-acetylglucosamine (These are found associated with band 3 and band 4.5 proteins.).
2. Asparagine-linked oligosaccharides
3. Serine-linked tetrasaccharides

Both of the latter types are present in glycophorin.

MEMBRANE CHANGES DURING ONTOGENY

During human development, the first red blood cells are formed by about 3 weeks of gestation.[8] These are derived from embryonic progenitor cells located in the yolk sac. The major site of erythropoiesis changes to the fetal liver by about the end of the first trimester and then shifts to the bone marrow during the second trimester and remains there thereafter.[9] Throughout development, it is likely that the process of red blood cell differentiation may be achieved by a similar, if not identical, sequence of changes starting from the stem cells and finally forming mature erythrocytes. A schematic representation of erythropoiesis during development is shown.

ONTOGENY

DIFFERENTIATION	Primordial stem cells (early embryo)	→ Embryonic stem cells (yolk sac)	→ Fetal stem cells (fetal liver)	→ Neonatal stem cells (bone marrow)	→ Adult stem cells (bone marrow)
		↓	↓	↓	↓
		Embryonic erythroblasts	Fetal erythroblasts	Neonatal erythroblasts	Adult erythroblasts
		↓	↓	↓	↓
		Embryonic erythrocytes	Fetal erythrocytes	Neonatal erythrocytes	Adult erythrocytes

The most notable developmental change in human red cells is in the composition of globin chains.[9] The epsilon and zeta globin chains are expressed in embryonic red blood cells and they form hemoglobins Gower I and Gower II and hemoglobin Portland. The alpha and gamma globin chains are next expressed to form fetal hemoglobin until about 7 months of gestation, at which time the switch from gamma globin (fetal hemoglobin) to beta globin (adult hemoglobin) expression begins. Developmental changes in other parameters such as red cell enzyme activity and metabolic rate have also been observed.[11]

Relatively little is known about the membranes of human embryonic and fetal erythrocytes because of obvious practical problems. To date, most of the ontogenic studies have been performed by comparing membranes of red blood cells from newborn humans with those from adults.[12] Some significant differences have been observed and these are summarized in Table 1 and discussed below.

Membrane Proteins

The general membrane protein composition of neonatal red cells is essentially identical to that of adult red cells.[13,14] Although spectrin (band 1) is less extractable from neonatal than from adult erythrocyte ghosts,[15] they appear to be the same protein[16] and are phosphorylated to the same extent.[17] The membrane skeleton (triton shells) of neonatal red cells is more resistant to urea solubilization,[17] but shows no significant difference in mechanical stability.[12]

Lipids

There are considerable differences in lipid composition and distribution between neonatal and adult erythrocytes. Neonatal cells have 20% more lipids per unit surface area than cells.[19,20] However, a part of this excess may be due to higher numbers of intracellular endocytotic vessicles.[21]

The cholesterol-to-phospholipid ratio is fairly constant during development. On the other hand, there appear to be relatively more sphingomyelin and less phosphatidylcholine in neonatal red cell membranes as compared to that of the adult.[19] There are also relatively higher levels of phosphatidylinositol as well as saturated fatty acid in neonatal cells. Despite these variations, there appears to be no difference in membrane viscosity between neonatal and adult red cell membranes,[22] and the significance of these lipid differences is not yet evident.

Carbohydrates

Red cell membrane carbohydrates are predominately associated with glycoproteins and to a lesser extent with glycolipids. Several interesting developmental changes have been studied and the nature of these changes has been delineated.[7,23]

Fetal and neonatal red cells express a cell surface antigen (i) which is immunologically distinct from that of the adult erythrocyte (I).[7,11] The developmental change from i to I antigen occurs in parallel with the switch from fetal to adult hemoglobin during the first year of human life.[24] These antigens have been identified to be carbohydrate chains associated with the major membrane glycoprotein band 3,[25,26] as well as with membrane glycolipids.[27-29] The main difference between the i and I antigens is that the former is a linear polylactosamine structure, whereas the latter is branched.[30,31] Also, it was shown that the fetal i antigenic moiety contains a NeuNAcα2 $\rightarrow$ 8 NeuNAcα2 $\rightarrow$ 3 Gal structure which is absent in the adult antigen.[31]

The progressive branching of carbohydrates during development has been observed not only for the Ii antigens, but for cell surface glycoconjugates in general.[29,31-33] In addition, the level of sialation of membrane carbohydrates appears to be higher in newborn than in adult erythrocytes.[34]

Blood Groups and Surface Antigens

During development, the red blood cell surface antigens change both quantitatively as well as qualitatively. In the previous section, the Ii antigens have already been mentioned. Other surface antigens that differ between neonatal and adult erythrocytes include the Lutheran, S, Bg, A, and B antigens which are present at lower levels in neonatal red cells[24,35-37] and the Lewis antigens which are not detectable in neonatal erythrocytes.[38] A progressive branching of the carbohydrates associated with glycolipids during development may partially account for the increase of ABH blood groups with age.[33]

Enzymes Associated with Membranes

Acetylcholinesterase

Acetylcholinesterase is an erythrocyte enzyme that is externally disposed on the membrane. Neonatal red cells have only about half the acetylcholinesterase activity as compared with adult erythrocytes.[39-41] Consistent with this developmental change, blood cells from premature infants have less activity than full-term newborns.[42] A change in the isoelectric point of the enzyme from 4.6 to 4.8 parallels fetal to adult development.[43]

Cytochrome-b_5 Reductase

This enzyme exists in both soluble and membrane-bound forms. The membrane-bound form is a precursor which is further processed to the soluble enzyme.[44] Neonatal red cells have reduced cytochrome-b_5 reductase activity (by about 50%) and this is primarily due to a lower level of the soluble form of the enzyme.[45] The deficiency of this enzyme in newborns is correlated to their greater susceptibility to methemoglobinemia.

Glycolytic Enzymes

Aldolase, glyceraldehyde-3-phosphate dehydrogenase (GAPD), and phosphoglycerate kinase (PGK) are three of the glycolytic enzymes that associate with the red cell membrane to varying degrees.[46,47] All three show greater activity in fetal and neonatal red cells than in the adult.[48,49]

ATPase

Total Mg^{2+}-dependent ATPase activity is higher in neonatal red cell ghosts than that in the adult.[11,50] Whether any differences exist with regard to Na^+,K^+-ATPase is still unclear.

MEMBRANE CHANGES DURING ERYTHROID DIFFERENTIATION

Significant membrane differences have been detected in erythroid cells at various stages of maturation. Until recently, studies of erythroid precursor cells have been limited. However, with the development of *in vitro* cultures of erythroid cells as well as more sensitive and sophisticated techniques, it has become possible to identify and even characterize some of the early precursor cells such as colony forming units (CFU-E) and blast forming units (BFU-E). Using a variety of direct and indirect approaches, the changing membrane and cell surface properties of differentiating erythroid cells have been followed from stem cell to mature erythrocyte.

Membrane Skeleton

The major red cell membrane skeletal component, spectrin, becomes first detectable at the proerythroblast stage and subsequently accumulates throughout erythroid differentiation.[51-55] The final, mature membrane skeleton is not formed until the erythrocyte stage is reached. The process of enucleation during reticulocyte formation serves to concentrate spectrin in the membrane since the portion of the membrane that surrounds the extruded nucleus is preferentially spectrin-free.[56] In addition, this spectrin enrichment process is continued during maturation of the reticulocyte by means of endocytosis of membrane vessicles which are not lined with spectrin. Other membrane skeletal elements such as band 4.1 and ankyrin presumably follow a similar pattern, although direct observations have yet to be made. On the other hand, actin, being also a cytoskeletal element, is likely present even in the earliest stem cells, but probably does not become part of the membrane skeleton until sufficient spectrin molecules have accumulated in the early erythroblasts.

Integral Membrane Proteins
Band 3

The major membrane glycoprotein, band 3, appears by the basophilic erythroblast stage.[7,23,57] It also accumulates during maturation until synthetic activity in the reticulocytes ceases. Band 3 functions as the anion transporter[58] and therefore gains physiologic importance only when erythrocytes enter the circulation. It is currently hypothesized that band 3 may act as a regulatory element in the assembly of the red cell membrane skeleton since it provides an anchorage point (membrane receptor) for the unstable spectrin-ankyrin network.[59]

Band 4.5

Band 4.5 represents a group of membrane glycopeptides that range in molecular weight from about 45,000 to about 80,000 Da.[7,23] Included in this group is the glucose transporter, which is a transmembranous integral glycoprotein.[4,23,60,61] Band 4.5 may also include proteolytic digestion products from larger membrane proteins such as band 3 and band 4.1.[4,7,62] Bands 3 and 4.5 are readily labeled by the galactose oxidase/NaB[3]$_4$ procedure[63] and these two bands are similar in terms of time of their appearance and accumulation during erythroid differentiation.[10]

Glycoprotein

The major membrane sialoglycoprotein, glycophorin, is a red cell-specific membrane protein. It is first detectable in proerythroblasts of normal human bone marrow and accumulates during maturation.[64-68] Also, in a line of human erythroleukemic cells, K562 cells, which is in many respects representative of the proerythroblast stage, glycophorin is present at significant levels, whereas band 3 is absent.[7,10,68-70]

Gp 105 and Gp 95

Two glycoproteins, Gp 105 and Gp 95, are preferentially present on immature red blood cells and are gradually lost during differentiation.[7,10,23] These two glycoproteins are also detected to be present in K562 cells.[10,71] Thus far, no known function has been assigned to either glycophorin or Gp 105/Gp 95. It has been hypothesized that these components may serve as "receptors" for endogenous lectins[72] or as hormones that may be important in differentiation.[7,66,68]

Carbohydrates
Lactosaminoglycans

In the section entitled Carbohydrates, the ontogenic changes in the carbohydrates associated with membrane glycoproteins and glycolipids have been discussed. There are apparently similar changes in this regard during erythroid differentiation.[7,10,23] Immature cells display low levels of linear lactosaminoglycans, whereas mature erythrocytes contain high levels of branched lactosaminoglycans. This increase in lactosaminoglycans parallels the increase of bands 3 and 4.5 glycoproteins and may be the basis for the change in Ii antigen expression during erythroid differentiation (see below).

Glycosylation of Glycophorin

Another carbohydrate change associated with erythroid differentiation is the level of O-glycosylation of glycophorin.[67] Glycophorin isolated from mature erythrocytes contains 1 N-glycosidic and 15 O-glycosidic oligosaccharides. Similar analyses of immature erythroblasts from bone marrow as well as K562 cells show significantly lower levels of O-glycosylation in these cells. Thus, during erythroid differentiation, there is a progressive increase of O-glycosylation of the glycophorin molecules.

Sialation of Glycoconjugates

It has been observed that sialation of glycoproteins and glycolipids exists at higher levels in K562 cells than in normal erythrocytes. The ratio of gangliosides to neutral glycolipids in K562 cells is 1:1, whereas it is much lower, 1:15, in mature red cells.[74] Since K562 cells are considered to represent an early erythroblast stage of differentiation, it is interpreted that sialation of glycoconjugates decreases significantly during erythroid maturation.[23]

Blood Groups and Antigens
I/i

The ontogenic change in the expression of Ii antigens on erythrocytes has already been discussed. It is of interest that changes involving the Ii antigens occur during erythroid maturation as well.[7,10,23,65,75-77] The general picture that emerges from numerous studies of both cultured bone marrow and K562 cells indicates that (1) there is a transient expression of the i antigen in early erythroblasts which then diminishes until it is no longer detectable in mature erythrocytes, and (2) the I antigen is readily detectable even in BFU-E and their expression increases with erythroid differentiation until it is maximal in mature red blood cells.

ABO, RhD

It has been reported that the A or B antigens are present on a small percent of erythroid progenitors which are at an even earlier stage than BFU-E namely, the BFU-GEMM (erythroid burst forming unit in bone marrow mixed colony forming assay).[78] The proportion of progenitor cells expressing the A or B antigens increases subsequently in BFU-E and CFU-E and essentially all erythroblasts through erythrocytes carry one of these antigens.[65,78,79] The RhD antigen follows a similar differentiation pattern as the ABH antigens; it increases in expression between the erythroblast and erythrocyte stages.[80]

The progressive gain in the above-mentioned blood group antigens (Ii, ABH, and RhD) correlates with a similar increase in the amount of band 3 glycoprotein as well as with an increase in the branching of cell surface lactosaminoglycan. It is possible that the latter two parameters form the bases for increased blood group antigen expression during erythroid maturation.[7,23]

HLA-ABC, HLA-DR (Ia), OKT-10 Antigens

These antigens represent a group of erythroid (and myeloid) cell surface antigens whose expression diminishes with maturation.[65] Some of these may be expressed transiently. The HLA-DR or Ia antigens are present in a large percentage of BFU-E, but are present in only a small proportion of CFU-E.[64,65] They are also reported to be absent in pluripotent stem cells.[81] The HLA-ABC antigens diminish more gradually and may be present even in reticulocytes, but are absent in mature erythrocytes.[64,65,82]

The OKT-10 antigen is associated with a glycoprotein of 40,000-Da molecular weight.[83] It is present on erythroid progenitor cells and gradually diminishes during maturation.[65] It is not present on mature erythrocytes.

Receptors

Transferrin receptor — The cell surface receptor for transferrin plays an important role in the transport of iron into cells. This receptor has been extensively studied and a large store of information has accumulated.[84-86] Since iron is required for oxidative metabolism, the transferrin receptors are present in almost every cell type. Exceptionally high levels of these receptors are found in rapidly proliferating cells,[87,88] immature red blood cells,[89] and placental syncytiotrophoblasts.[90,91] More than 50% of BFU-E and almost all CFU-E have been found to have transferrin receptors.[65] Erythroblasts have the highest concentration of transferrin receptors which gradually decline with further maturation.[65,92-94] Reticulocytes still express significant levels of transferrin receptors, but none are detectable in mature erythrocytes.[95,96] The mechanism for the progressive loss of transferrin receptors has been shown in sheep reticulocytes to be a preferential exocytosis of transferrin receptors contained in membrane vessicles.[97] This may represent a general mechanism by which cellular membranes are remodeled during differentiation.

Insulin receptors — Specific insulin receptors have been identified on human erythrocytes.[98,99] It was observed that the density of insulin receptors on erythroid cells decreases with maturation in normal[100-103] as well as in erythroleukemic cells.[104]

ACKNOWLEDGMENTS

I am indebted to Dr. M. Fukuda whose reprints and preprints were invaluable in the preparation of this review.

REFERENCES

1. **Cohen, C. M.**, The molecular organization of the red cell membrane skeleton, *Semin. Hematol.*, 20, 141, 1983.
2. **Marchesi, V. T.**, The red cell membrane skeleton: recent progress, *Blood*, 61, 1, 1983.
3. **Tanner, M. J.**, Erythrocyte membrane structure and function, *Ciba Found. Symp.*, 94, 3, 1983.
4. **Schrier, S. L.**, Red cell membrane biology, *Clin. Haematol.*, 14, 1, 1985.
5. **Schwartz, R. S., Chiu, D. T.-Y, and Lubin, B.**, Plasma membrane phospholipid organization in human erythrocytes, *Curr. Top. Hematol.*, 5, 63, 1985.

6. **Steck, T. L.,** The organization of proteins in the human red blood cell membrane, *J. Cell Biol.*, 62, 1, 1974.
7. **Fukuda, M. and Fukuda, M. M.,** Cell surface glycoproteins and carbohydrate antigens in development and differentiation of human erythroid cells, in the *Biology of Glycoproteins*, Ivatt, R. J., Eds., Plenum Press, New York, 1984, 183.
8. **Moore, K. L.,** *The Developing Human*, W. B. Saunders, Philadelphia, 1974, 49.
9. **Wood, W. G., Clegg, J. B., and Weatherall, D. J.,** Developmental biology of human hemoglobins, *Prog. Hematol.*, 10, 43, 1977.
10. **Fukuda, M., Fukuda, M. N., Papayannopoulou, T., and Hakamori, S.,** Membrane differentiation in human erythroid cells: unique profiles of cell surface glycoproteins expressed in erythroblasts *in vitro* from three ontogenic stages, *Proc. Natl. Acad. Sci. U. S. A.*, 77, 3474, 1980.
11. **Matthay, K. K. and Mentzer, W. C.,** Erythrocyte enzymopathies in the newborn, *Clin. Haematol.*, 10, 31, 1981.
12. **Matovcik, L. M. and Mentzer, W. C.,** The membrane of the human neonatal red cell, *Clin. Haematol.*, 14, 203, 1985.
13. **Shapiro, D. L. and Pasqualini, P.,** Erythrocyte membrane proteins of premature and full-term newborn infants, *Pediatr. Res.*, 12, 176, 1978.
14. **Rosenblum, B. B.,** Two-dimensional gel electrophoresis of erythrocyte membrane proteins, *Prog. Clin. Biol. Res.*, 56, 251, 1981.
15. **Landaw, S. A., Rathbun, S. C., and Guancial, R. L.,** Evidence for a tightly linked spectrin lattice in red blood cells of newborn man, *Pediatr. Res.*, 16, 208A, 1981.
16. **Lawler, J., Liu, S. C., Palek, J., and Prchal, J.,** Molecular defect of spectrin in a subset of patients with hereditary elliptocytosis: alterations in the subunit domain involved in spectrin self-association, *J. Clin. Invest.*, 73, 1688, 1984.
17. **Shapiro, D. L.,** Spectrin phosphorylation in neonatal and adult erythrocyte membranes, *Arch. Biochem. Biophys.*, 193, 264, 1979.
18. **Landaw, S. A., Rathbun, S. C., and Guanicial, R. L.,** The effect of spectrin cross linking on red blood cell membrane properties and survival, *Pediatr. Res.*, 14, 536, 1980.
19. **Neerhout, R. C.,** Erythrocyte lipids in the neonate, *Pediatr. Res.*, 2, 172, 1968.
20. **Neerhout, R. C.,** Erythrocyte lipids in infants with low birth weights, *Pediatr. Res.*, 5, 101, 1971.
21. **Matovcik, L. M., Lubin, B. H., and Schrier, S. L.,** The extreme heterogeneity of neonatal RBC is reflected in measurements of their plasma membrane surface area, *Blood*, 62, 34a, 1983.
22. **Kehry, M., Yguerabide, J., and Singer, S. J.,** Fluidity in the membrane of adult and neonatal human erythrocytes, *Science*, 195, 486, 1977.
23. **Fukuda, M.,** Cell surface glycoconjugates as oncodifferentiation markers in hematopoietic cells, *Biochim. Biophys. Acta*, 780, 119, 1985.
24. **Marsh, W. L.,** Anti-i: cold antibody defining Ii relationship in human red cells, *Br. J. Haematol.*, 7, 20, 1961.
25. **Fukuda, M. N., Fukuda, M., Watanabe, K., and Hakamori, S.,** Modification of cell surface antigenicity by endogalactosidase of *E. freundii*, *Fed. Proc.*, 37, 1601, 1978.
26. **Childs, R. A., Feizi, T., Fukuda, M., and Hakamori, S.,** Blood group I activity associated with Band 3, the major intrinsic membrane protein of human erythrocytes, *Biochem. J.*, 173, 333, 1978.
27. **Watanabe, K., Hakamori, S., Childs, R. A., and Feizi, T.,** Characterization of a blood group I-active ganglioside: structural requirements for I and i specificities, *J. Biol. Chem.*, 254, 3221, 1979.
28. **Feizi, T., Childs, R. A., Watanabe, K., and Hakamori, S.,** Three types of blood groups I specificity among monoclonal anti-I autoantibodies revealed by analogues of a branched erythrocyte glycolipid, *J. Exp. Med.*, 149, 975, 1979.
29. **Nieman, H., Watanabe, K., Hakamori, S., Childs, R. A., and Feizi, T.,** Blood group i and I activities of "lacto-N-norhex aosylceramide" and its analogues: the structural requirements for i-specificities, *Biochem. Biophys. Res. Commun.*, 81, 1286, 1978.
30. **Fukuda, M., Fukuda, M. N., and Hakamori, S.,** Developmental change and genetic defect in the carbohydrate structure of Band 3 glycoprotein of human erythrocyte membrane, *J. Biol. Chem.*, 254, 3700, 1979.
31. **Koschielak, J., Zdebska, E., Wilczynska, Z., Miller-Podraza, H., and Dzierzkowa-Borodej, W.,** Immunochemistry of Ii-active glycosphingolipids of erythrocytes, *Eur. J. Biochem.*, 96, 331, 1979.
32. **Fukuda, M., Dell, A., and Fukuda, M. N.,** Structure of fetal lactosaminoglycan, the carbohydrate moiety of Band 3 isolated from human umbilical cord erythrocytes, *J. Biol. Chem.*, 259, 4782, 1984.
33. **Watanabe, K. and Hakamori, S.,** Status of blood group carbohydrate chains in ontogenesis and in oncogenesis, *J. Exp. Med.*, 144, 664, 1976.
34. **Fukuda, M. N. and Levery, S. B.,** Glycolipids of fetal, newborn, and adult erythrocytes: glycolipid pattern and structural study of H_3-glycolipid from newborn erythrocytes, *Biochemistry*, 22, 5034, 1983.

35. **Larson, G. and Samuelsson, B. E.**, Blood group type glycosphingolipids of human cord blood erythrocytes, *J. Biochem.*, 88, 647, 1980.
36. **Vulpis, N.**, Studies on the antigens of human red cell ghosts, *Acta Haematol.*, 30, 229, 1963.
37. **Marsh, W. L. and Allen, F. H.**, Erythrocyte blood groups in humans, in *Hematology of Infancy and Childhood*, Nathan, D. G. and Oski, F. A., Eds., W. B. Saunders, Philadelphia, 1981, 1411.
38. **Andresen, P. H.**, Blood group with characteristic phenotypical aspects, *Acta Pathol. Microbiol. Scand.*, 24, 616, 1947.
39. **Jones, P. E. H. and McCance, R. A.**, Enzyme activities in the blood of infants and adults, *Biochem. J.*, 45, 464, 1949.
40. **Herz, F. and Kaplan, E.**, A review: human erythrocyte acetylcholinesterase, *Pediatr. Res.*, 7, 204, 1973.
41. **Garre, C., Ravazzolo, R., Ajmar, F., and Bruzzone, G.**, Electrophoretic differences between fetal and adult acetylcholinesterase of human red blood cell membranes, *Cell Differ.*, 9, 165, 1980.
42. **Burman, D.**, Red cell cholinesterase in infancy and childhood, *Arch. Dis. Child.*, 36, 362, 1961.
43. **Ravazzolo, R., Garre, C., Bruzzone, G., and Ajmor, F.**, Variability of acetylcholinesterase in adult and fetal red cell membranes, *Prog. Clin. Biol. Res.*, 55, 439, 1981.
44. **Choury, D., Wajcman, H., Boissel, J. P., and Kaplan, J. C.**, Evidence for endogenous proteolytic solubilization of human red cell membrane NADH cytochrome b_5 reductase, *FEBS Lett.*, 126, 172, 1981.
45. **Choury, D., Reghis, A., Pichard, A.-L., and Kaplan, J. C.**, Endogenous proteolysis of membrane-bound red cell cytochrome-b_5 reductase in adults and newborns: its possible relevance to the generation of the soluble "methemoglobin reductase," *Blood*, 61, 894, 1983.
46. **Schrier, S. L.**, Studies of the metabolism of human erythrocyte membranes, *J. Clin. Invest.*, 42, 757, 1963.
47. **Duchon, G. and Collier, H. B.**, Enzyme activities of human erythrocyte ghosts: effects of various treatments, *J. Membr. Biol.*, 6, 138, 1971.
48. **Lestas, A. N., Rodeck, C. H., and White, J. M.**, Normal activities of glycolytic enzymes in fetal erythrocytes, *Br. J. Haematol.*, 50, 439, 1982.
49. **Oski, F. A. and Komazawa, M.**, Metabolism of the erythrocytes of the newborn infant, *Semin. Hematol.*, 12, 209, 1975.
50. **Sigstrom, L., Waldenstrom, J., and Karlberg, P.**, Characteristics of active sodium and potassium transport in erythrocytes of healthy infants and children, *Acta Paediatr. Scan.*, 70, 374, 1981.
51. **Chang, H., Langer, P. J., and Lodish, H. F.**, Asynchronous synthesis of erythrocyte membrane proteins, *Proc. Natl. Acad. Sci. U. S. A.*, 73, 3206, 1976.
52. **Eisen, H., Bach, R., and Embery, R.**, Induction of spectrin in erythroleukemic cells transformed by Friend virus, *Proc. Natl. Acad. Sci. U. S. A.*, 74, 3898, 1977.
53. **Fehlmann, M., Lafleur, L., and Marceau, N.**, Surface membrane differentiation of hemopoietic cells as observed by radioactive labeling, *J. Cell. Physiol.*, 90, 455, 1977.
54. **Geiduschek, S. B. and Singer, S. J.**, Molecular changes in the membranes of mouse erythroid cells accompanying differentiation, *Cell*, 16, 149, 1979.
55. **Tong, B. D. and Goldwasser, E.**, The formation of erythrocyte membrane proteins during erythropoietin-induced differentiation, *J. Biol. Chem.*, 256, 12,666, 1981.
56. **Zweig, S. E., Tokuyasu, K. T., and Singer, S. J.**, Membrane-associated changes during erythropoiesis. On the mechanism of maturation of reticulocytes to erythrocytes, *J. Supramol. Struct. Cell Biochem.*, 17, 163, 1981.
57. **Foxwell, B. M. and Tanner, M. J.**, Synthesis of the erythrocyte anion-transport protein: immunochemical study of its incorporation into the plasma membrane of erythroid cells, *Biochem. J.*, 195, 129, 1981.
58. **Cabantchik, Z. I. and Rothstein, A.**, The nature of the membrane sites controlling anion permeability of human red blood cells as determined by studies with disulfonic stilbene derivatives, *J. Membr. Biol.*, 15, 207, 1972.
59. **Moon, R. T. and Lazarides, E.**, Biogenesis of the avian erythroid membrane skeleton: receptor-mediated assembly and stabilization of ankyrin (goblin) and spectrin, *J. Cell Biol.*, 98, 1899, 1984.
60. **Sogin, D. C. and Hinkle, P. C.**, Characterization of the glucose transporter from human erythrocytes, *J. Supramol. Struct.*, 8, 447, 1978.
61. **Sogin, D. C. and Hinkle, P. C.**, Immunological identification of the human erythrocyte glucose transporter, *Proc. Natl. Acad. Sci. U. S. A.*, 77, 5725, 1980.
62. **Tarone, G., Hamasaki, N., Fukuda, M., and Marchesi, V. T.**, Proteolytic degradation of human erythrocyte Band 3 by membrane-associated protease activity, *J. Membr. Biol.*, 48, 1, 1979.
63. **Gahmberg, C. G.**, External labeling of human erythrocyte glycoproteins: studies with galactose oxidase and fluorography, *J. Biol. Chem.*, 251, 510, 1976.
64. **Robinson, J., Sieff, C., Delia, D., Edwards, P. A. W., and Greaves, M.**, Expression of cell-surface HLA-DR, HLA-ABC and glycophorin during erythroid differentiation, *Nature (London)*, 289, 68, 1981.

65. Sieff, C., Bicknell, D., Caine, G., Robinson, J., Lam, G., and Greaves, M. F., Changes in cell surface antigen expression during hemopoietic differentiation, *Blood,* 60, 703, 1982.
66. Gahmberg, C. G., Jokinen, M., and Andersson, L. C., Expression of the major sialoglycoprotein (glycophorin) on erythroid cells in human bone marrow cells, *Blood,* 52, 379, 1978.
67. Gahmberg, C. G., Ekblom, M., and Andersson, L. C., Differentiation of human erythroid cells is associated with increased O-glycosylation of the major sialoglycoprotein, glycophorin A, *Proc. Natl. Acad. Sci. U. S. A.,* 81, 6752, 1984.
68. Yurcheno, P. D. and Furthmayr, H., Expression of red cell membrane proteins in erythroid precursor cells, *J. Supramol. Structr.,* 13, 255, 1980.
69. Horton, M. A., Cedar, S. H., and Edwards, P. A., Expression of red cell specific determinants during differentiation in the K562 erythroleukaemia cell line, *Scand. J. Haematol.,* 27, 231, 1981.
70. Gahmberg, C. G., Jokinen, M., and Anderson, L. C., Expression of the major red cell sialoglycoprotein, glycophorin A, in the human leukemia cell line K562, *J. Biol. Chem.,* 254, 7442, 1979.
71. Turco, S. J., Rush, J. S., and Laine, R. A., Presence of erythroglycan on human K562 chronic myelogeneous leukemia derived cells, *J. Biol. Chem.,* 255, 3266, 1980.
72. Harrison, F. L. and Chesterton, C. J., Erythroid developmental agglutinin is a protein lectin mediating specific cell-cell adhesion between differentiating rabbit erythroblasts, *Nature (London),* 286, 502, 1980.
73. Fukuda, M., K562 human leukaemic cells express fetal type (i) antigen on different glycoproteins from circulating erythrocytes, *Nature (London),* 285, 405, 1980.
74. Suzuki, A., Karol, R. B., Kundu, S. K., and Marcus, D. M., Glycosphingolipids of K562 cells: a chemical and immunological analysis, *Int. J. Cancer,* 28, 271, 1981.
75. O'Hara, C. J., Shumak, K. H., and Prince, G. B., The i antigen on human myeloid progenitors, *Clin. Immunol. Immunopathol.,* 10, 420, 1978.
76. Vainchenker, W., Testa, U., Guichard, J., Titeux, M., and Breton-Gorius, J., Heterogeneity in the cellular commitment of a human leukemic cell line: K562, *Blood Cells,* 7, 357, 1981.
77. Vainchenker, W., Teston, U., Rochant, H., Titeux, M., Henri, A., Bouguet, J., and Breton-Gorius, J., Cellular regulation of i and I antigen expression in human erythroblasts grown *in vitro, Stem Cells,* 1, 97, 1981.
78. Blacklock, H. A., Katz, F., Michalevicz, R., and Hazlehurst, G. R. P., A and B blood group antigen expression on mixed colony cells and erythroid precursors: relevance for human allogenic bone marrow transplantation, *Br. J. Haematol.,* 58, 267, 1984.
79. Karhi, K. K., Andersson, L. C., Vuopio, P., and Gahmberg, C. G., Expression of blood group A antigens in bone marrow cells, *Blood,* 57, 147, 1981.
80. Rearden, A. and Masonredis, S. P., Blood group D antigen content of nucleated red cell precursors, *Blood,* 50, 981, 1977.
81. Moore, M. A., Broxmeyer, H. E., Sheridan, A. P., Meyers, P. A., Jacobson, N., and Winchester, R. J., Continuous bone marrow cells: Ia antigen characterization of probable pluripotential stem cells, *Blood,* 55, 682, 1980.
82. Brown, G., Biberfeld, P., Christenson, B., and Mason, D. Y., The distribution of HLA on human lymphoid, bone marrow and peripheral blood cells, *Eur. J. Immunol.,* 9, 272, 1979.
83. Terhorst, C., Van Agthoven, A., LeClair, K., Snow, P., Reinhertz, E., and Scheossman, S., Biochemical studies of the human thymocyte on cell-surface antigens T6, T9 and T10, *Cell,* 23, 771, 1981.
84. Newman, R., Schneider, C., Sutherland, R., Vodinelich, L., and Greaves, M., The transferrin receptor, *TIBS,* 7, 397, 1982.
85. Seligman, P. A., Structure and function of the transferrin receptor, *Prog. Hematol.,* 13, 131, 1983.
86. Testa, U., Transferrin receptors: structure and function, *Curr. Top. Hematol.,* 5, 127, 1985.
87. Sutherland, R., Delia, D., Schneider, C., Newman, R., Kemshead, J., and Greaves, M., Ubiquitous cell-surface glycoprotein on tumor cells in proliferation-associated receptor for transferrin, *Proc. Natl. Acad. Sci. U. S. A.,* 78, 4515, 1981.
88. Trowbridge, I. S. and Omary, M. B., Human cell surface glycoprotein related to cell proliferation is receptor for transferrin, *Proc. Natl. Acad. Sci. U. S. A.,* 78, 3039, 1981.
89. Jandle, J. H., Inman, J. K., Simons, R. L., and Allen, D. W., Transfer of iron from serum iron-binding protein to human reticulocytes, *J. Clin. Invest.,* 38, 161, 1959.
90. Seligman, P. A., Schneider, R. B., and Allen, R. H., Isolation and characterization of the transferrin receptor from human placenta, *J. Biol. Chem.,* 254, 9943, 1979.
91. Wada, H. G., Hass, P. E., and Sussman, H. H., Transferrin receptor in human placental brush border membranes, *J. Biol. Chem.,* 254, 12,629, 1979.
92. Nunez, M. T., Glass, J., Fischer, S., Lavidor, L. M., Lenk, E. M., and Robinson, S. H., Transferrin receptors in developing murine erythroid cells, *Br. J. Haematol.,* 36, 519, 1977.
93. Horton, M. A., Expression of transferrin receptors during erythroid maturation, *Exp. Cell Res.,* 144, 301, 1983.

94. **Iacopetta, B. J., Morgan, E. H., and Yeoh, G. C.,** Transferrin receptors and iron uptake during erythroid cell development, *Biochim. Biophys. Acta,* 687, 204, 1982.
95. **Jandle, J. L. and Katz, J. H.,** The plasma to cell cycle of transferrin, *J. Clin. Invest.,* 42, 314, 1963.
96. **Frazier, J., Caskey, J., Yoffe, M., and Seligman, P.,** Studies of the transferrin receptor on both human reticulocytes and nucleated human cells in culture: comparison of factors regulating receptor density, *J. Clin. Invest.,* 69, 853, 1982.
97. **Pan, B. and Johnstone, R. M.,** Fate of the transferrin receptor during maturation of sheep reticulocytes *in vitro*: selective externalization of the receptor, *Cell,* 33, 967, 1983.
98. **Gambhir, K. K., Archer, J. A., and Carter, L.,** Insulin radioreceptor assay for human erythrocytes, *Clin. Chem.,* 23, 1590, 1977.
99. **Gambhir, K. K., Archer, J. A., and Bradley, C. J.,** Characteristics of human erythrocyte insulin receptors, *Diabetes,* 27, 701, 1978.
100. **Thomopoulos, P., Berthellier, M., and Laudat, M.-H.,** Loss of insulin receptors on maturation of reticulocytes, *Biochem. Biophys. Res. Commun.,* 85, 1460, 1978.
101. **Eng, J., Lee, L., and Yalow, R. S.,** Influence of the age of erythrocytes on their insulin receptors, *Diabetes,* 29, 164, 1980.
102. **Kosmakos, F. C., Nagulesparan, M., and Bennett, P. H.,** Insulin binding to erythrocytes: a negative correlation with red cell age, *J. Clin. Endocrinol. Metab.,* 51, 46, 1980.
103. **Polychronakos, C., Ruggere, M. D., and Benjamin, A.,** The role of cell age in the difference in insulin binding between adult and cord erythrocytes, *J. Clin. Endocrinol. Metab.,* 55, 290, 1982.
104. **Ginsberg, B. H., Brown, T., and Raizada, M.,** Decrease in insulin receptors during Friend erythroleukemia cell differentiation, *Diabetes,* 28, 823, 1979.

DEVELOPMENT OF IMMUNITY AND IMMUNE SYSTEM

DEVELOPMENT OF LYMPHOID TISSUES

Ilan Bleiberg and Myra Small

INTRODUCTION

Hemopoietic stem cells have the potential to differentiate to a variety of mature cells which are found in the blood. Blood cells as diverse as erythrocytes, macrophages, lymphocytes, polymorphonuclear cells, and thrombocytes are believed to be derived from a single stem cell.[1] One class of these cells which is involved in defense, the lymphocytes, can be divided into many subclasses of specialized cells, each with its own particular function. The lymphoid system, which is programmed to react against foreign antigens over the whole area of the body, is unique in its organization into many discrete organs which are dispersed in strategic locations throughout the body. Functional lymphocytes which are found in these highly organized tissues, and in the blood and lymph circulations that link these diffuse organs, will be considered in detail elsewhere in this book. Here we will consider the pathway of development leading to activity of these diverse cells which occurs as the lymphocytes pass through successive dispersed specialized environments. Stem cells from the hemopoietic tissues undergo early lymphoid differentiation in primary lymphoid organs. There, early development of each main lymphoid lineage occurs in its own microenvironment before the cells migrate to the peripheral lymphoid tissues. Development of the lymphoid lineage, unlike other hemopoietic lineages, is thus a multitissue process that involves migration in a unidirectional flow. The first stage takes place in adult bone marrow (or liver and spleen in early embryonic life); the next stage takes place in either the bone marrow (embryonic liver) or thymus; and the mature cells then migrate to the spleen, lymph nodes, and scattered peripheral lymphoid tissues where they are ready to react. At each stage it seems clear that interaction between the mobile lymphocytes and the fixed cells of the lymphoid organs is a characteristic feature involved in directing both lymphocyte development and lymphocyte activity. A pictorial example of such an interaction can be seen in Figure 1, which illustrates the spatial relation between lymphocytes and stromal cells in the thymus. Although we believe that supporting cells are part of specialized microenvironments, or often of many adjoining submicroenvironments differing in function, we are still far from understanding the nature and mechanism of activity of these units.

As in differentiation of other tissues, development of lymphocyte classes and subclasses from pluripotent stem cells consists of successive narrowing of the possibilities in terms of both structure and function of the cells. At this time, we have very little idea of the nuclear changes which the cells undergo at each step. Our knowledge (also far from complete) is more detailed concerning the corresponding changes in surface properties of the differentiating lymphocytes and the tissue locations at which these changes occur. It is the aim of this chapter to summarize the information on the process of lymphocyte development from the vantage point of 1985, without attempting to include all of the numerous and interesting investigations in print. We begin with a brief ontogenic survey of the sites of lymphoid development. Then we will consider the development of T- and B-lymphocytes which occurs within these organs during fetal and adult life.

ONTOGENY OF LYMPHOID TISSUES

Development of the lymphoid system in humans as in all mammals begins around the first trimester of pregnancy, although not all lymphatic organs develop simultaneously. Activity of these organs also begins at different times. While some tissues are active during

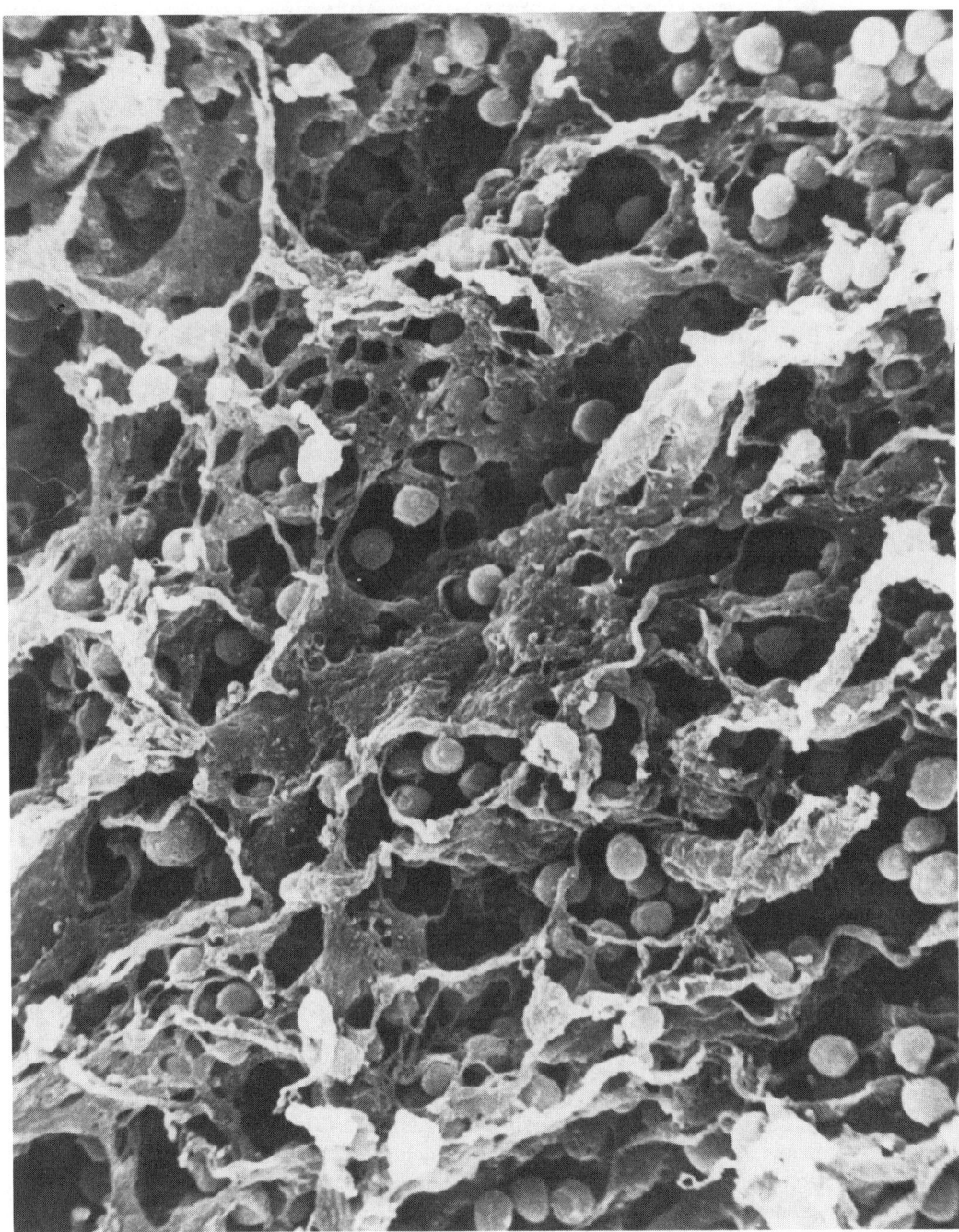

FIGURE 1. Scanning electron micrograph of the thymic cortex. Smooth-surfaced thymocytes are interposed in an interlacing meshwork of epithelial cells. (Magnification × 10,000.) (From van Ewijk, W., *Am J. Anat.*, 170, 311, 1984. With permission.)

the fetal period, others such as lymphoid nodules reach full activity only after birth. Lymphocyte precursors are detected only after establishment of the circulatory system is complete, and it was first shown by Moore and Owen[2] that embryonic lymphoid tissue is colonized by hemopoietic stem cells which enter from the circulating blood. In mammals, it is generally accepted that the yolk sac is the source of the earliest hemopoietic stem cells,[3] although the extent of this contribution to the lymphoid lineage of adults remains a matter of controversy. In birds there is evidence that early lymphocyte precursors are derived from intraembryonic blood foci[4] and a similar phenomenon has been demonstrated in amphibians.[5] In the light of this evidence, perhaps it is necessary to reexamine the question in mammals as well. Mature lymphocytes are found in human fetuses from the third month onward, a month later than granulocytes.[6] Between the fourth and fifth months, monocytes are found in the spleen and lymph nodes. Although the lymph nodes usually become reactive only at birth with exposure to external immunologic challenge, embryonic lymphoid tissues are capable of lymphopoiesis in response to a stimulus *in vitro*. In short, it seems that already in early embryogenesis there is activity in the lymphoid organs which is mainly involved in establishment of the complex mechanisms of the lymphoid system that will function after birth. Before appearance of lymphocytes in the various organs, establishment and organization of the fixed constituents of these tissues take place and seem to be prerequisite.

Fetal Liver

Around the time of midgestation, the fetal liver becomes the main hemopoietic organ in the human embryo. Lymphoid precursors are among the stem cells which are produced in this site, and transplantation of human fetal liver to an infant with severe combined immunodeficiency disease resulted in development of immunocompetent T- and B-cells of donor origin.[7] During this period, development of B-cells from their precursors occurs in this organ as well.

The liver starts to develop during the first trimester of gestation from the endodermal epithelia of the duodenum which is later covered by mesodermal tissue and highly vascularized. Since the liver becomes primarily an endocrine and exocrine gland, morphological heterogeneity of the tissue makes investigation of stem cell development more difficult.

The ability of the fetal liver to facilitate development of B-cells is well established.[8-10] A wave of development of B-cells from pre-B occurs toward the end of the first trimester and continues into the second.[11] Gathings et al. showed that 0.2% of human nucleated fetal liver cells at 7.5 weeks of age are pre-B.[12] This increases to 1.1% of the cells in the 11-week fetal liver, whereas these cells are not found in the bone marrow before the 14th week. The first mature B-cells appear in the fetal liver around 10 to 11 weeks (0.6% of the nucleated cells). A similar pattern was found in mice[13] and in rabbits.[14]

The presence of T-cell progenitors in mouse fetal liver has been demonstrated, perhaps most convincingly by transfer experiments to irradiated recipients.[15] In addition it has been suggested that hemopoietic cells from the yolk sac might undergo obligatory development in the fetal liver before colonizing the thymus.[16] The possibility has also been raised that potentially active T-cells are present in the liver, but that liver suppressor cells might interfere with detection of their activity.[17] In short, the role of this organ in supporting early development and determination of T-cells is controversial and requires further investigation.

Thymus

The thymus has stimulated much research in the past two decades, but in many respects it still remains an enigma. This mysterious organ was without a known function for years since its removal from adult animals did not appear to result in decrease of any activity. Today it seems clear that the thymus plays a central role in establishment of the cellular immune response as well as in development of regulatory lymphocytes. The variety of T-cells involved in these functions, such as helper, suppressor, and cytotoxic T-lymphocytes, all develop in the thymus.[18] These heterogeneous T-cells also differ in the surface properties of both the mature cells and the intermediates at successive stages of development.[19] Explanations must be found for this heterogeneous development, perhaps in terms of cell programming and/or microenvironmental differences within the thymus. In contrast to other lymphoid tissues, the fixed components of the thymus (which we shall call stroma) consist of epithelial cells. These epithelial cells are derived from the pharyngeal pouches where they become surrounded by mesenchymal tissue. The latter develops into connective tissue and blood vessels and is also a necessary inductive stimulus for development of the epithelial thymus.[20-22]

While there is no doubt that an endodermal contribution is required for the thymus to develop, the importance of the ectoderm is a matter of controversy. On the one hand, Le Douarin et al. have demonstrated development of the avian thymus *in vitro* from pharyngeal endoderm and mesenchymal elements only.[23] On the other hand, others[24,25] have claimed that a full-sized and functional thymus develops in mice only when ectodermal cells also contribute to the stroma. At any rate, the thymic stroma consists mainly of a network of epithelial cells which, as we will see later, are heterogeneous both in function and in structure and which can be identified histologically about the sixth week of gestation. Precursors of the antigen-presenting cells arrive in the thymus later and become incorporated into the stromal network.[26,27] The first lymphocytes appear in the human thymus at about 8 to 9 weeks of gestation, and the cortex can be distinguished from the medulla at about 12 weeks.[28] Since (in birds) thymic stem cells are present some days before thymic colonization occurs,[29] it seems that the onset of colonization is not limited by availability of the precursor cells, but more likely depends on changes in the thymic primordium.

Spleen

From the primary lymphoid tissues, the T- and B-cells migrate to the peripheral lymphoid organs, the largest of which is the spleen. This organ also plays a transient hemopoietic role during embryonic life. The spleen begins to develop in the first trimester and at first its function is limited to hemopoiesis and only the red pulp is active. Development of the white pulp begins at about 14 weeks with growth of cells from the blood vessel wall into the periarterial area. At first they are concentrated close to the blood vessels. Gradually reticular cells appear and are integrated into larger areas around the arteries, especially the smaller arteries, in which lymphoid cells will settle, to form the periarterial lymphatic sheaths and the marginal zone surrounding the white pulp.[30] At this stage,[31] the follicular dendritic cells (FDC) arrive in the white pulp. The interdigitating cells (IDC) also arrive with the circulating blood and settle mainly in the periarterial regions. Their processes form invaginations within which T-lymphocytes later accumulate.[32,33] These two antigen presenting cells thus may direct the development of B-cell and T-cell areas in the developing spleen. A similar process appears to occur in the embryonic lymph node, and the same cells may also be involved in maintenance of the two separate domains in adult life. The lymphocytes which begin to accumulate in the spleen about the middle of pregnancy are derived at first from the hematogenic fetal liver. At a later date, the bone marrow becomes an hemopoietic organ and takes over as the source of the lymphoid stem cells for the remainder of the life of the organism. There is no evidence that hemopoiesis of the splenic red pulp contributes directly to the settlement of lymphoid cells in the white pulp. It is more likely that such a process takes place indirectly via the blood circulation, perhaps involving development in an intervening organ.

Bone Marrow

In humans the bone marrow takes over as the main hemopoietic organ during the last part of the fetal period and continues in this role during all of adult life. This tissue is clearly detected in human bone during the second trimester of gestation and is closely interrelated with processes occurring in the bone tissue and with development of blood vessels in the bone. It is also possible to reproduce and study the initiation of fetal bone marrow development in postembryonic experiments by the approach developed by Reddi and Huggins.[34] Implantation of demineralized bone matrix subcutaneously results in a cascade of events leading to the development of endochondral bone, penetration of blood vessels into it, and the development of bone marrow supportive tissue. Here as in other hemopoietic tissues, only after establishment of the stroma do hemopoietic stem cells settle and differentiate. The mechanisms leading to cessation of function of the fetal liver and later the spleen as hemopoietic organs are entirely unknown. It is likewise unclear whether this loss of hemopoietic function involves changes in the stroma or in the stem cells which bring an end to the production of lymphoid precursor cells in these organs. There is no doubt that bone marrow of the adult organism is the main source of precursor cells for both the T-cell and B-cell lineages. In addition, human bone marrow appears to be the site of subsequent development of active B-cells (a function which in birds is linked to the bursa of Fabricius). It is still not clear, however, which bone marrow stromal elements are inductive for any of the hemopoietic cells and especially for those that are lymphocytic.

Lymph Nodes

Lymph nodes and nodules are the last of the lymphoid organs to reach functional maturity even though their development begins at the end of the first trimester. They organize in various regions along the lymphatic vascular system which developed previously. Cells migrating from the wall of the blood vessels also contribute to the structure and function of these lymphatic organs. In general reticular cells of mesenchymal origin appear, secrete an extracellular matrix, and form a loose supporting tissue. Gradually macrophages and other accessory cells as well as lymphocytes accumulate. As in the spleen, primary lymphoid follicles develop in the areas around the FDC. In human lymph nodes, follicles are found from 17 weeks onward and contain mainly B-cells.[31]

Distinct areas of cortex, paracortex, and medulla appear only after birth. Germinal centers also generally appear after birth when antigenic stimulation reaches the lymphatic tissues. However, the capacity of these organs to react against foreign antigens matures during the embryonic period.[6] In the lymphoid nodules (which are not surrounded by a connective tissue capsule), the whole structure appears in response to antigenic challenge and the tissue often disappears after the challenge ceases.

In short, the functional tissue of the secondary lymphoid organs such as spleen and lymph nodes is derived from specialized microenvironments in other organs. These primary organs, where the supportive tissue dictates the development of either T-cells (thymus) or B-cells (bone marrow, fetal liver, or bursa of Fabricius in birds), in turn receive less differentiated hemopoietic precursors from other sites which were oriented for earlier stages of proliferation and development of precursor cells. During ontogeny this hemopoietic function is transferred with time from the liver to the spleen to the bone marrow. In each case, migration of cells can occur only after both the migratory cells develop in the primary organ and the environment in which they are to settle is prepared. The sequence of events of these developmental changes both in humans and in mice has been summarized in Table 1.

MATURATION OF T- AND B-LYMPHOCYTES IN SEPARATE MICROENVIRONMENTS

On this descriptive background of appearance of the lymphoid tissues, we can consider some of the experimental evidence pertaining to the step-wise development of functional T- and B-cells and their relationship with the fixed stromal cells. In the hemopoietic system,

Table 1
DEVELOPMENT OF LYMPHO-HEMATOPOIESIS IN MICE AND HUMANS

	Time of appearance in gestation (d)	
Event	Mouse	Human
Gestation period	19—21	270
Hematopoiesis		
Yolk sac, etc.	6—8	21—28
Liver	10—12	42—45
Spleen	15—16	90—100
Bone marrow	17—18	140—150
Thymus anlage	10—11	40—42
Lymphocytes		
Thymus	12—14	70—72
Thymus (T)	15—16	70—72
Thymus (cortex-medulla)	17—18	70—75
Liver (B)	15—16	65—68
Spleen	16—18	140—150
Spleen (B)	17—19	140—150
Spleen (T)	18—19	
Lymph nodes	18—19	70—130
Blood	17—19	120—130
Peyer's patches	19-birth	120?

Note: "Hematopoiesis" and "thymic anlage" are self-explanatory. "Lymphocytes" implies the first appearance of cells with lymphoid morphology as well as the appearance of T- (defined by surface markers such as Thy-1 in mice) and B-cells (defined by Ig in surface or cytoplasm). For references to the table, see the text.

From Stutman, O., *Clin. Immunol. Allergy,* 5, 191, 1985. With permission.

the self-renewal and differentiation that occur during embryogenesis continue throughout the life of the animal, and the sequential association of developing lymphocytes with differing microenvironments is also important throughout life. Because there is clear separation of the microenvironments involved in production of T-cells or B-cells, we will consider lymphocyte development in each microenvironment separately. Before this, however, we can consider the commitment of lymphocytes to each of these lineages and the commitment of precursor cells to the lymphocyte pathway in general. There is little doubt that pluripotent stem cells are the precursors of all the cells which circulate in the blood.[35-37] Our first question is thus at what stage does commitment to the lymphocyte pathway occur? Some clinical and experimental evidence indicates that at a later stage of differentiation T- and B-lymphocytes and myeloid cells may be derived from a common more restricted progenitor. In mice, cells with radiation-induced chromosome markers were transplanted to mutant recipients and in some cases a single chromosome marker appeared in myeloid, T- and B-cells.[38] Also, B-cells have been reported in multipotential *in vitro* colonies.[39] In humans, enzyme markers and chromosome markers have appeared in both lymphocytes and myeloid cells of patients, suggesting mutation of a common lymphohemopoietic cell.[39,40] However, from this minimal evidence, it is still not yet clear whether or not there is a common precursor for T- and B-cells only[39] or to what extent the precursors in each lineage are restricted.[40]

Development of T-Cells

Does commitment to the T- or B-lineage occur in the bone marrow or does random arrival in the thymus confer direction on unprogrammed T-cells? Although the evidence is still inconclusive, several experiments point to precommitment of adult bone marrow cells before they reach the thymus. Experiments of Abramson et al. had suggested (although the evidence was somewhat meager) that the adult bone marrow contains cells committed to the T-cell lineage.[38] Further evidence in this direction has been obtained by Dyer and Hunt, who found that T-cell reconstitution of irradiated rats could be achieved without parallel B-cell reconstitution,[41] and by Lepault et al., who reported that enrichment of adult mouse bone marrow for pre-B-cells resulted in parallel depletion of cells homing to the thymus when these cells were injected into irradiated mice.[42] Commitment of thymic-homing cells does not seem to be absolute, however, since colony forming units-spleen (CFU-S) activity was also found in the cells which repopulated the thymus of irradiated mice.[43] In embryos, on the other hand, there are at least two suggestions that some of the cells which home to the bursa of Fabricius can subsequently develop along the T-cell line.[44,45] Such cells are found in the bursa up to 12 d of gestation[45] and subsequently their numbers decrease, reaching zero by the day of hatching. In addition, transient myeloid and erythroid activity has been noted in the embryonic bursa.[4] Thus, it may be that during embryogenesis the precursor cells undergo progressive restriction, and after birth cells home to the thymus or the bursa (or the bursal-analogue in the bone marrow) according to previous programming rather than at random.

The elegant experiments of Le Douarin and co-workers[4] have shown that both the thymus and the bursa are receptive to colonizing precursor cells at particular and separate times during embryonic development. Data from these detailed and comprehensive experiments on chick-quail chimeras are summarized in Figure 2, where it can be seen that the thymus receives three successive waves of lymphohemopoietic precursors. These cells were shown to originate within the embryo and it was reported that although yolk sac cells have the potential to colonize the avian thymus, they do not do so. The receptivity of the thymus is proposed to be mediated by intermittant production of a chemoattractive factor acting together with another product purported to cause circulating precursor cells to stick to the blood vessel endothelia.[4] Precursors of accessory cells of hemopoietic origin were also found to colonize the thymus with the first wave of lymphocytes. It appears, however, that lymphocytes of the first wave differ from those of the second wave (at least in terms of the terminal transferase enzyme which is expressed within the thymus).[46]

As we have discussed above, precursors of lymphocytes as well as other blood cells are found transiently in fetal liver and spleen, and after birth they are produced mainly in the bone marrow. Do the thymic precursors express T-cell markers? Apparently they do not. After mouse bone marrow cells are labeled with fluorescein isothiocyanate and injected into irradiated hosts, it is possible to detect those precursors which home to the thymus.[42] Thy-1, Ly-1, and Ly-2 were not detected on the precursor cells. When we ask how many precursor cells can colonize the thymus, the surprising answer from several laboratories is less than ten and perhaps as few as one or two.[21,47,48] This being the case, it is difficult to imagine that the T-cell repertoire existed in the precursor cells. Rather it seems more likely that diversity develops within the thymus.

This brings us to questions of central importance in T-cell development. How does the body establish a T-cell population that is diverse in its ability to react against a variety of foreign antigens and at the same time is prevented from reacting against self-antigens? There had been many suggestions of a role for the thymus in distinguishing between self and not-self which were confirmed and expanded by the experiments of Zinkernagel.[49] The results of his investigations, and work of others,[50,51] indicated that in terms of H-2 restriction,*

---

* H-2 or MHC restriction — a requirement for identity at the major histocompatibility locus (MHC) between antigen presenting cells, T-cells involved in the immune response, and their target cells in order for the response to occur.

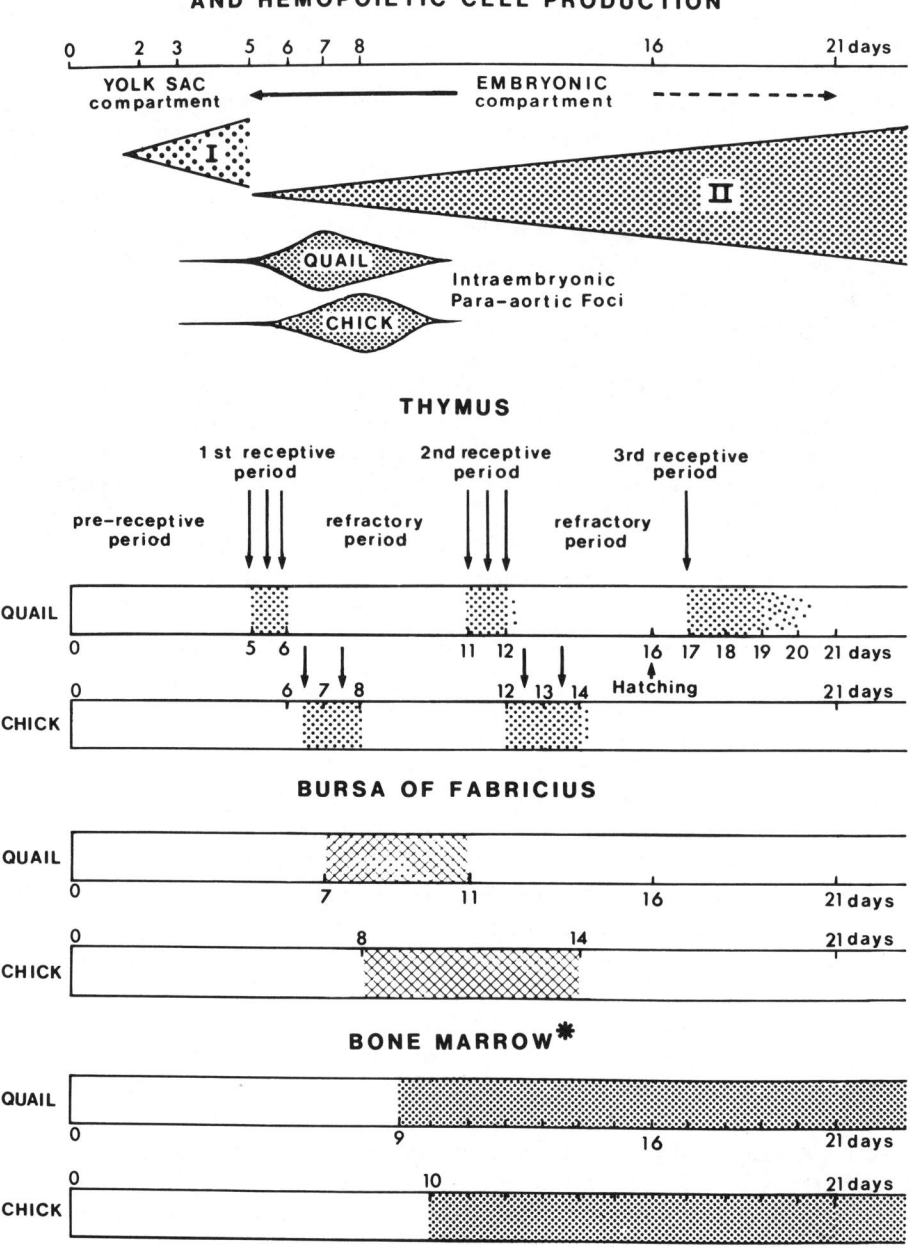

FIGURE 2. Timetable of the major events in the ontogeny of the hemopoietic system in quail and chick embryos. The periods when the thymus and bursa of Fabricius are receptive to incoming lymphocyte precursors can be compared with the period of production of precursors in the bone marrow. (From Le Douarin, N. M., Dieterlen-Lievre, F., and Oliver, P. D., *Am. J. Anat.*, 170, 261, 1984, Alan R. Liss. With permission.)

bone marrow cells which matured in a nonsyngenic thymus later recognized as self the H-2 type of the thymus in which they developed (rather than the H-2 type of their bone marrow origin). Since H-2 restriction determines the ability of the T-cells to communicate with the rest of the immune system, it appeared that the thymic microenvironment confers identity upon the developing precursor cells which determines their subsequent ability to react. It does not necessarily follow, however, that a single mechanism confers MHC restriction, self-tolerance, and establishment of T-cell diversity as well. Since the nonlymphoid component of the thymus consists mainly of epithelial cells, it was originally assumed that the inductive effects of the thymic microenvironment are linked to their activity,[49,52] although the importance of cells of bone marrow origin was also asserted by other investigators.[53] In the case of self-tolerance, it now appears that the mechanism may involve activity of thymic dendritic cells (when the term is used broadly without attempting to distinguish between FDC, IDC, and macrophages) rather than the epithelia.[21,54] These cells are known for their efficiency in presentation of foreign antigens to mature T-cells[55] and we should now ask what is the result of their presentation of self-antigen to immature T-cells. In other words, the possibility exists that tolerance results from the exposure of immature thymocytes to antigen[56] and that the lymphocytes encounter self-antigen presented by accessory cells very early in thymic maturation.

What is the current view of differentiation of the lymphocytes within the thymus? We will first briefly describe the stromal elements which appear to contribute to this process. Epithelial cells make up the bulk of the thymic stroma. This characteristic distinguishes the thymus from other mammalian lymphoid organs where the supportive framework consists of reticular connective tissue cells. For this reason, it was originally suggested that thymic stromal cells secrete a hormone-like substance.[57] Today it is clear that such a peptide or peptides are produced in the thymus,[58,59] by a minority of the stromal epithelial cells.[60-62] At the present time, further morphological distinction can be made between epithelial cells of the medulla, the cortex, and the outer cortex (the subcapsular region). In the outer cortical region, lymphocytes are found within or strongly affixed to very large epithelial cells.[63,64] These are called nurse cells because when isolated they encircle the lymphocytes. There is some evidence that *in vivo* as well the lymphocytes are enveloped by the nurse cells[65,66] and thus an isolated microenvironment is formed. Spidery epithelial cells joined by their long processes form a lacy network[27] that serves as the scaffolding for the inner cortical thymocytes. Most medullary epithelial cells appear to be more spatulate in shape with fewer cytoplasmic projections.[27] Although all of these cells are considered as fixed elements, mitotic figures have been reported,[67] the potential for cell division is expressed in culture,[68] and one investigation has proposed that turnover of epithelial cells might be analogous to that occurring in the skin.[69] In addition, the medulla of human thymus contains Hassall's corpuscles, circular epithelial structures whose function is still unclear.

The notion that different types of epithelial stromal cells might be involved in various stages of thymocyte development prompted investigation of surface differences in these cells. Differences have so far been reported in lectin binding,[70] in expression of MHC antigens (which differs between cortex and medulla in mice and in humans as well),[52,71-73] in expression of antigens bound by some monoclonal antibodies, and in the fetal age when such antigens have been detected.[69,74,75] Receptors for steroid hormones also appear on these cells and this might be related to involution of the thymus which begins about the age of puberty.[76-78] Macrophages of bone marrow origin are concentrated in the cortical-medullary junction and in the medulla and are scattered in the cortex.[79,80] Bone marrow-derived interdigitating cells (probably the same as dendritic accessory cells in the terminology of some investigators) are also concentrated in the medulla and cortical-medullary junction.[79,80] Ia antigens are found on most epithelial cells of both the cortex and the medulla,[52,71-73] on some of the macrophages,[79] and on many IDC (at least on those termed dendritic cells).[54,81] Ia is detected in the embryonic thymus about the time when the incoming lymphocytes begin to proliferate.[82] It is not found on other embryonic tissues of pharyngeal origin. However, this antigen does not appear to be important for seeding of the thymus with the first lymphocytes during ontogeny since these cells enter 2 d before Ia antigens are detected in the mouse.[73]

On the other hand, the presence of the Ia antigens on sessile cells is believed to be involved in the development of T-helper cells since helper cell development could be inhibited by anti-Ia administered to neonatal mice.[83,84] Since activation of mature helper cells (which regulate responses of the whole immune system) involves their receptors for Ia antigens, it has been suggested that in this way they obviate the need to develop tolerance to all of the individual organ-specific antigens on syngeneic tissues.[85]

The site of entry of incoming thymic-homing cells is believed to be the blood vessels of the cortical-medullary junction[79,86] and from there they appear to develop in at least two separate lineages, one cortical and the other medullary.[47,87] Nothing more will be said here regarding the medullary cells since they remain an enigma at this time. The developing cortical lymphocytes are found in intimate contact with a succession of fixed stromal cells. In irradiated mice, lymphocytes which enter the thymus are found in association first with macrophages and 2 d later with both IDC (again called dendritic cells) and nurse cells.[81] Thus, from the cortical-medullary junction the lymphocytes appear to migrate to the outer cortex, where they are then found in the environment created by the subcapsular nurse cells.[64,65] There the thymocytes proliferate rapidly[88] and may thus be amplified in clonal fashion. Although the signal for this proliferation has not been identified (IL-2 has been mentioned, but does not seem to account for all divisions), a large excess of thymocytes is thus created. A barrier between the blood and the outer cortex[89] creates an environment which is not usually penetrated by foreign antigens. In other words, the immature thymocytes are exposed almost exclusively to self at this stage. Since only about 1% of the thymocytes are exported as mature T-cells,[90] the majority of the lymphocytes die within the confines of the thymus. It is thus possible that the migration of developing cells, which proceeds from the subcapsular region to the inner cortex,[88,91] is part of a selection process and/or involves maturation of subclasses of T-cells guided by the epithelial cells of the framework. The antigens characteristic of T-cell subclasses can be detected at successive time points in mice.[92] In humans, individual mid-cortical cells first express antigens common to both cytotoxic and helper T-cells; then one of these surface structures disappears as the cells reach full maturity and migrate.[93] Convincing evidence has been presented that the migrant cells depart from the cortex and do not pass through the medulla when identified in terms of receptors used by the T-cells to enter peripheral lymph nodes.[94] Surprisingly, these purportedly mature cells are dispersed throughout the cortex,[75,94] suggesting that they may emigrate from diverse regions of the cortex.

What are the biochemical events which underlie these developmental changes leading to T-cell activity? It is obvious that knowledge of molecular mechanisms should clarify our understanding of the process. At the present time, the worldwide immunologic community has made a beginning on this problem, but knowledge is still fragmentary and a clear summary is far from our capabilities. It is apparent that T-cell development depends upon changes in surface structures of the lymphocytes. The best characterized of these so far is the T-cell antigen receptor which seems to be a transmembrane glycoprotein, usually consisting of two polypeptide chains called α and β. In addition, a γ-gene has been found.[95,99] It appears that like the DNA for immunoglobulin, the DNA which codes these peptides must be rearranged before the receptor is expressed. Thus, rearrangement of DNA, transcription of the appropriate mRNA, and appearance of receptor protein and associated structures determine a stepwise process. It seems that both in man and in mice, rearrangement of the β-chain DNA is an intrathymic event,[93,97] indicating that the thymic environment triggers the process. Messenger RNA for β and γ is relatively abundant in immature thymocytes and the latter falls to low levels in mature T-cells while the former stays roughly constant.[98] Messenger RNA for the α-subunit increases later in ontogeny and increases as T-cells mature. Three additional proteins called the T_3 complex have been found associated with the receptor in humans.[95] The T_3 molecule also is expressed on stage III thymocytes. The task now remains to correlate expression of these surface structures with the stromal microenvironments in which they occur.

For the immune system, 5 years (the time that has elapsed between preparation of this review and its publication) is long enough for considerable progress. During the past 5 years, studies of T-cell differentiation have been so numerous that we feel compelled to direct the reader to at least some of the important publications. Isolation of pluripotential stem cells[122] and characterization of their potential for thymic colonization imply that there is a lack of comitment of the prethymocytes. This finding also illustrates the progress in a single year over the knowledge summarized in a comprehensive review from the same lab.[123] The complexities of intrathymic events and developmental lineages are also described in Scollay et al.[124] Work on the T-cell receptor can be found in Mak[125] or Wilson et al.[126] Fowlkes and Pardoll[127] have also summarized knowledge on intrathymic events. These include the demonstration that clonal deletion can occur when thymocytes interact with bone marrow-derived thymic stromal cells.[128] Finally, we return to ontogeny of human T-cell development in Haynes et al.[129]

Early Development of B-Lymphocytes

In contrast to the situation with T-cells, we have considerable information on the molecular mechanisms leading to functional B-lymphocytes, but we know very little about the locations where these changes occur and the microenvironmental signals involved. In a recent article,[100] Jerne has described the B-cell complement of an imaginary mouse. At 10 weeks of age, this mouse was endowed with 10^8 mature B-cells, half of which are short lived and replaced with a renewal rate in the bone marrow of 6×10^5 cells per hour. Of the 10^8 marrow lymphocytes produced daily (about 80% are of B-cell lineage), some may die in the marrow while many are continuously released to the bloodstream and home to the peripheral lymphoid tissue.[101] By 3 d after their production, virtually all of the newly formed lymphocytes have left the marrow.[102] The development of B-lymphocytes from their precursors is an antigen-independent event that occurs, in adult mammals, in the bone marrow. Subsequent changes in B-cell structure are antigen mediated, occur in peripheral lymphoid tissue, and will not be considered here. The ability of these B-cells (numbering 10^{12} in man) to respond to an almost unlimited number of foreign antigenic determinants depends on the ability of each B-cell to recognize an antigen which is complementary to the immunoglobulin that it carries on its cell membrane. Diversity of the B-cells is thus a function of the surface receptors which these lymphocytes develop during differentiation. The complexity of the B-cell surface is shown schematically in Figure 3.

How do these cells acquire millions of different receptors? Since the receptors are proteins, they are encoded by DNA which undergoes rearrangement during development of the B-cell to create unlimited possibilities. We will summarize this unusual process briefly and recommend several review articles.[103-106] An immunoglobulin molecule consists of four polypeptide chains (two sets of identical pairs called heavy and light chains). Each of these polypeptide chains contains a region that is constant (and encoded by one or a few gene segments) and a region that is variable (and encoded by many gene segments). Before B-cell differentiation occurs, these gene segments are separated from each other in different parts of the genome and they must be joined in order to be expressed. During differentiation of a B-lymphocyte precursor, any one of the variable genes can be translocated so that it lies next to a gene for the constant region. This process first involves joining of two gene segments to form the variable region gene for the light chain and recombination of three gene segments to assemble the variable region gene of the heavy chain. Each of these genes then recombines with the DNA for its constant region, the information in additional DNA segments is removed by RNA splicing, and the message is translated into surface immunoglobulin of the B-cell.

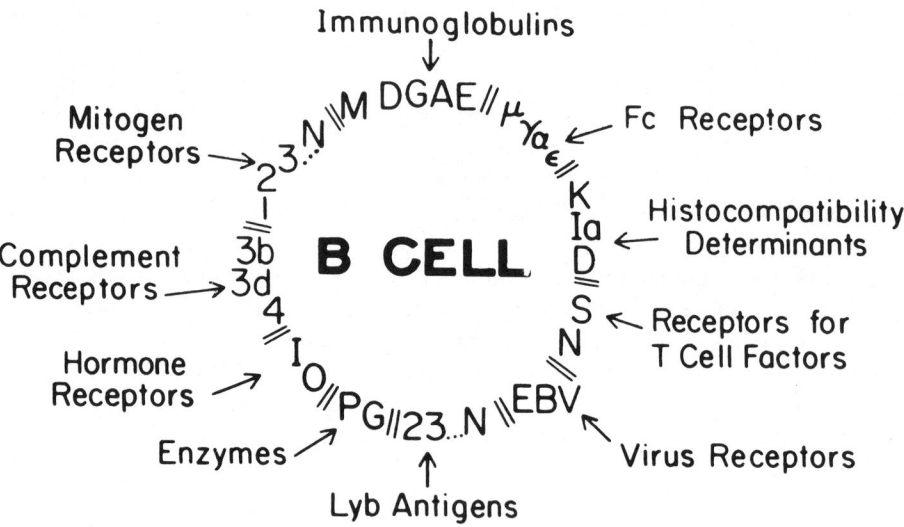

FIGURE 3. Diagram of some membrane-bound molecules which are expressed on mature B-lymphocytes. These include single or multiple immunoglobulin isotypes, receptors for the Fc portion of exogenously produced immunoglobulin molecules, receptors for antigen specific and nonspecific T-cell factors, receptors for viruses, antigens that serve as differentiation markers of B-lymphocytes, glycosidases, and phosphorylating enzymes, receptors for insulin and other hormones, receptors for components of complement, and receptors for mitogens. (From Cooper, M. D., Kearney, J., and Scher, I., in *Fundamental Immunology*, Paul, W. E., Ed., Raven Press, New York, 1984, 43. With permission.)

Where does this differentiation occur? In order to follow such a process *in situ*, markers are necessary to distinguish the precursors from early cells of other differentiation pathways and to identify B-cells at various stages of development. Since the site of B-cell differentiation in mammals is at first the liver (in embryos) and later the bone marrow (in adults), the first difficulty is to distinguish these cells from all the other hemopoietic cells being formed. Identification of early precursors of B-cells has also posed a problem that has only recently been attacked. B-cells are characterized by surface immunoglobulin (sIg), and their precursors can be detected by the presence of the μ-immunoglobulin heavy chain in the cytoplasm, but not on the cell membrane. However, this intracellular protein can be detected only on fixed cells which cannot then proceed to develop. By preparing monoclonal antibodies against surface antigens on pre-B cells, several laboratories have now produced reagents which can be used to identify and isolate precursors of B-lymphocytes.[107-110] Using this approach, Coffman[107] has distinguished between two types of pre-B cells: one which contains rearranged DNA for the immunoglobulin heavy chain only and a smaller more mature cell which has an additional identifiable surface marker and is currently undergoing the process of light chain DNA rearrangement. These latter cells are believed to mature into sIg$^+$ B-cells, and the total process corresponds to a sequence of events that takes 3 to 4 d *in vivo*.[111] This promising type of approach has also been taken by other investigators to identify pre-B-cells in mice[9,112] and in humans.[31,113]

Is B-cell development in the bone marrow localized in particular areas? There have been several attempts to map the bone marrow from the point of view of orientation of the stromal cells and conditions necessary for differentiation into each hemopoietic lineage.[114,115] The localization for lymphocyte development is even less clear than that for other developing cell types. Osmond[116] describes the perivascular spaces in the bone marrow as areas of continuous proliferation of lymphoid cells. Nossal and Pike[10] found a multifocal distribution of the lymphopoietic microenvironment during ontogenesis. Osmond et al.[101] showed that in transverse sections of diaphyseal marrow, small lymphocytes labeled with ^{3}H-thymidine are found first in the subendosteal regions and then they migrate inward together with a decrease in proliferation. In this mapping, no foci of concentrated lymphocytes were found. Rather, a sort of gradient of maturation progressing toward the center of the bone marrow was found. Although these authors believe in the existence of a lymphopoietic microenvironment in the area of the cortical bone, the nature and exact function of such a microenvironment in the production of B-cells are unknown. The extensive kinetic experiments of Osmond and co-workers have recently utilized perfusion of radiolabeled antibodies to detect sIg$^+$ cells in the bone marrow of young mice, using electron microscopic preparations.[102] Osmond et al.[102] describe the structure of the bone marrow which is organized around its blood circulation as consisting of an extravascular compartment between the sinusoids where hemopoiesis occurs and an intravascular compartment within the sinusoids where newly formed cells enter the circulation. An irregular network of reticular and macrophage stroma surrounds the hemopoietic cells. Many sIg$^+$ small lymphocytes were found throughout the hemopoietic compartment showing that sIg is expressed before the cells are released into the sinusoids. No area of the marrow appeared to be an exclusive B-cell microenvironment. Also in fetal liver, B-cells were found to be dispersed and were found among the erythroid and myeloid elements in extrasinusoidal areas next to the liver parenchymal cells.[117] However, the location of marrow sIg$^+$ cells was not random. They were frequently associated with reticular cells or reticular cell processes and were generally concentrated near sinusoid walls in areas rich in other mature cells. There also appeared to be an increase in sIg density on cells in the intrasinusoidal areas, suggesting that maturation may continue in this compartment.[102]

Investigation of the stromal cells necessary for differentiation of B-lymphocytes has also been approached *in vitro*. Mouse B-cell precursors in 12-d fetal liver were cloned in agar and generated cultures that secrete antibodies.[118] In this way, it was possible to evaluate the effects of fetal liver and bone marrow stromal cells and several soluble products produced by other tissues on B-cell development. The soluble products known to support growth of cells in other hemopoietic lineages also seemed to be important here. In addition a difference was found in the ability of stromal cells from fetal liver or from adult bone marrow to support the development of mature B-cells. Long-term cultures of B-lymphocytes and their precursors have been established[119] and have been shown to facilitate the development of mature B-cells *in vivo* as well as *in vitro*.[120] It thus became possible to compare in morphological terms the cells appearing in these B-lymphocyte cultures with those of long-term myeloid cultures.[121] When the three-dimensional stromal cell layer of the B-cell cultures was characterized, numerous lymphoid cells were found within the membrane infoldings of large mononuclear stromal cells which appeared to be critical for B-cell differentiation. Classification of these stromal cells was less certain and both macrophages and epithelial cells were suggested as candidates for further study. In the chicken, the early stages of B-cell differentiation occur in the isolated environment of the bursa of Fabricius, where the dominant stromal cells are epithelial. The process of DNA rearrangement leading to B-cell heterogeneity occurs in this microenvironment.[11] Perhaps comparison of the stromal layer of B-cell cultures with the stromal cells of this purely B-cell environment could shed light on the inductive environment involved in the early stages of B-cell development.

Later stages of B-cell differentiation to synthesize different classes of sIg and eventually to become plasma cells occur after the cells have migrated to the peripheral lymphoid organs and are triggered by exposure to foreign antigens. As antigen-driven lymphocyte development is outside the scope of this article, we recommend recent communications[31-33,79] for these and other aspects of lymphocyte activity in peripheral organs.

In summary, the development of the lymphoid organs and the cells that differentiate within them is a sequential process which begins early in embryonic life. As described above, the developing lymphoid cells undergo successive stages of differentiation in dispersed specialized environments. At each stage, interaction occurs when both the mobile lymphoid cells and the stationary stromal cells have reached the necessary degree of development. It seems that the development of each organ is dependent on the function of an earlier site of activity. As migration proceeds from the hemopoietic tissues to the central lymphoid organs, and from there to the periphery, the potential of the lymphoid cells to differentiate is successively restricted. The factors involved in this restriction await further elucidation as do the factors which control the streams of migration from one site to the next. Investigators are only beginning to approach the question of the specific contribution of each microenvironment to the cells developing in its midst. In conclusion, it seems that understanding of the role of specific microenvironments in the step-by-step development of lymphoid cells is far from complete. We predict that this will be a major field of investigation in the years to come.

ACKNOWLEDGMENT

We wish to express our thanks to Drs. A. Globerson, M. Aronson, L. Barr-Nea, O. Stutman, W. Van Ewijk, N. M. Le Douarin, M. D. Cooper, and A. J. P. Veerman for kind advice and contributions. Special thanks go to Shoshana Dvir for skilled secretarial assistance and unlimited patience.

REFERENCES

1. **Edwards, G. E., Miller, R. G., and Phillips, R. A.,** Differentiation of rosette-forming cells from myeloid stem cells, *J. Immunol.*, 105, 719, 1970.
2. **Moore, M. A. S. and Owen, J. J. T.,** Chromosome marker studies on the development of hemopoietic system in the chick embryo, *Nature (London)*, 208, 956, 1965.
3. **Metcalf, D. and Moore, M. A. S.,** Haemopoietic cells, in *Frontiers in Biology,* Newberger, A. and Tatum, E. L., Eds., North-Holland, Amsterdam, 1971, 183.
4. **Le Douarin, N. M., Dieterlen-Lievre, F., and Oliver, P. D.,** Ontogeny of primary lymphoid organs and lymphoid stem cells, *Am. J. Anat.*, 170, 261, 1984.
5. **Turpen, J. B., Cohen, N., Deparis, P., Joylet, A., Tompkins, R., and Volp, E. S.,** Ontogeny of amphibian hemopoietic cells, in *The Reticuloendothelial System,* Vol 3, Cohen, N. and Sigel, M., Eds., Plenum Press, New York, 1982, 569.
6. **Sterzl, J. and Silverstein, A. M.,** Developmental aspects of immunity, *Adv. Immunol.*, 6, 337, 1967.
7. **Keightley, R. G., Lawton, A. R., and Cooper, M. D.,** Successful fetal liver transformation in a child with severe combined immunodeficiency, *Lancet*, 2, 850, 1975.
8. **Owen, J. J. T.,** Ontogeny of mammalian lymphocytes, in *The Reticuloendothelial System,* Vol. 3, Cohen, N. and Sigel, M., Eds., Plenum Press, New York, 1982, 617.
9. **Kincade, P. W.,** Formation of B lymphocytes in fetal and adult life, *Adv. Immunol.*, 31, 177, 1981.
10. **Nossal, G. J. V. and Pike, B. L.,** Studies on differentiation of B lymphocytes in the mouse, *Immunology*, 25, 33, 1975.
11. **Cooper, M. D., Kearney, J., and Scher, I.,** B lymphocytes, in *Fundamental Immunology,* Paul, W. E., Ed., Raven Press, New York, 1982, 43.
12. **Gathings, W. E., Lawton, A. R., and Cooper, M. D.,** Immunofluorescent studies of the development of pre B cells, B lymphocytes and immunoglobulin isotype diversity in humans, *Eur. J. Immunol.*, 7, 804, 1977.
13. **Owen, J. J. T., Wright, D. E., Habu, S., Raff, M. C., and Cooper, M. D.,** Studies on the generation of B lymphocytes in fetal liver and bone marrow, *J. Immunol.*, 118, 2067, 1977.
14. **Hayward, A. R. and Lydyard, P. M.,** Suppression of B lymphocyte differentiation by newborn T lymphocytes with an Fc receptor for IgM, *Clin. Exp. Immunol.*, 34, 374, 1978.
15. **Rabinowich, H., Umiel, T., and Globerson, A.,** T cell progenitors in the mouse fetal liver, *Transplantation*, 35, 40, 1983.

16. **Stutman, O.**, Intrathymic and extrathymic T cell maturation, *Immunol. Rev.*, 42, 138, 1978.
17. **Auerbach, R., Globerson, A., and Umiel, T.**, Ontogeny of cellular immune reactivity in the mouse, in *The Reticuloendothelial System*, Vol. 3, Cohen, N. and Sigel, M., Eds., Plenum Press, New York 1982, 687.
18. **Miller, J. F. A. P.**, Experimental thymology has come of age, *Thymus*, 1, 3, 1979.
19. **Scollay, R. and Shortman, K.**, Thymocyte subpopulations: an experimental review, *Thymus*, 5, 245, 1983.
20. **Haynes, B. F., Scearce, R. M., Lobach, D. F., and Hensley, L. L.**, Phenotypic characterization and ontogeny of mesodermal-derived and endocrine epithelial components of the human thymic microenvironment, *J. Exp. Med.*, 159, 1149, 1984.
21. **Owen, J. J. T. and Jenkinson, E. J.**, Early events in T lymphocyte genesis in the fetal thymus, *Am. J. Anat.*, 170, 301, 1984.
22. **Auerbach, R.**, Some aspects of tissue interaction in-vitro, in *Epithelial-Mesenchymal Interactions*, Fleischmajer, R., Ed., William & Wilkins, Baltimore, 1968, 200.
23. **Le Douarin, N., Jotereau, F., Houssaint, E., Martin, C., and Dieterlen-Lievre, F.**, Ontogeny of avian lymphocytes, in *The Reticuloendothelial System*, Vol. 3, Cohen, N. and Sigel, M., Eds., Plenum Press, New York 1982, 589.
24. **Hammond, W. S.**, Origin of thymus in the chick embryo, *J. Morphol.*, 95, 501, 1954.
25. **Cordier, A. C., and Heremans, J. F.**, Nude mouse embryo. Ectodermal nature of the primordial thymic defect, *Scand. J. Immunol.*, 4, 193, 1975.
26. **Beller, D. I. and Unanue, E. R.**, Ia antigens and antigen-presenting function of thymic macrophages, *J. Immunol.*, 124, 1443, 1980.
27. **Rouse, R. V. and Weissman, I. L.**, Microanatomy of the thymus: its relationship to T-cell differentiation in microenvironment, in *Microenvironments in Haemopoietic and Lymphoid Differentiation*, Ciba Found Symp. 84, Porter, R. and Whelan, J., Eds., Pitman, London, 1981, 161.
28. **Hayward, A. R. and Soothill, J. F.**, Reaction to antigen by human fetal thymus lymphocytes, in *Ontogeny of Acquired Immunity*, Ciba Found Symp. 5, Porter, R. and Knight, J., Eds., Elsevier, Amsterdam, 1972, 261.
29. **Le Douarin, N. and Jotereau, F.**, Tracing of cells of the avian thymus through embryonic life in interspecific chimera, *J. Exp. Med.*, 142, 17, 1975.
30. **Weiss, L. and Chen, L.**, The differentiation of white pulp and red pulp in the spleen of human fetuses (72-145 mm crown-rump length), *Am. J. Anat.*, 141, 393, 1974.
31. **Bofill, M., Janossy, G., Janossa, M., Burford, G. D., Seymour, G. J., Wernet, P., and Kelemen, E.**, Human B cell development. II. Subpopulations in the human fetus, *J. Immunol.*, 134, 1531, 1985.
32. **Veerman, A. J. P. and van Ewijk, W.**, White pulp compartments in the spleen of rats and mice, *Cell Tissue Res.*, 156, 417, 1975.
33. **Rouse, R. V., Reichert, R. A., Gallatin, W. M., Weissman, I. L., and Butcher, E. G.**, Localization of lymphocyte subpopulations in peripheral lymphoid organs: directed lymphocyte migration and segregation into specific microenvironments, *Am. J. Anat.*, 170, 391, 1984.
34. **Reddi, A. H. and Huggins, C. B.**, The formation of bone-marrow in fibroblast transformation ossicles, *Proc. Natl. Acad. Sci. U. S. A.*, 72, 2212, 1975.
35. **Micklem, H. S., Ford, C. E., Evans, E. P., and Gray, J.**, Interrelationships of myeloid and lymphoid cells, *Proc. R. Soc. London Ser. B*, 165, 78, 1966.
36. **Metcalf, D. and Moore, M. A. S.**, Haemopoietic cells, in *Frontiers of Biology*, Neuberger, A. and Tatum, T. L., Eds., Elsevier/North-Holland, Amsterdam, 1971.
37. **Mintz, B., Anthony, K., and Litwin, S.**, Monoclonal derivation of mouse myeloid and lymphoid lineages from totipotent hematopoietic stem cells experimentally engrafted in fetal hosts, *Proc. Natl. Acad. Sci. U. S. A.*, 81, 7835, 1984.
38. **Abramson, S., Miller, R. G., and Phillips, R. A.**, The identification in adult bone marrow of pluripotent and restricted stem cells of the myeloid and lymphoid system, *J. Exp. Med.*, 145, 1567, 1977.
39. **Ogawa, M., Porter, P. N., and Nakahata, T.**, Renewal and commitment to differentiation of hemopoietic stem cells (an interpretive review), *Blood*, 61, 823, 1983.
40. **Till, J. E. and McCulloch, E. A.**, Hemopoietic stem cell differentiation, *Biochim. Biophys. Acta*, 605, 431, 1980.
41. **Dyer, M. J. S. and Hunt, S. V.**, Committed T lymphocyte stem cells of rats. Characterization by surface W3/13 antigen and radiosensitivity, *J. Exp. Med.*, 154, 1164, 1981.
42. **Lepault, F., Coffman, R. L., and Weissman, I. L.**, Characteristics of thymus — homing bone marrow cells, *J. Immunol.*, 131, 64, 1983.
43. **Kraal, G., Geldof, A. A., and Boden, D.**, T-cell differentiation in lethally irradiated and reconstituted mice, *Cell. Immunol.*, 49, 110, 1980.
44. **Jotereau, F. V., Houssaint, E., and Le Douarin, N. M.**, Lymphoid stem cell homing to the early thymic primordium of the avian embryo, *Eur. J. Immunol.*, 10, 620, 1980.

45. **Brand, A., Gatton, J., and Gilmour, D. G.,** Committed precursors of B and T lymphocytes in chick embryo bursa of Fabricius, thymus and bone marrow, *Eur. J. Immunol.*, 13, 449, 1983.
46. **Penit, C., Jotereau, F., and Gelabert, M. J.,** Relationships between terminal transferase expression stem cell colonization and thymic maturation in the avian embryo: studies in thymic chimeras resulting from homospecific and heterospecific grafts, *J. Immunol.*, 134, 2149, 1985.
47. **Ezine, S., Weissman, I. L., and Rouse, R. V.,** Bone marrow cells give rise to distinct clones within the thymus, *Nature (London)*, 309, 629, 1984.
48. **Wallis, V., Leuchars, E., Chwalinski, S., and Davies, A. J. S.,** On the sparse seeding of bone marrow and thymus in radiation chimeras, *Transplantation*, 19, 2, 1975.
49. **Zinkernagel, R. M., Callahan, G. N., Althage, A., Cooper, S., Klein, P. A., and Klein, J.,** On the thymus in the differentiation of "H-2 self-recognition" by T cells: evidence for dual recognition?, *J. Exp. Med.*, 147, 882, 1978.
50. **Bevan, M. J. and Fink, P. J.,** The influence of thymus H-2 antigens on the specificity of maturing killer and helper cells, *Immunol. Rev.*, 42, 3, 1978.
51. **Von Boehmer, H., Haas, W., and Jerne, N. K.,** Major histocompatibility complex linked immune responsiveness is acquired by lymphocytes of low responder mice differentiation in thymus of high responder mice, *Proc. Natl. Acad. Sci. U. S. A.*, 75, 2439, 1978.
52. **Rouse, R. V., Van Ewijk, W., Jones, P. J., and Weissman, I. L.,** Expression of MHC antigens by mouse thymic dendritic cells, *J. Immunol.*, 122, 2508, 1979.
53. **Longo, D. and Schwarz, R. H.,** T cell specifying for H-2 and Ir gene phenotype correlates with the phenotype of thymic antigen-presenting cells, *Nature (London)*, 287, 44, 1980.
54. **Jenkinson, E. J., Jhittay, P., Kingston, R., and Owen, J. J. T.,** Studies of the role of the thymic environment in the induction of tolerance to MHC antigens, *Transplantation*, 39, 331, 1985.
55. **Kyewski, B. A., Fathman, C. G., and Kaplan, H. S.,** Intrathymic presentation of circulating non-major histocompatibility complex antigens, *Nature (London)*, 308, 196, 1984.
56. **Small, M.,** Tumor enhancing T lymphocytes in mice: further studies on characteristics and mechanism of activity, *Int. J. Cancer*, 29, 465, 1982.
57. **Trainin, N.,** Thymic hormones and the immune response, *Physiol. Rev.*, 54, 272, 1974.
58. **Trainin, N., Small, M., Zipori, D., Umiel, T., Kook, A. I., and Rotter, V.,** Characteristics of THF, a thymic hormone, in *Biological Activity of Thymic Hormones*, Van Bekkum, D. W., Ed., Kooyker, Rotterdam, 1975, 117.
59. **Trainin, N., Small, M., and Kook, A. I.,** The role of thymic hormones in regulation of the lymphoid system, in *B and T Cells in Immune Recognition*, Loor, F. and Roelants, G. E., Eds., John Wiley & Sons, London, 1977, 83.
60. **Savino, W. and Dardenne M.,** Thymic hormone-containing cells. VI, *Eur. J. Immunol.*, 14, 987, 1984.
61. **Hirokawa, K., McClure, J. E., and Goldstein, A. L.,** Age-related changes in localization of thymosin in the human thymus, *Thymus*, 4, 19, 1982.
62. **Berrih, S., Savino, W., Azoulay, M., Dardenne, M., and Bach, J.-F.,** Production of anti-thymulin monoclonal antibodies by immunization against human thymic epithelial cells, *J. Histochem. Cytochem.*, 32, 432, 1984.
63. **Wekerle, H., Ketelsen, U.-P., and Ernst, M.,** Thymic nurse cells: lymphoepithelial complexes in murine thymuses: morphological and serological characterization, *J. Exp. Med.*, 151, 925, 1980.
64. **Kyewski, B. A. and Kaplan, H. S.,** Lymphoepithelial interactions in the mouse thymus: phenotypic and kinetic studies on thymic nurse cells, *J. Immunol.*, 128, 2287, 1982.
65. **Andrews, P. and Boyd, R.,** The murine thymic nurse cell: an isolated thymic microenvironment, *Eur. J. Immunol.*, 15, 36, 1985.
66. **Ritter, M. A., Sauvage, C. A., and Cotmore, S. F.,** The human thymus microenvironment: in vivo identification of thymic nurse cells and other antigenically-distinct subpopulations of epithelial cells, *Immunology*, 44, 439, 1981.
67. **Mandel, T.,** Differentiation of epithelial cells in the mouse thymus, *Z. Zellforsch. Mikrosk. Anat.*, 106, 498, 1970.
68. **Small, M., Barr-Nea, L., and Aronson, M.,** Culture of thymic epithelial cells from mice and age-related studies on the growing cells, *Eur. J. Immunol.*, 14, 936, 1984.
69. **Haynes, B. F.,** The human thymic microenvironment, *Adv. Immunol.*, 36, 87, 1984.
70. **Raedler, A., Raedler, E., Scholz, K.-U., Arndt, R., and Thiele, H.-G.,** The intrathymic microenvironment: expression of lectin receptors and lectin-like molecules of differentiation antigens and MHC gene products, *Thymus*, 5, 311, 1983.
71. **Rouse, R. V., Parham, P., Grumet, F. C., and Weissman, I. L.,** Expression of HLA antigens by human thymic epithelial cells, *Hum. Immunol.*, 5, 21, 1982.
72. **Van Ewijk, W., Rouse, R. V., and Weissman, I. L.,** Distribution of H-2 microenvironments in the mouse thymus. Immunoelectron microscopic identification of I-A and H-2K bearing cells, *J. Histochem. Cytochem.*, 28, 1089, 1980.

73. **Jenkinson, E. J., Van Ewijk, W., and Owen, J. J. T.**, Major histocompatibility complex antigen, expression on the epithelium of the developing thymus in normal and nude mice, *J. Exp. Med.*, 153, 280, 1981.
74. **Van Vliet, E., Melis, M., and van Ewijk, W.**, Monoclonal antibodies to stromal cell types of the mouse thymus, *Eur. J. Immunol.*, 14, 524, 1984.
75. **Van Ewijk, W.**, Immunohistology of lymphoid and non-lymphoid cells in the thymus in relation to T lymphocyte differentiation, *Am. J. Anat.*, 170, 311, 1984.
76. **Harrison, D. E., Archer, J. R., and Astle, C. M.**, The effect of hypophysectomy on thymic aging in mice, *J. Immunol.* 129, 2673, 1982.
77. **Luster, M. I., Hayes, H., Korach, K., Tucker, A. N., Dean, J. H., Greenlee, W. F., and Boorman, G. A.**, Estrogen immunosuppression is regulated through estrogenic responses in the thymus, *J. Immunol.*, 133, 110, 1984.
78. **Barr, I. G., Pyke, K. W., Pearce, P., Toh, B.-H., and Funder, J. W.**, Thymic sensitivity to sex hormones develops post-natally; an in vivo and an in vitro study, *J. Immunol.*, 132, 1095, 1984.
79. **Butcher, E. C., and Weissman, I. L.**, Lymphoid tissues and organs in *Fundamental Immunology*, Paul, W. E., Ed., Raven Press, New York, 1984, 109.
80. **Duijvestijn, A. M., and Hoefsmit, E. C. M.**, Ultrastructure of rat thymus: the micro-environment of T-lymphocyte maturation, *Cell Tissue Res.*, 218, 279, 1981.
81. **Kyewski, B. A., Rouse, R. V., and Kaplan, H. S.**, Thymocyte rosettes: multicellular complexes of lymphocytes and bone-marrow-derived stromal cells, *Proc. Natl. Acad. Sci. U.S.A.*, 79, 5646, 1982.
82. **Jenkinson, E. J., Owen, J. J. T., and Aspinall, R.**, Lymphocyte differentiation and major histocompatibility complex antigen expression in the embryonic thymus, *Nature (London)*, 284, 177, 1980.
83. **Kruisbeek, A. M., Fultz, M. J., Sharrow, S. O., Singer, A., and Mond, J. J.**, Early development of the T cell repertoire. In vivo treatment of neonatal mice with anti-Ia antibodies interferes with differentiation of I restricted T cells but not K/D restricted T cells, *J. Exp. Med.*, 157, 1932, 1983.
84. **Sprent, J.**, Effects of blocking helper T induction in vivo with anti-Ia antibodies, *J. Exp. Med.*, 152, 996, 1980.
85. **Cowing, C.**, Does T-cell restriction to Ia limit the need for self-tolerance?, *Immunol. Today*, 6, 72, 1985.
86. **Weissman, I. L., Rouse, R. V., Kyewski, B. A., Lepault, F., Butcher, E. C., Kaplan, H. S., and Scollay, R. G.**, Thymic lymphocyte maturation in the thymic microenvironment, in *Behring Institute Mittelung*, Seiler, F. R. and Schwick, H. G., Eds., Medizinische Verlagsgesellschaft, Marburg, Germany, 1982, 242.
87. **Shortman, K., von Boehmer, H., Lipp, J., and Hopper, K.**, Subpopulations of T-lymphocytes, *Transplant. Rev.*, 25, 163, 1975.
88. **Weissman, I. L.**, Thymus cell migration, *J. Exp. Med.*, 126, 291, 1967.
89. **Raviola, E. and Karnovsky, M. J.**, Evidence for a blood thymus barrier using electron-opaque tracers, *J. Exp. Med.*, 136, 466, 1972.
90. **Scollay, R. G., Butcher, E. C., and Weissman, I. L.**, Thymus cell migration, quantitative aspects of cellular traffic from the thymus to the periphery in mice, *Eur. J. Immunol.*, 10, 210, 1980.
91. **Fathman, C. G., Small, M., Herzenberg, L. A., and Weissman, I. L.**, Thymus cell maturation. II. Differentiation of three mature subclasses in vivo, *Cell. Immunol.*, 15, 109, 1975.
92. **Lepault, F. and Weissman, I. L.**, An in vivo assay for thymus-homing bone marrow cells, *Nature (London)*, 293, 151, 1981.
93. **Royer, H. D., Acuto, O., Fabbi, M., Tizard, R., Ramachandran, K., Smart, J. E., and Reinherz, E. L.**, Genes encoding the Tiβ subunit of the antigen/MHC receptor undergo rearrangement during intrathymic ontogeny prior to T_3-Ti expression, *Cell*, 39, 261, 1984.
94. **Reichert, R. A., Gallatin, W. M., Butcher, E. C., and Weissman, I. L.**, A homing receptor-bearing cortical thymocyte subset: implications for thymus cell migration and the nature of cortisone-resistant thymocytes, *Cell*, 38, 89, 1984.
95. **Reinherz, E. L.**, A molecular basis for thymic selection: regulation of T_{11} induced thymocyte expansion by the T_3-Ti antigen/MHC receptor pathway, *Immunol. Today*, 6, 75, 1985.
96. **Hood, L., Kronenberg, M., and Hunkapiller, T.**, T cell antigen receptors and the immunoglolubin supergene family, *Cell*, 40, 225, 1985.
97. **Trowbridge, I. S., Lesley, J., Trotter, J., and Hyman, R.**, Thymocyte subpopulation enriched for progenitors with an unrearranged T-cell receptor β chain gene, *Nature (London)*, 315, 666, 1985.
98. **Raulet, D. H., Garman, R. D., Saito, H., and Tonegawa, S.**, Developmental regulation of T-cell receptor gene expression, *Nature (London)*, 314, 103, 1985.
99. **Rupp, F., Acha-Orbea, H., Hengartner, H., Zinkernagel, R., and Joho, R.**, Identical $V_β$ T-cell receptor genes used in alloreactive cytotoxic and antigen plus I-A specific helper T cells, *Nature, (London)*, 315, 425, 1985.
100. **Jerne, N. K.**, Idiotypic networks and other preconceived ideas, *Immunol. Rev.* 79, 5, 1984.

101. **Osmond, D. G., Fahlman, M. T. E., Fulop, G. M., and Rahel, D. M.**, Regulation and localization of lymphocyte production in the bone marrow, in *Microenvironments in Haemopoietic and Lymphoid Differentiation,* Ciba Found. Symp. 84., Porter, R. and Whelan, J., Eds., Pitman, London, 1981, 68.
102. **Osmond, D. G. and Batten, S. J.**, Genesis of B lymphocytes in the bone marrow, *Am. J. Anat.,* 170, 349, 1984.
103. **Adams, J. M.**, The organization and expression of immunoglolubin genes, *Immunol. Today.,* 1, 10, 1980.
104. **Honjo, T., Nakai, S., Nishida, Y., Kataoka, T., Yamawaki-Kataoka, Y., Takahashi, N., Obata, M., Shimiza, A., Yaoita, Y., Nikaido, T., and Ishida, N.**, Rearrangements of Immunoglobin genes during differentiation and evolution, *Immunol. Rev.,* 59, 33, 1981.
105. **Leder, P.**, The genetics of antibody diversity, *Sci. Am.,* 246(5), 72, 1982.
106. **Joho, R., Nottenburg, C., Coffman, R. L., and Weissman, I. L.**, Immunoglobulin gene rearrangement and expression during lymphocyte development, in *Current Topics in Developmental Biology,* Vol. 18, Moscona, A. A. and Monroy, A., Eds., Academic Press, New York, 1983, 16.
107. **Coffman, R. L.**, Surface antigen expression and immunoglolubin gene rearrangement during mouse pre-B cell development, *Immunol. Rev.,* 69, 5, 1982.
108. **Shinefeld, L. A., Sato, V. L., and Rosenberg, N. E.**, Monoclonal anti-mouse brain antibody detects Abelson murine leukemia virus target cells in mouse bone marrow, *Cell,* 20, 11, 1980.
109. **Kincade, P. W., Grace, L., Takishi, W., Leslie, S., and Schied, M. P.**, Antigens displayed on murine B lymphocyte precursors, *J. Immunol.,* 127, 2262, 1981.
110. **Dessner, D. S. and Loken, M. R.**, DNL 1.9: a monoclonal antibody which specifically detects all murine B lineage cells, *Eur. J. Immunol.,* 11, 282, 1981.
111. **Osmond, D. G. and Nossal, G. J. V.**, Differentiation of lymphocytes in mouse bone marrow, *Cell. Immunol.,* 13, 132, 1974.
112. **Velardi, A. and Cooper, M. D.**, An immunofluorescence analysis of the ontogeny of myeloid, T and B lineage cells in mouse hemopoietic tissues, *J. Immunol.,* 133, 672, 1984.
113. **Hofman, F. M., Danilovs, J., Husmann, L., and Taylor, C. R.**, Ontogeny of B cell markers in the human fetal liver, *J. Immunol.,* 133, 1197, 1984.
114. **Lambertsen, R. H. and Weiss, L.**, A model of intramedullary hemopoietic microenvironments based on stereologic study of the distribution of enclonned marrow colonies, *Blood,* 63, 287, 1984.
115. **Lichtman, M. A.**, The ultrastructure of hemopoietic environment of the marrow: a review, *Exp. Hematol.,* 9, 391, 1981.
116. **Osmond, D. G.**, Formation and maturation of bone marrow lymphocytes, *J. Reticuloendothel. Soc.,* 17, 97, 1975.
117. **Kamps, W. A. and Cooper, M. D.**, Microenvironmental studies of pre-B and B cell development in human and mouse fetuses, *J. Immunol.,* 129, 526, 1982.
118. **Paige, C. J., Gisler, R. H., McKearn, J. P., and Iscove, N.**, Differentiation of murine B cell precursors in agar culture, *Eur. J. Immunol.,* 14, 979, 1984.
119. **Whitlock, C. A. and Witte, O. N.**, Long-term culture of B lymphocytes and their precursors from murine bone marrow, *Proc. Natl. Acad. Sci. U. S. A.,* 79, 3608, 1982.
120. **Kurland, J. I., Ziegler, S. F., and Witte, O. N.**, Long-term cultured B-lymphocytes and their precursors reconstitute the B-lymphocyte lineage in vivo, *Proc. Natl. Acad. Sci. U. S. A.,* 81, 7554, 1984.
121. **Dorshkind, K., Schouest, L., and Fletcher, W. H.**, Morphologic analysis of long-term bone marrow cultures that support B-lymphopoiesis or myelopoiesis, *Cell Tissue Res.,* 239, 375, 1985.
122. **Spangrude, G. J., Heimfeld, S., and Weissman, I. L.**, Purification and characterization of mouse hemopoeitic stem cells, *Science,* 241, 58, 1988.
123. **Adkins, B., Mueller, C., Okada, C. Y., Reichert, R. A., Weissman, I. L., and Spangrude, G. J.**, Early events in T-cell maturation, *Annu. Rev. Immunol.,* 5, 325, 1987.
124. **Scollay, R., Wilson, A., D'Amico, A., Kelly, K., Edgerton, M., Pearse, M., Li, W., and Shortman, K.**, Developmental status and reconstitution potential of subpopulations of murine thymocytes, *Immunol. Rev.,* 104, 81, 1988.
125. **Mak, T. W., Ed.**, *The T-Cell Receptors,* Plenum Press, New York, 1988.
126. **Wilson, R. K., Lai, E., Concannon, P., Barth, R. K., and Hood, L. E.**, Structure organization and polymorphism of murine and human T-cell receptor α and β chain genes, *Immunol. Rev.,* 101, 149, 1988.
127. **Fowlkes, B. J. and Pardoll, D.**, Molecular and cellular events of T cell development, *Adv. Immunol.,* 44, 207, 1989.
128. **Marrack, P., Lo, D., Brinster, R., Palmiter, R., Burkly, L., Flavell, R. H., and Kappler, J.**, The effect of thymus environment on T cell development and tolerance, *Cell,* 53, 627, 1988.
129. **Haynes, B. F., Denning, S. M., Singer, K. H., and Kurtzberg, J.**, Ontogeny of T-cell precursors, *Immunol. Today,* 10, 87, 1989.

IMMUNITY DEVELOPMENT

Paul Leibson and Anthony Hayward

INTRODUCTION

Growth and development, in the context of the immune system, take place in two different ways. One aspect of growth is the production, without specific antigen stimulation, of cells and proteins encompassing a range of antimicrobial activities. This process starts early in fetal life and, by birth, the nonspecific effector mechanisms which these cells and products mediate are sufficiently mature to protect the infant from some but by no means all infections. The other type of growth follows the exposure of the infant to environmental antigens such as microbes, food, and inhaled allergens. These stimuli elicit proliferation by lymphocytes (which mediate specific immunity) and they may also activate nonspecific immune mechanisms such as complement and phagocytosis. The description of immunity development in this chapter is divided into sections corresponding to the major effector mechanisms of immunity.

DEVELOPMENT OF NONSPECIFIC IMMUNITY

Nonspecific immunity is mediated by phagocytes (polymorphonuclear neutrophils, monocytes and fixed cells of the mononuclear phagocyte system), natural killer (NK) cells, complement, and other soluble antimicrobial substances such as lysozyme and lactoferrin. Since the development of these effectors does not require prior antigen exposure, their function is at least partly independent of active T- or B-cell immunity. However, lymphokines released by lymphocytes in the course of specific immune responses can amplify some functions, such as killing by macrophages and NK cells.

Phagocytic Cells

The activities of polymorphonuclear neutrophils (PMNs) and monocytes/macrophages include phagocytosis, microbial killing, antigen processing and presentation, and the production of soluble mediators (e.g., interleukin-1 and tumor necrosis factor). In order for phagocytic cells to mediate these functions in local inflammatory sites, they must first adhere to vascular endothelium adjacent to the inflammatory site (pavementing) and then pass through the junction between endothelial cells (diapedesis). These motile cells then move toward chemotactic signals in the inflammatory site by receptor-facilitated migration. Upon arrival in the inflammatory focus, phagocytic cells can attach to opsonized particles (e.g., antibody- or complement-coated microbes), ingest these materials, and process them (which involves an intracellular respiratory burst, intracellular enzymatic degradation, and extracellular enzyme release). The development of phagocytic function is reviewed in general terms in the following paragraphs; for more detail see Johnston[25] and Gordon et al.[16]

PMNs and macrophages develop initially from common pluripotential stem cells. Myelopoiesis begins in the fetal liver at about 6 weeks of gestation and later involves the spleen and bone marrow. After birth, the bone marrow is the site of continuing myelopoiesis. Differentiation of precursor cells along a myeloid pathway is influenced by glycoprotein hormones termed colony-stimulating factors (e.g., G-CSF, GM-CSF, interleukin-3). The subsequent morphologically distinguishable cellular stages include myeloblasts, promyelocytes, and the final release of mature PMNs into the circulation. Circulating PMNs survive for only a few days and large numbers (10^9 PMNs per kilogram per day) are released from the bone marrow to maintain the circulating pool.

Mononuclear phagocytes develop in the fetal liver, spleen, and bone marrow and later in the adult bone marrow. The development from monocytic precursors to mature monocytes occurs under the influence of CSFs (e.g. M-CSF, GM-CSF, interleukin-3) and takes 6 to 7 d. Mature monocytes released from the bone marrow circulate with a half-life of approximately 3 d, after which a portion of the monocytes leaves the vascular compartment to participate either as local acute inflammatory cells or as fixed macrophages in the liver, spleen, and other tissues.

Homing and Adhesion

During the differentiation of phagocytic cells, there is a complex and dynamic pattern of adhesive interactions between different cell types and between cells and the surrounding extracellular matrix components (for review see Wawryk et al.[52]). Early in development, these interactions keep the developing cells in the bone marrow until the mature cells are ready to migrate into the circulation. After release from the bone marrow, circulating phagocytic cells can enter localized inflammatory sites or take residence in a specific tissue only after adherence to vascular endothelium. This cellular interaction can be influenced by distinct adhesive molecules, such as the LFA-1/ICAM-1 family of molecules.

After phagocytic cells adhere to vascular endothelium and pass into the adjacent tissue, chemotactic signals, such as chemotaxins and activated complement components, direct their migration to localized sites. Newborn PMNs have an impaired response to chemotactic signals. *In vitro* testing using either modified Boyden chambers or agarose gels shows that migration in response to chemoattractants (i.e., adult sera activated by bacteria, endotoxin, or antibody-antigen complexes) is decreased for premature or term newborn PMNs as compared to adult controls. Studies also show decreased receptor density on newborn PMNs for a specific chemotactic peptide (*N*-formyl-methonyl-leucyl-phenylalanine). Quantitative reduction in the receptors able to respond to chemotactic signals could provide a partial explanation for the impaired chemotaxis of newborn MNCs.

Phagocytosis and Microbial Killing

Many investigations have focused on the capacity of newborn PMNs and monocytes to ingest and kill a broad range of bacterial targets. The data suggest that with optimum culture conditions, cells from healthy neonates can perform these functions as well as adult cells. However, under "stress conditions" (i.e., increased bacteria-to-cell ratio, decreased serum concentration, or using cells from infants with sepsis, meconium aspiration, respiratory distress syndrome, or premature rupture of membranes), phagocytosis and bactericidal activity are decreased as compared to adult controls. This impaired function may be due, in part, to the incomplete development of intracellular components necessary for the generation of reactive oxygen metabolites and cytotoxic enzymes. However, recent data suggest that defective phagocyte function in newborns may also be secondary to impaired γ-interferon production by neonatal T-cells. The activation of phagocytic cells by γ-interferon is required for optimal cellular function.

Antigen Processing and Presentation

In addition to directly ingesting and killing infectious agents, macrophages play an important role in processing antigens for presentation to T-cells. Lymphokines released by the activated T-cell act both to amplify the inflammatory response and to increase the expression of major histocompatibility antigens (HLA in man) on monocytes. The antigen processing function of mononuclear phagocytes is important because T-cells belonging to the CD4 subset recognize small antigenic fragments (actually peptides of 10-20 amino acids) which have bound in the groove formed on the cell surface by the α-helices of the class II major histocompatability antigens (HLA-DP, DQ, and DR in humans; IA and IE in mice).

Several studies of newborn rats and mice indicate impaired presentation of antigen by neonatal macrophages. This correlates with reduced Ia antigen expression on the neonatal cells. The reduced Ia expression appears to reflect local *in vivo* production in the developing fetus of prostaglandin E and α-feto-protein. The relevance of these observations in rodents to events in man (where the gestation period is much longer) is uncertain. Although decreased HLA-DR expression has been described on human newborn macrophages, experiments designed to test *in vitro* antigen presentation (e.g., tetanus toxoid, herpes simplex virus) by neonatal macrophages have shown normal function.

Natural Killer Cells

Natural killer (NK) cells are a subpopulation of lymphocytes that, without prior sensitization, lyse neoplastic and virus-infected cells. Since the initial descriptions of this cell type, NK cells have been reported to be participants in other immunologic functions, including resistance against microbial infections; regulation of T- and B-cell immunity; cytokine production; regulation of hematopoietic stem cell differentiation and growth; and control of malignant growth.[48] After development from bone marrow progenitors, phenotypically mature ($CD16^+$, $CD3^-$, $CD56^+$) NK cells can be found in peripheral blood (average = 15% of peripheral blood lymphocytes) and spleens (3 to 4% of lymphocytes). NK cells are absent from the lymph nodes and thymus of healthy individuals. Morphologically, NK cells are large granular lymphocytes, i.e., have a high cytoplasm-nucleus ratio, round or indented nuclei, and large numbers of lysosomal organelles. One of the membrane-bound receptors on NK cells is a low affinity receptor for the Fc portion of immunoglobulin G (FcR or CD16). Through their FcR, NK cells can recognize antibody coated target cells and initiate antibody-dependent cell mediated cytotoxicity.

NK Cell Development

The determination of the differentiation pathway of NK cells was initially hampered by the need to identify NK cells by function (i.e., their ability to kill tumor cells in a non-MHC restricted manner). Since many hematopoietic cell types are capable of mediating this activity, conclusions were often conflicting. However, since defining mature NK cells by their distinct phenotype ($CD16^+$, $CD3^-$, $CD56^+$), it has been established that (1) NK cells are dependent on intact bone marrow and not on the thymus for differentiation and (2) mature NK cells are a distinct population that is separate from T-cells, B-cells, and myeloid cells.

Precise information on the fetal development of NK cells has been difficult to obtain because of the lack of known surface markers on NK cell precursors. Although "NK activity" is first detectable in the fetal liver at 8 to 11 weeks of gestation, the phenotype and subsequent lineage of this effector population remains to be determined. At birth, mature human $CD16^+$ NK cells are present in normal numbers, but their capability to mediate cytotoxicity against neoplastic cells is diminished. Although augmentation of this newborn NK cell activity can be elicited by exposure to either interleukin-2 or α-interferon, it is usually not raised to the level of activity exhibited by lymphokine stimulated adult NK cells[31] During childhood, NK cells reach full cytotoxic competency, and subsequently the proportion of $CD16^+$ NK cells in the peripheral blood remains relatively constant throughout life.

DEVELOPMENT OF COMPLEMENT

The complement system consists of a family of plasma proteins that can be activated to lyse cells. The activation pathway which was described first (and so is known as the classical pathway) is initiated by the binding of an IgG or an immunoglobulin M (IgM) antibody to an antigen. This exposes a site on the immunoglobulin to which C1q binds and which results in the activation of C4, C2, and C3. Direct activation of C3 also follows contact with bacterial endotoxins and is mediated by an alternative pathway using Properdin factor B. Complement activation contributes to immunity through the opsonization of microorganisms, regulation of the inflammatory response (i.e., chemotaxis and vasodilation), solubilization of immune complexes, and cellular lysis. In man, the genes for some complement proteins (i.e., C4, C2, and factor B) have been mapped to the major histocompatability complex on chromosome 6. Both serum complement components and locally produced complement proteins participate in the inflammatory response. Therefore, studies of the ontogeny of the complement system have included assessment of serum complement proteins and the regulation of local complement biosynthesis (for review, see Cole and Colten[8] and Frank and Fries[13]).

Serum Complement

Serum complement reaches detectable levels as early as 5 to 6 weeks of gestation, and gradually increases during development *in utero*. By 26 to 28 weeks of gestation, serum C1 and C3 are approximately two thirds of adult levels. From 28 weeks to term, most serum complement component levels rise with increasing gestational age. Term infants have total serum complement activity (CH_{50}) and individual complement component concentrations (C1, C4, C3, C7, and C9) between 35% and 100%. of adult levels (for review see Johnston et al.[26]). The reduced neonatal complement levels reflect lower fetal biosynthesis since transplacental passage of complement has not been detected (analysis using genetically distinct complement proteins) and since complement consumption is not greater in the healthy newborn (as determined by measuring complement split products). Serum complement levels then begin to rise during the first week of life (stimulus unknown) and subsequently increase at a rate that varies for different complement components and for each sex.

Complement Biosynthesis by Fetal Tissue

The primary source of serum complement proteins is the liver. In addition mononuclear phagocytes synthesize significant amounts of complement components. *In vitro* synthesis of complement by fibroblast and epithelial cells has also been detected, but its physiologic importance has not been determined.

In vitro studies of fetal tissue using labeled amino acids have shown complement synthesis (C3 and C1 inhibitor) by fetal liver as early as 4 weeks of gestation, with C2, C3 and C4 synthesis by 8 weeks and C5 synthesis by 9 weeks. Human peritoneal and alveolar cells from 14 weeks of gestation fetuses produce C3 and C4. The production of these components by fetal cells quantitatively increases throughout gestation. However, at term certain components still have reduced rates of *in vitro* synthesis (which correlates with reduced *in vivo* serum levels).

Functional Complement Activity

The activity of both the classical and alternative pathways is quantitatively reduced, on average, in the serum of premature and term infants (some term infants do have levels comparable to adult controls). Reduced activity of the alternative pathway is more common than decreased activity of the classical pathway. This functional defect results in decreased opsonization of microorganisms which requires the alternative pathway for clearance when antibody is not present (e.g., endotoxin containing bacteria such as *Escherichia coli*). Reduced activation, by either pathway, of complement in the serum of premature and term infants also results in the defective generation of fragments (C3a and C5a) with chemotactic activity for PMNs. These functional impairments in complement activity may contribute to the susceptibility of neonates to severe infections.

DEVELOPMENT OF SPECIFIC IMMUNITY

Specific immunity is conventionally divided into antibody-mediated (humoral immunity) and cell-mediated immunity. Antibodies are made by plasma cells, which are the descendants of B-lymphocytes. The stimuli required to initiate proliferation by B-cells and their differentiation into plasma cells include binding of antigen to B-cell surface antibody receptors and simultaneous exposure to two or more soluble factors derived from activated T-cells (these factors are currently thought to include interleukins 2, 4, 5, 6, and 7). The specificity of an antibody response results from the fact that B-cells need to bind antigen through their own cell surface antibody receptors in order to become activated and susceptible to the T-cell-derived factors. The plasma cells derived from each activated B-cell make antibody of the same specificity as the antibody previously expressed on the B-cell surface. As antibody responses mature, the progeny of cells which previously made IgM antibody switch to making IgG or IgA antibodies. T-lymphocytes recognize antigen in a different way from B-cells in that their receptors bind simultaneously to HLA determinants and to antigen. When stimulated, T-lymphocytes divide and release mediators such as interferon and interleukin-2 which recruit other cells into an inflammatory response.

Both antibody and T-cell receptors recognize antigen through specific receptors comprised of two polypeptide chains with variable amino acid sequences in the first 200 to 300 residues of the N terminus. This part of the antigen receptor is called the variable region to contrast it with the remaining, more constant, part of the receptor molecule. The specificity of the receptor-antigen interaction is dictated by the amino acid sequence of the variable region that results mainly from different recombinations of the germ line genes.

In describing the development of specific immunity in the fetus in the following paragraphs the main emphasis is on the development of lymphocytes. This is because both B- and T-lymphocytes are necessary, though not sufficient, for specific immune responses.

Development of B-Cells

B-cells are derived from pluripotential stem cells which develop first in blood islands in the yolk sac and subsequently migrate to the fetal liver. Pre-B-cells, which have immunoglobulin µ chains in their cytoplasm, but lack cell surface immunoglobulin, are the first morphologically detectable members of the B-cell series.[14] They appear in the fetal liver at about 8 weeks of gestation and B-cells are found 1 to 2 weeks later. The proportion of pre-B-cells in the liver falls as the bone marrow develops and most B-cell production appears to take place in the marrow after 12 weeks of gestation. Occasional pre-B-cells are present in the spleen throughout gestation: these cells are small and it seems probable that they are in transit, rather than originating in the spleen itself.[5]

Pre-B-cells divide spontaneously and their progeny are small pre-B-cells and immature B-cells. Studies of virus transformed mouse pre-B-cell lines suggest that the different progeny of an individual pre-B-cells have different rearrangements of their immunoglobulin heavy chain variable region genes. Commitment to a particular VDJ rearrangement for the heavy chain genes therefore appears to be made at the point of pre-B-cell division. Similarly, human B lymphocyte precursors from 14- to 18-week gestation bone marrow are transformed by Epstain Barr virus into cell lines expressing each of the five immunoglobulin classes.[9] The rearrangements of immunoglobulin genes on chromosome 14 which characterize the different developmental stages of B-cells are summarized in Table 1.

Maturation from the pre-B to the B-cell stage is accompanied by rearrangement of light chain genes and the expression of immunoglobulin on the cell surface. The κ light chain genes on chromosome 2 are rearranged first, and, if a transcribable light chain gene does not result, then the λ genes on chromosome 22 are rearranged. B-cells appear in increasing numbers in the fetal liver, bone marrow and spleen between 8 and 20 weeks of gestation. Before 12 weeks, these cells are mostly surface $\mu^+\delta^-$. Cells bearing IgD appear in increasing numbers from the 14th week of gestation in the spleen, lymph nodes, and blood. $\mu^+\delta^-$-cells outnumber δ^+ cells in the liver and bone marrow throughout gestation. This most probably results from the continued production of new, immature B-cells at these sites.

As immature B-cells mature they start to express a range of other surface antigens and receptors. These include receptors for complement components and for the Fc region of IgG and

Table 1
STAGES OF B-CELL MATURATION

Cell	Ig gene arrangement	Ig synthesis
Stem	D to J on chromosome 4	None
Pre-B	V to JD to μ	Cytoplasmic μ
Immature B	V to J to κ on 2 or λ on 22	Complete Ig monomer with light and heavy chain
Mature B	VDJ to μ or δ, γ, α, ε genes	Surface Ig with heavy chains
Plasma cell	As mature B-cell, but with intervening sequences deleted	High rate and secreted

antigens such as HLA-DR determinants. B-cells with surface IgG or IgA appear first in the fetal liver and by 14 weeks they are present in the spleen. Synthesis of IgG and IgA requires a further rearrangement of immunoglobulin genes on chromosome 14 to bring the VDJ sequence adjacent to a downstream γ or α gene. Most of the γ- or α-positive cells also bear surface μ and δ. This simultaneous expression of μ and δ with γ or α-heavy chains is rarely found on the B-lymphocytes of adults. The difference may result from the lack of exogenous antigen stimulus in the developing fetus. B-cells of newborns express other antigens associated with immaturity such as the early common acute lymphoblastic leukemia antigen (CALLA).[42]

In vitro studies of B-cell development are generally in agreement with the cell phenotyping studies described above. Incorporation of radiolabeled amino acids into IgM by fetal liver suspensions starts as early as 9 weeks of gestation and synthesis of both IgG and IgA was detected a few weeks later.[15] The cells responsible for this synthesis are not clearly identified. Plasma cells are not found in the tissues of uninfected fetuses until around 20 weeks of gestation, and at that time they mainly contain IgM. IgG and IgA plasma cells occur in very low numbers after 32 weeks.

Functional Studies of B-Cell Responses

For a B-cell to mature into a plasma cell and make large amounts of antibody, it must first be stimulated by antigen and, in most cases, receive soluble factors from a T-cell which promote cell proliferation and differentiation. B-cells bind antigen through their surface immunoglobulin and retain intact and catabolized fragments of the antigen on their surface. This antigen, and class II HLA antigens, are then available for recognition by the T-cells which supply antigen-specific help to the B-cell.[30] Since T-cells primed by defined antigens are not present in uninfected fetuses and newborns, functional testing of their antibody responses is difficult. The best evidence that fetal B-cells have a potential for antigen-specific responses comes from the study of fetuses with a congenital infection. An IgM antibody which has been detected in newborn sera appears to have specificity for the receptors on a subset of maternal T-cells which recognize HLA antigens which the infant has inherited from the father.[37] This is an intriguing observation because the antibody might protect the fetus from attack by maternal T-cells which had specificity for paternal HLA antigens. Leakage of maternal T-cells across the placenta and into the fetus would presumably be required to stimulate the production of this antibody. The maternal T-cells could themselves provide T-cell help under these circumstances, as a result of recognizing foreign HLA determinants.(This is called allospecific help.) The study certainly suggests that newborns can make antibody responses in circumstances when an antigen stimulus and T-cell help are both available. The report of anti-influenza antibody production following *in vitro* stimulation of B-cells of newborns in the presence of T cells of unrelated donors (as a source of allospecific help) supports this view.[55]

The response of fetal and newborn B-cells to nonspecific activation by mitogens gives additional information on B-cell development. Pokeweed mitogen-stimulated cultures of newborn cells make little immunoglobulin unless the T-cells are first irradiated to overcome their suppressor activity. Following irradiation of the T-cells, most of the immunoglobulin made is IgM and production of IgG and IgA is either delayed or absent. Stimulation with Epstein Barr virus (EBV) also elicits an immature response with predominance of IgM and little IgG or IgA. The change to an adult distribution of immunoglobulin classes in the response occurs over the first 2 years of life, and little IgG_2 is made before 2 years of age.

Maternal Antibody Transfer

IgG antibodies start to cross the placenta at about 16 weeks of gestation and an active transport mechanism develops after 22 weeks.[14a] In consequence the fetal IgG levels continue to rise and usually reach 1000 to 1500 g/dl by birth. The transport of IgG_3 is slightly less efficient than the other 3 subclasses but there is no selection for antibodies of a particular specificity. Maternal antibodies to HLA antigens inherited from the father do not appear in the fetal circulation though they are bound in the placenta. With this exception, newborns have antibodies of the same range of specificity as their mothers.

IgM does not cross the placenta and healthy newborns have levels below 20 mg/dL (they are higher following congenital infection). IgA levels are also very low in the serum and secretions of healthy newborns; maternal IgA does not cross the placenta. Large amounts of IgA are present in breast milk though this does not enter the infant's circulation.

Antibody Responses in Infancy and Childhood

Most of the available information concerning the ability of newborns and infants to make antibody responses derives from attempts at therapeutic immunization. Antigens to which antibody has been transplacentally acquired from the mother elicit substantially weaker responses by newborns than by older infants. The difference may in part be overcome by adding an adjuvant and it seems likely that the pertussis component of the diptheria, pertussis, and tetanus vaccine (DPT) routinely given at 2, 4, and 6 months of age has this effect. Other antigens, such as hepatitis B, elicit antibody responses following immunization started in the first month of life, whether or not maternal antibody was present.[6] Immunization with unmodified bacterial polysaccharides of infants under 18 months of age has generally been unsuccessful. Most antibodies to polysaccharides in adults are of the IgG_2 subclass but production of this subclass is normally delayed until 18 to 24 months of age.[4] Recent studies indicate that antipolysaccharide antibody can be made earlier (between 6 and 9 months of age) if the antigen is given with an adjuvant or linked to a protein.[17]

The need for infants to be able to initiate antibody responses in the presence of maternal antibody has further implications in relation to the regulation of specific immunity. Antibody, by promoting ingestion of antigen by mononuclear phagocytes, could promote antigen processing and representation to T-cells. There is also a theoretical possibility that antibody has an instructive role in the development of the immune repertoire. This is because the immune system has the potential to recognize the antigen binding site of an antibody as foreign, as well as the antigen to which the antibody binds. Mature animals make antibodies to the antigen binding sites of their own antibodies following immunization with purified antibody of a single binding specificity. These antibodies are called anti-idiotypes and under different experimental conditions they may boost or suppress an immune response to their corresponding antigen. Neonatal administration of large amounts of anti-idiotype antibodies suppresses the development of B-cells bearing the corresponding idiotype.[3] Whether the titer of an anti-idiotype antibody transferred across the placenta could ever be great enough to have a comparable effect seems unlikely. An essential stimulatory role for anti-idiotype antibodies also seems unlikely since the infants of agammaglobulinemic mothers usually make normal antibody responses. However, response to an anti-idiotype antibody has been proposed as the mechanism to explain the production of secretory anti-*E. coli* and polio virus antibodies by infants who have apparently never encountered this antigen[36] and the issue is not as yet resolved.

Early immunization alters the subsequent response to reimmunization. This effect has been interpreted as evidence against early immunization, though several studies suggest that infants who receive measles vaccination before 10 months make effective antibody responses.[45]

Immunoglobulin Development in Infancy and Childhood

Serum IgG levels fall for the first 3 to 4 months of life as maternal IgG is catabolized and diluted. The infant's own IgG synthesis probably starts in the first weeks of life, but insufficient amounts are made to maintain the serum IgG levels. Adult levels of IgG are usually reached by 3 years of age. IgM synthesis develops faster and adult levels are normally reached by 9 months of age. Serum IgA is the last immunoglobulin to reach adult levels: this normally occurs between the ages of 9 and 12 years. IgA and IgA antibodies appear in secretions (tears, saliva) much earlier (between 3 and 6 weeks of age). Much of the secreted IgA belongs to the IgA_2 subclass.

Development of the Thymus

Thymic development is described separately from other lymphoid development because the thymus is a primary lymphoid organ where the rate of cell division and lymphocyte production is independent of antigen stimulation. The thymic anlage is formed between 5 and 6 weeks of gestation by the invagination of the third and fourth pharyngeal pouches into the neck, followed by their growth into the mediastinum. The first lymphocytes appear between 9 and 10 weeks, and although direct data in man are lacking, observations in birds suggest that the thymus is populated by committed stem cells which develop in the yolk sac and which migrate via the blood to the thymus. A densely packed lymphoid medulla is visible by 12 weeks and Hassals corpuscles are present by 14 weeks of gestation. The earliest lymphoid cells to arrive in the thymus express the CD7 antigen and the express CD2 shortly afterward. In mice there is evidence that CD8 is expressed on the cell surface, without CD3 or CD4, at about the time that the β chain of the T-cell receptor is rearranged. By 10 weeks of gestation, cells simultaneously expressing CD1, CD4, and CD8 are present in the thymic cortex while cells expressing high levels of CD3 do not appear until about 12 weeks.[33]

As the stem cell matures in the thymus, it first rearranges the V, D, and J genes on chromosome 7 to make the complete variable region of the β chain of the T-cell receptor. This step is followed by rearrangement of the variable region genes for the α chain on chromosome 14 and expression of the complete heterodimer on the cell surface. The phenotype of the cells which first express T-cell receptors is $CD4^+$, $CD8^+$, $CD1^+$, $CD3^{low}$. It is at this point that the cell becomes susceptible to the selection process exerted by thymic epithelial cells and macrophages (see below).

The frequency of T-cells of different subsets in the fetal thymus by 18 weeks is generally similar to that of adults.[42] There is little direct information on the early emigration of fetal T-cells from the thymus into the periphery. The proliferative responses of spleen cell suspensions at 15 weeks of gestation to PHA is weak, which suggests that few T-cells have arrived in the spleen by this time. Fetal blood samples obtained between 16 and 20 weeks of gestation have percentages of $CD3^+$, $CD4^+$, and $CD8^+$ cells very similar to adults: this allows infants with congenital severe immunodeficiencies to be identified early enough in gestation for termination to be considered.[11] Differences reported between newborn and adult T-cells include the expression of the CD10 antigen on newborn $CD3^+$ T-cells, the occasional presence of increased numbers of $CD1^+$ cells,[54] the presence of PNA receptors on $CD8^+$ cells,[34] and the presence of some cells expressing both CD4 and CD8.[18]

Selection and Development of the T-Cell Repertoire

It is clear that random rearrangement of V, D, and J genes for the α and β chains of the T-cell receptor can give rise to a very large number of binding specificities. T-cells normally respond only to antigen which is seen in association with an antigen of the major histocompatibility complex (MHC). This phenomenon is called MHC-restriction, and the MHC antigens by which T-cell responses are restricted in man belong to the HLA series. Responses by CD4+ cells are restricted to HLA class II antigens (DP, DQ, and DR), while the CD8+ cells are restricted to the class I (A, B, and C) antigens. Recent studies using transgenic mice suggest that the thymus provides two selective pressures. One of these (termed positive selection) promotes the development of T-cells whose receptors have the potential for interaction with the individual's own MHC. This step can be selectively interfered with by injecting developing mice with antibodies to their own class I or class II MHC antigens, in which case the development of CD8 or CD4 T-cells is selectively suppressed. The other (negative selection) eliminates T-cells whose receptors would result in damaging interactions with self-antigens. Positive selection[6a] appears to require an interaction between a developing T-cells and thymic epithelial cells and it occurs at the developmental stage associated with expression of CD4 + CD8 + CD1. It is currently unclear whether positive selection precedes, follows, or is simultaneous with negative selection.

Negative selection in the thymus is thought to be linked to interactions between developing T-cells and thymic macrophages. The latter are derived from the bone marrow and express both class I and class II MHC antigens. In mice, large changes in the frequency of expression of a Vβ family can result from the presence of a particular MHC antigen. An example is the low number of Vβ5 expressing cells in mice which have the IE antigen. Other antigens can also result in the elimination of cells during their thymic development, such as the reduction in Vβ6 expressing cells in mice with the Mls[a] antigen. Recent results in man suggest that the subset of thymus cells which is destined to die within the thymus (presumably because the selection criteria were not met) can be identified by virtue of expression of the CD45R0 antigen.

There may be additional mechanisms by which the developing T-cell repertoire could be modified. One such is the production of antibodies to the antigen-binding portion (or idiotype) of developing T cells. Indirect evidence to support the potential for anti-idiotypes comes from the alteration of developing mouse T-cell repertoire by injections of anti-IgM antibody.[35]

Tolerance

A major factor in the development of T- and B-cell repertoires must be the avoidance of the production of activated self-reactive cells. Studies in developing mice and cattle showed that it was much easier to induce immunologic unresponsiveness (tolerance) to injections of foreign cells in fetal or newborn animals than in adult animals. Several mechanisms appear to operate in the induction of tolerance. The most simply understood is the clonal inactivation model, in which lymphocytes coming into contact with antigen at a critical stage in their development respond by becoming quiescent instead of continuing their differentiation. Immature B-cells pass through this stage when they have surface μ but not γ immunoglobulin and they are exposed to antigen in the absence of T-cell derived growth factors.[27] Experimental examples of this phenomenon include the suppression of the entire B-cell repertoire development by anti-μ antibodies and the more selective suppression which is achieved by anti-idiotype antibodies.[3] Pre-B-cells, which do not have surface immunoglobulin receptors for antigen, are not suppressed by anti-immunoglobulin antibodies. Abortion of potential self-reactive B-cells by antigen contact during their early development is most likely to be important for widely distributed antigens such as serum proteins. B-cell production does, however, appear to permit the release from the marrow of cells capable of making the range of autoantibodies which occur in autoimmune diseases, such as to thyroglobulin, red cell antigens, and IgG(Fc) determinants. Although B-cells with these specificities may come into contact with antigen, it is thought that they do not normally mature to make antibody because they do not receive T-cell help for these antigens.

In addition to the negative selection pressures which eliminate autoreactive cells in the thymus described above, mechanisms also exist to induce T-cell unresponsiveness (or tolerance) in the periphery, as follows the intravenous injection of antigens in cross-linked form or attached to cells.[24a] Models to account for these phenomena are generally based on the view that antigen binds to the receptor of resting T-cells in the absence of an essential lymphokine such as IL1. For example, mature T-cells maintained in continuous proliferation by antigen stimulation lose the ability to respond to the antigen if the monocyte derived factor, interleukin 1, is not present in the cultures.[28] This observation raises the possibility that developing self-reactive T-cells might be inactivated by encountering the antigen-HLA combination for which they had specificity in the absence of specialized antigen presenting cells. An attractive feature of this model is that it does not require a critical period for tolerance induction and it provides a mechanism for effective fetal immune responses to viral or bacterial antigens, which could be expected to be processed via monocytes and macrophages.

T-Cell Suppression

Adult blood lymphocytes make immunoglobulin in cultures stimulated by pokeweed mitogen (PWM), but when newborn lymphocytes are added the immunoglobulin production falls.[20] The negative influence of the cord blood cells is described as suppression and it is abolished by irradiating the cells or adding cortisol to the cultures. Both CD4 and CD8 cells have been implicated as necessary for the suppression and the effect may be mediated by soluble factors. Suppression is easy to detect in the laboratory,[40] and persists for several months after birth,[51] but it may be an artifact of mitogen stimulation and of little biological significance. For example, newborn T-cells only suppress EBV-induced immunoglobulin production by cultures of adult blood when PWM is also added. Furthermore, available data suggest that newborns are indeed capable of making antibody responses following appropriate immunization.

Development of Mature Peripheral T-Cell Populations

Fetal T-cells which are mature as judged by their proliferative responses to mitogen and alloantigen (in mixed lymphocyte culture) are present in the thymus from 14 weeks of gestation. The amount of proliferation by PHA stimulated thymus cells tends to diminish after 24 weeks of gestation and the fall continues into the first year of life. The reduction is in part due to the dilution of mature $CD3^+$ cells (which is the subset which responds to PHA) by increasing numbers of $CD1^+$ $CD4^+$ $CD8^+$ $CD3^{low}$ cortical thymocytes. The presence of large proportions of mature T-cells in the thymus early in fetal life is not fully explained. Delay in the peripheralization of developing T-cells is a possible contributing factor as very few mature T-cells are present before 15 weeks in the spleen and before 20 weeks in the lymph nodes.

Fewer than 5% of T-cells in newborns express the CD45R0 form of the common leukocyte antigen which is associated with memory cell status.[26a] This percentage increases to adult values after 3 to 4 years and is accompanied by a reduction in frequency of cells expressing the reciprocal CD45RA form of the common leukocyte antigen.[19] CD45RA bearing cells make less γ-interferon than CD45R0 cells and this difference probably underlies the lag in *in vitro* production of γ-interferon by mitogen stimulated newborn T-cells.[32] This difference is of potential importance in circumstances where a rapid immune response would be beneficial, as in the response to HSV. There is also evidence that CD45RA T cells require more cofactors to secure maximal proliferation than the more mature CD45R0 subset.[44a]

Immune Responses to Congenital and Perinatal Infection

The presence of plasma cells in the spleen, liver, and other tissues of fetuses aborted with congenital syphilis at 20 or more weeks of gestation provided the first direct evidence that immune responses could be made before birth. Other pathogens which can cross the placenta include rubella virus, cytomegalovirus, toxoplasmosis, and occasionally varicella zoster and herpes simplex viruses. In contrast to syphilis, these infections do not usually cause abortion and affected newborns have characteristic findings such as cardiac defects, hepatosplenomegaly, and thrombocytopenia. Immunologic features include the presence of raised (>20 mg/dl) serum IgM levels; IgA is often detectable and there are germinal centers in the lymph nodes and spleen. The immune responses are appropriate in that the immunoglobulin which is made has antibody activity for the pathogen. However, the newborn immune responses appear to eliminate the pathogen less effectively than the immune responses of adults. For example, infants with congenital rubella continue to excrete the virus in their urine for the first years of life. CMV also is excreted in the urine of babies infected congenitally or neonatally for many years. This suggests that the response to infection acquired early is inefficient. Our own studies have dealt primarily with immune responses to HSV and CMV following perinatal infection and they show that these infections are followed by a smaller increase in numbers of circulating virus specific T-cells than are primary HSV infections in adults. The reasons for this difference are incompletely understood, but they do not appear to involve any defect of antigen processing by monocytes. The survivors of neonatal HSV infections are clearly not immunologically tolerant of the virus since they make anti-HSV antibody. Cells with the potential to proliferate in response to the virus are not eliminated either, as shown by the production of low levels of soluble mediators following stimulation of the cells of infants with viral antigen.[19]

An increased incidence of immunodeficiency and autoimmunity is another adverse consequence of congenital rubella infection, but apparently not of other congenital infections. The immunodeficiency is most often selective IgA deficiency (occasionally IgG is deficient, too) and the best documented autoimmune association is with insulin dependent (type 1) diabetes mellitus (IDDM). Selective IgA deficiency sometimes occurs with auto-immune disorders so a single mechanism could account for both associations. The observation that congenital rubella infants who develop IDDM have the same HLA DR 3 and 4 antigens as other IDDM patients suggests that the congenital infection increases autoantibody production only in genetically susceptible individuals.

Maternal Tolerance for the Fetus

While the uterine environment may protect the fetus from most pathogens, it does entail exposure to the mother's own immune responses. Fetal major histocompatibility (MHC) antigens of the HLA class I (A,B, and C) and class II (DP, DQ, and DR) series inherited from the father elicit antibody production by the mother and the titers are increased by multiple pregnancies. These antibodies are of the IgG class and cross the placenta, but they are not found free in the fetal circulation, probably because most are bound in the subtrophoblastic layers of the placenta. The mechanisms by which maternal T-cells with specificity for paternal MHC class II antigens are prevented from attacking the fetus are incompletely understood. During the implantation stage, the blastocyst appears to express surface MHC antigens and consequently is a potential target for attack. However, implantation may involve so much proteolytic activity that antigen recognition is prevented. The developing trophoblast appears to have very little surface MHC class II antigen, which would reduce the potential for immunologic recognition of the fetus, and in mice there is evidence for the development of a specific suppressor mechanism localized to the decidua. A range of other nonspecific immunosuppressive factors which appear in increased concentration in the plasma or serum of pregnant women is described.[12a] The significance of nonspecific factors is uncertain because women do not appear to be much more susceptible to common environmental microorganisms during pregnancy.

CONCLUSIONS

Cells and humoral factors with nonspecific immune effector functions appear early in fetal life and their activity increases with gestation, although adult levels are not reached until the first year of life. Contact with microorganisms starts at birth and immaturities of nonspecific immunity which are likely to limit the newborn response to infection include:

1. Slow response of newborn neutrophils to chemotactic signals
2. Reduced phagocytosis and bacterial killing under adverse conditions
3. A low reserve of neutrophils in the marrow
4. Reduced cytotoxicity by newborn natural killer cells towards virus infected cells
5. Reduced rates of complement synthesis, resulting in lower plasma complement levels

The fetal development of specific immune mechanisms extends to the production of T- and B-cells with diverse repertoires for antigen. The precursors of lymphocytes arise in the yolk sac and develop first in the liver and later in the bone marrow. Their diversity is achieved by rearrangements, in lymphocyte precursors, of the germ line genes which code for the components of the antigen receptor. T- and B-cells appear in the blood from 12 weeks of gestation onward and reach adult numbers by 30 weeks of gestation. The antigen specificities which the T-cells of fetuses and newborns can recognize are described as the expressed repertoire and is shaped by positive and negative selection within the thymus. Lymphocytes which bind antigen in the presence of appropriate helper factors are triggered to divide as well as to release soluble mediators. It is the antigen-driven expansion of clones of T- and B-cells which converts the naive cells of the newborn into the mature memory lymphocyte populations of the adult.

REFERENCES

1. **Abo, T., Miller, C. A., Gartland, G. L., and Balch., C. M.,** Differentiation stages of human natural killer cells in lymphoid tissues from fetal to adult life, *J. Exp. Med.,* 157, 273, 1983.
2. **Abruzzo, L. V. and Rowley, D. A.,** Homeostasis of the antibody response: immunoregulation by NK cells, *Science,* 222, 581, 1983.
3. **Accolla, R. S., Gearhart, P. J., Sigal, N. H., Cancro, M. P., and Klinman, N. R.,** Idiotype specific neonatal suppression of phosphorylcholine responsive B cells, *Eur. J. Immunol.,* 7, 876, 1977.
4. **Andersson, U., Bird, A. G., Britton, S., and Palacios, R.,** Humoral and cellular immunity in humans studied at the cell level from birth to two years of age, *Immunol. Rev.,* 57, 5, 1981.
5. **Asma, G. E. M, Langlois, R., and Vossen, J. M.,** Development of pre-B and B lymphocytes in the human fetus, *Clin. Exp. Immunol.,* 56, 407, 1984.
6. **Barin, F., Goudeau, A., Denis, F., Yvonnet, B. Chiron, J. P., Coursaget, P., and Diop Mar, I.,** Immune response in neonates to hepatitis B vaccine, *Lancet,* i, 251, 1982.
6a. **Benoist, C. and Mathis, D.,** Positive selection of the T cell repertoire: where and when does it occur?, *Cell,* 58, 1027, 1989.
7. **Chilmonczyk, B., Levin, M. J., McDuffie, R., and Hayward, A. R.,** Characterization of the newborn response to herpesvirus antigens, *J. Immunol.,* 134, 4184, 1985.
8. **Cole, F. S. and Colten, H. R.,** Complement, in *Neonatal Infections: Nutritional and Immunological Interactions,* Olgra, P. L., Ed., Grune & Stratton, Orlando, FL, 1984, 37.
9. **Dosch, H. M., Lam, P., Hui, M. F., and Hibi, T.,** Concerted generation of Ig isotype diversity in human fetal bone marrow, *J. Immunol.,* 143, 2470, 1989.
10. **Drew, J. H. and Arroyave, C. M.,** The complement system of the newborn infant, *Biol. Neonate,* 37, 209, 1980.
11. **Durandy, A., Oury, C., Griscelli, C., Dumez, Y., Oury, J. F., and Henrion, R.,** Prenatal testing for inherited immune deficiencies by fetal blood sampling, *Prenatal Diagn.,* 2, 109, 1982.
12. **Eibl, M., Zielinski, Ch. C., Ahmad, R., Steurer, F., and Rockenschraub, A.,** Plaque forming cells in human cord blood: studies on T and B cell function, *Clin. Exp. Immunol.,* 41, 176, 1980.
12a. **Faulk, W. P. and McIntyre, J. A.,** Immunological studies of human trophoblast: markers, subsets, and functions, *Immunol. Rev.,* 75, 139, 1983.
13. **Frank, M. M. and Fries, L. F.,** Complement, in *Fundamental Immunology,* Paul, W. E., Ed., Raven Press, NY, 1989, 679.
14. **Gathings, W. E., Lawton, A. R., and Cooper, M. D.,** Immunofluorescent studies of the development of pre-B cells, B lymphocytes and immunoglobulin isotype diversity in humans, *Eur. J. Immunol.,* 7, 804, 1977.
14a. **Gitlin, D.,** Development and metabolism of immunoglobulins, in *Immunologic Incompetence,* Kargan, B. M. and Stiehm, R., Eds., Year Book Medical Publishers, Chicago, 1971.

15. **Gitlin, D. and Biasucci, A.,** Development of G, A, M, B1C/B1A, C11 esterase inhibitor, ceruloplasmin, transferrin, hemopexin, haptoglobin, fibrinogen, plasminogen, alpha-1-antitrypsin, orosomucoid, B lipoprotein, alpha-2-macroglobulin, and prealbumin in the human conceptus, *J. Clin. Invest.,* 48, 1433, 1969.
16. **Gordon, S., Keshav, S., and Chung, L. P.,** Mononuclear phagocytes: tissue distribution and functional heterogeneity, *Curr. Opinions Immunol.,* 1, 26, 1988.
17. **Granoff, D. M. and Cates, K. L.,** Haemophilus influenzae type b polysaccharide vaccines, *J. Pediatr.,* 107, 330, 1985.
18. **Griffiths-Chu, S., Patterson, J. A. K., Berger, C. L., Edelson, R. L., and Chu, A. C.,** Characterization of immature T cell subpopulations in neonatal blood, *Blood,* 64, 296, 1984.
19. **Hayward, A. R., Leibson, P., and Arvin, A.,** Development of lymphocyte responses to herpes simplex virus following neonatal infection, in *Immunology of the Neonate,* Burgio, G. R., Eds., Springer, New York, 1986, 112.
20. **Hayward, A. R. and Lawton, A. R.,** Induction of plasma cell differentiation of human fetal lymphocytes: evidence for functional immaturity of T and B cells, *J. Immunol.,* 119, 1213, 1977.
21. **Hayward, A. R. and Malmberg, S.,** Response of human newborn lymphocytes to alloantigen: lack of evidence for suppression induction, *Pediatr. Res.,* 18, 414, 1984.
22. **Herberman, R. B.,** *Natural Cell Mediated Immunity Against Tumors,* Academic Press, New York, 1980.
23. **Hoffman, A., Hayward, A. R., Kurnick, J. T., Defreitas, E. C., McGregor, J., and Harbeck, R. J.,** Presentation of antigen by human newborn monocytes to maternal tetanus toxoid-specific T cell blasts, *J. Clin. Immunol.,* 1 217, 1981.
24. **Holmberg, L. A., Miller, B. A., and Ault, K. A.,** The effect of natural killer cells on the development of syngeneic hematopoietic progenitors, *J. Immunol.,* 133, 2933, 1984.
24a. **Jenkins, M. K. and Schwartz, R. H.,** Antigen presentation by chemically modified splenocytes induces antigen-specific T cell unresponsiveness in vitro and in vivo, *J. Exp. Med.,* 165, 302, 1987.
25. **Johnston, R. B.,** Monocytes and macrophages, *N Engl. J. Med.,* 318, 747 1988.
26. **Johnston, R. B., Altenburger, K. M., Atkinson, A. W., and Curry, R. H.,** Complement in the newborn infant, *Pediatrics,* 64(Suppl), 781, 1979.
26a. **Kingsley, G., Pitzalis, C., Waugh, A. P., and Panayi, G. S.,** Correlation of immunoregulatory function with cell phenotype in cord blood lymphocytes, *Clin. Exp. Immunol.,* 73, 40, 1988.
27. **Klinman, N. R., Wylie, D. E., and Teale, J. M.,** B cell development, *Immunol. Today,* 2, 212, 1981.
28. **Lamb, J. R. and Feldmann, M.,** Essential requirement for major histocompatibility complex recognition in T cell tolerance induction, *Nature,* 308, 72, 1984.
29. **Lanier, L. L., Cwirla, S., Federspiel, N., and Phillips, J. H.,** Human natural killer cells isolated from peripheral blood do not rearrange T cell antigen receptor beta chain genes, *J. Exp. Med.,* 163, 209, 1986.
30. **Lanzavecchia, A.,** Antigen specific interaction between T and B cells, *Nature,* 314, 537, 1985.
31. **Leibson, P. J. Hunter-Laszlo, M., Douvas, G. S., and Hayward, A. R.,** Impaired neonatal natural killer cell activity to herpes simplex virus: decreased inhibition of viral replication and altered response to lymphokines, *J. Clin. Immunol.,* 6, 216, 1986.
32. **Lewis, D. B., Larsen, A., and Wilson, C. B.,** Reduced interferon-gamma mRNA levels in human neonates, *J. Exp. Med.,* 163, 1018, 1986.
33. **Lobach, D. F., Hensley, L. L., Ho, W., and Haynes, B. F.,** Human T cell antigen expression during the early stages of fetal thymic maturation, *J. Immunol.,* 135, 1752, 1985.
34. **Maccario, R., Nespoli, L., Mingrat, G., Vitello, A., Ugazio, A. G., and Burgio, G. R.,** Lymphocyte subpopulations in the neonate, *J. Immunol.,* 130, 1129, 1983.
35. **Martinez, A, C., Bernabe, R. R., de la Herta, A., Pereira, P., Cazenave, P.-A., and Couthino, A.,** Establishment of idiotypic helper T cell repertoires early in life, *Nature,* 317, 721, 1985.
36. **Mellander, L., Carlsson, B., and Hanson, L. A.,** Secretory IgA and IgM antibodies to *E. coli* O and poliovirus type 1 antigens occur in amniotic fluid, meconium and saliva from newborns, *Clin. Exp. Immunol.,* 63, 555, 1986.
37. **Miyagawa, Y., Komiyama, A., Akabane, T., Uehara, Y., and Yano, A.,** Cord IgM antibody specific for human killer cells: T lymphocytotoxic human fatal antibody recognizing maternal killer T cells proliferating in the presence of interleukin 2, *J. Immunol.,* 129, 1993, 1982.
38. **Notarangelo, L. D., Chirico, G., Chiari, A., Colombo, A., Rondini, G., Plebani, A., Martini, A., and Ugazio, A. G.,** Activity of classical and alternative pathways of complement in pre-term and small for gestational age infants, *Pediatr. Res.,* 18, 251, 1984.
39. **Nunoi, H., Endo, F., Chikazawa, S., Namikawa, T., and Matsuda, I.,** Chemotactic receptor of cord blood granulocytes to the synthesized chemotactic peptide N-formyl-methionyl-leucyl-phenylalanine, *Pediatr Res.,* 17, 57, 1983.
40. **Olding, L. B. and Oldstone, M. B. A.,** Thymus derived peripheral lymphocytes from human newborns inhibit division of their mothers' lymphocytes, *J. Immunol.,* 116, 682-6, 1976.
41. **Perussia, B., Trinchieri, G., Jackson, A., Warner, N. L., Faust, J., Rumpold, H., Kraft, D., and Lanier, L. L.,** The Fc receptor for IgG on human natural killer cells: phenotypic, functional, and comparative studies with monoclonal antibodies, *J. Immunol.,* 133, 180, 1984.

42. **Rosenthal, P., Rimm, I. J., Umiel, T., et al.,** Ontogeny of human hemopoietic cells: analysis using monoclonal antibodies, *J. Immunol.,* 31, 232, 1983.
43. **Sato, T., Fuse, A., and Kuwata, T.,** Enhancement by interferon of natural cytotoxic activities of lymphocytes from human cord blood and peripheral blood of aged persons, *Cell. Immunol.,* 45, 458, 1979.
44. **Segal, A. W. and Soothill, J. F.,** Phagocytes, in *Paediatric Immunology,* Soothill, J. F., Hayward, A. R., and Wood, C. B. S., Eds., Blackwell Scientific, Oxford, 1983, 37.
44a. **Shaw, S.,** Lymphocyte differentiation, *Nature,* 338, 539, 1989.
45. **Stetler, H. C., Orenstein, W. A., Bernier, R. H., Herrmann, K. L., Sirotkin, B., Hopfensperger, D., Schuh, R., Albrecht, P., Lievens, A. W., and Brunell, P. A.,** Impact of revaccinating children who initially received measles vaccine before 10 months of age, *Pediatrics,* 77, 471, 1986.
46. **Tardieu, M., Hery, C., and Dupuy, J. M.,** Neonatal susceptibility to MHV infection in mice. II. Role of natural effector marrow cells in the transfer of resistance, *J. Immunol.,* 124, 418, 1980.
47. **Toivanen, P., Uksila, J., Leino, A., Lassila, O., Hirvonen, T., and Ruuskanen, O.,** Development of mitogen responding T cells and natural killer cells in the human fetus, *Immunol. Rev.,* 57, 89, 1981.
48. **Trinchieri, G.,** Biology of natural killer cells, *Adv. Immunol,* 47, 187, 1989.
49. **Trinchieri, G. and Perussia, B.,** Human natural killer cells: biologic and pathologic aspects, *Lab. Invest.,* 50, 489, 1984.
50. **Ueno, Y., Miyawaki, T., Seki, H., Matsuda, A., Taga, K., Sato, H., and Taniguchi, N.,** Differential effects of recombinant human gamma-interferon and interleukin 2 on natural killer cell activity of peripheral blood in early human development, *J. Immunol.,* 135(1), 180, 1985.
51. **Unander, A. M. and Olding, L. B.,** Ontogeny and postnatal persistence of a strong suppressor activity in man, *J. Immunol.,* 127, 1182, 1981.
52. **Wawryk, S. O., Novotny, J. R., Wicks, I. P., Wilkinson, D., Maher, D., Salvaris, E., Welch, K., Fecondo, J., and Boyd, A. W.,** The role of the LFA 1/ICAM 1 interaction in human leukocyte homing and adhesion, *Immunol. Rev.,* 108, 135, 1989.
53. **Welsh, R. M. and Zinkernagel, R. M.,** Heterospecific cytotoxic cell activity induced during the first three days of acute lymphocytic choriomeningitis virus infection in mice, *Nature,* 268, 646, 1977.
54. **Wilson, M.,** Immunology of the fetus and newborn: lymphocyte phenotype and function, *Clin. Immunol. Allergy,* 5, 271, 1985.
55. **Yarchoan, R. and Nelson, D. L.,** Specificity of *in vitro* anti-influenza antibody production by human lymphocytes: analysis of original antigenic sin by limiting dilution cultures, *J. Immunol.,* 312, 928, 1984.

THE HISTOCOMPATIBILITY SYSTEM IN HEALTH, DISEASE, AND AGING

Marilyn S. Pollack

INTRODUCTION

Histocompatibility determinants are cell-surface glycoprotein molecules that are involved in functional interactions between cells and in self-/nonself-recognition. Such histocompatibility systems are postulated to have originated in multicellular invertebrates, probably beginning with coelenterates.[1] It has been further postulated[2] that the system arose from the self-/nonself-recognition systems that prevent autofertilization in hermaphroditic invertebrates. However, while invertebrates allow fusion of heterozygous cells sharing either one or both "compatibility" alleles, the evolution of specific immune systems, which at the level of primitive chordates or fishes began to include regulatory immunoglobulins (with molecular homology to the histocompatibility determinants), provided the basis for vertebrate immunoregulatory networks capable of eliminating "foreign" cells carrying even one nonshared histocompatibility allele.

In humans, histocompatibility determinants are highly polymorphic and have significant alloimmunogenicity, affecting the transplantation of organs and transfusion of blood components from one individual to another. These determinants also play a major role in susceptibility or resistance to different diseases in an individual. They do so both as cell membrane components of antigen presenting cells that differentially bind processed auto- or foreign antigens to influence immune system responses and through their genetic association with other factors affecting immunocompetence or disease status. The following sections will briefly summarize the genetics of the human histocompatibility determinants and what is currently known about their possible role in human health, disease, and aging.

GENETICS, NOMENCLATURE, AND TECHNICAL FEATURES OF THE HLA SYSTEM

The designation "HLA" for the major histocompatibility (MHC) system of man derives from the initials for "human leukocyte locus A" because, although all nucleated cells have HLA antigens, the existence and immunogenicity of the system was first recognized by the study of leukocyte antibodies in transfusion patients.[4-6] The HLA system is determined by a series of highly polymorphic, closely linked genes on the short arm of chromosome 6. Three of these determinants code for serologically detected "class I" glycoprotein molecules that are expressed on the surfaces of all nucleated cells and platelets: the HLA-A, -B, and, -C antigens. These molecules each have one large subunit (molecular weight of approximately 44,000 Da) expressing alloantigen activity as the result of a few amino acid substitutions in specific "variable" regions[7] of the chromosome 6 gene product. The smaller subunit (11,000-Da molecular weight), coded for by a gene on chromosome 15, is referred to as B_2-microglobulin and apparently lacks alloimmunogenicity. The smaller and larger class I HLA subunits are inserted together into cell membranes to form single molecules lacking covalent bonds between the peptide chains (Figure 1). Recent studies utilizing X-ray crystallography techniques have demonstrated that a typical class I molecule has α-helical structures forming a rim on each side of a cleft, the bottom of which is formed by β-plated sheets (Figure 2).[8] This cleft could form an antigen binding and/or T-cell receptor binding site. Most of the amino acid substitutions that account for the differences between serologically defined class I alleles are located in the "rim" regions.

Class II HLA molecules consist of two, noncovalently linked, chromosome 6 coded

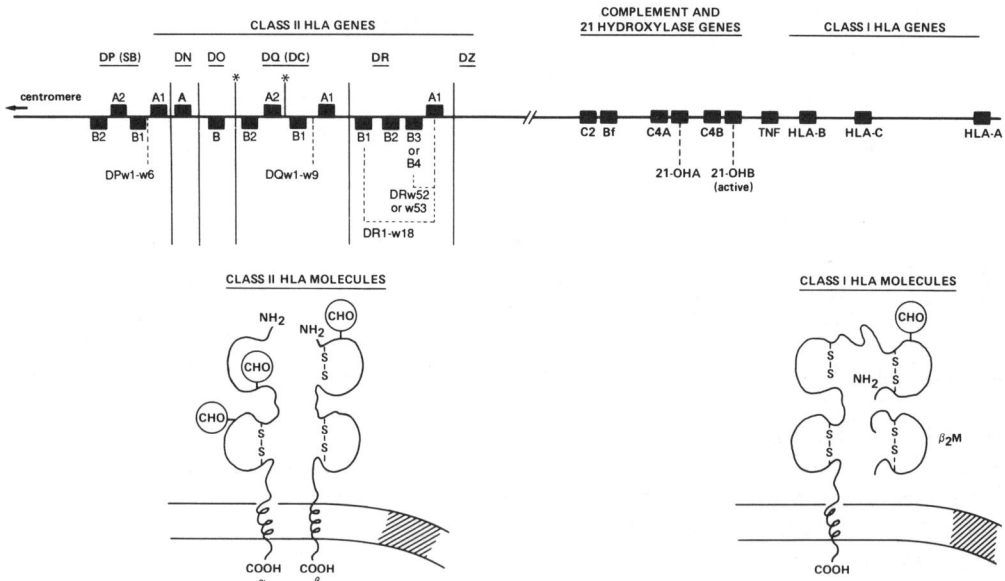

FIGURE 1. Major histocompatibility region genes and expressed class I and class II HLA molecules. Recombination "hot spots" are believed to exist at the points marked "*" within the class II HLA gene region, but there is significant positive linkage disequilibrium involving DQ, DR, complement, and class I region gene products. As illustrated, class I molecules contain one polypeptide coded for by a class I region gene and β-2-microglobulin, coded for by a gene on chromosome 15, while class II molecules consist of polypeptide chains coded for by functional α- and β-genes in the DR, DQ, or DP regions. The genes for each chromosome code for multiple, different HLA class II molecules and it is likely that *"trans"* molecules consisting of an α-gene product from one chromosome and a β-gene product from the other are also expressed. Most of the serologically defined allodeterminants on the DR molecules (see Table 1) are coded for by the first DR region gene (DR B1), while the DRw52 and w53 specificities are on DR B3 or DR B4 (no haplotype contains both) gene products, respectively. The expressed DR α-gene product is not polymorphic. Similarly, although in this case both gene products are polymorphic, the *serologically* defined DQ region polymorphisms are apparently coded for by the DQ β-genes. Different haplotypes express different numbers of gene products, but all have some "genes" that are nonexpressed pseudogenes, particularly, for example, the pseudo α-gene called "DN", the pseudo β-gene called DO, and the second β-gene of the DR region (DR B2).

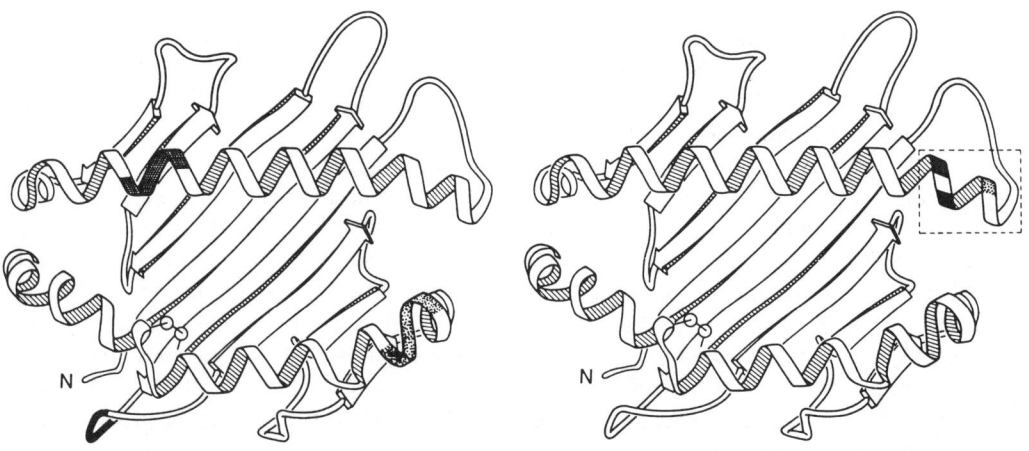

FIGURE 2. Three-dimensional views of class I molecules with dark and hatched areas at sites where amino acid changes distinguish A2, A25, and B17 molecules (left) and Bw4 vs. Bw6 molecules (right).

polypeptide chains and are constitutively expressed only on certain nucleated cells, notably on subpopulations of monocytes, the B-cell subpopulation of lymphocytes, Langerhans cells of the skin, and dendritic cells of the spleen and other organs.[9] They can also be expressed on other nucleated cells such as T-lymphocytes,[10] dermal fibroblasts,[11] and endothelial cells[12,13] that have been activated by stimulation with mitogens or γ-interferon. The small and large glycoprotein subunits are approximately 29,000 and 34,000 Da, respectively.[14] At least three biochemically and serologically distinguishable types of class II molecules, coded for by α- (large subunit) and β- (small subunit) genes in three separate regions of chromosome 6, are expressed on most class II positive cells: the DR, DQ (formerly MB or DS), and DP (formerly SB) molecules (see Figure 1). For most chromosomes, there are two different expressed DR β-gene products: those coded for by the DR B1 gene that carry the primary serologically defined DR alleles (e.g., DR1, DR2, etc.) and those coded for by the DR B3 gene (the DRw52 specificities) or the DR B4 gene (the DRw53 specificities).[15] With the notable exception of the α-chain subunit of DR molecules, both types of subunits of expressed class II molecules that have been studied to date are apparently polymorphic.[16] Genes coding for tumor necrosis factor (TNF) and the complement and 21-hydroxylase enzymes that have been mapped between the class I and class II MHC genes in humans[17-20] (see Figure 1) will be discussed further below.

Although no evidence exists that cell surface class II MHC molecules are formed by combinations of α- and β-genes from different chromosomal regions under physiological conditions, there is evidence that *"trans"* class II molecules are formed from α- and β-gene products from the same region of the two different #6 chromosomes[21] of an individual; the number of different polymorphic class II molecules that can thus theoretically be expressed on the surface of cells from a single individual is quite large. Whereas most well-characterized HLA class II antigen variants are conventionally detected by serological techniques (see below), additional polymorphic variants of the DQ molecules are primarily detected by biochemical techniques in which two-dimensional electrophoresis patterns of molecules precipitated by DQ specific monoclonal antibodies are analyzed.[21,22] Also, the polymorphic variants of the DP regions molecules have thus far been primarily detectable only by proliferative (or cytolytic) responses of primed or cloned T-cells.[23] The recent development of allele-specific oligonucleotide probes for use in conjunction with polymerase chain reaction (PCR) techniques for amplification of DP specific DNA now allows the products of this genetic region to also be rapidly identified.[24]

As noted above, most of the polymorphic DR class II variants and all well-characterized class I HLA-A, -B, and -C variants can be detected by (complement-dependent) serological reactions using human sera from donors immunized by pregnancy or (rarely) transplantation of blood transfusions. A list of the allelic class I and class II HLA variants that are currently well defined by these serological techniques or, in some cases, cellular techniques (DP region products and combined DR + DQ + DP ("HLA-D") products) is shown in Table 1. Restriction fragment length polymorphism (RFLP) techniques, which detect variations at the gene (DNA) level,[24a] define a vast number of additional polymorphisms, including those that do not affect expressed molecules. Analysis of actual DNA variations with RFLP techniques and also with synthetic oligonucleotide probes[25] or even DNA sequencing[26] has recently allowed for a much clearer perception of the actual basis for HLA associated disease susceptibility, as will be discussed further below.

THE HISTOCOMPATIBILITY SYSTEM IN HEALTH: TRANSFUSION, TRANSPLANTATION, AND RELATIONSHIP TESTING

The polymorphism of the HLA system has been a fundamental factor in its clinical significance and applications. Patients who have become sensitized to the HLA antigens of

Table 1
WHO RECOGNIZED HLA ALLOSPECIFICITIES[a] (1987)

Class I HLA alleles				"HLA-D"	Class II HLA alleles		
A	B	B'[b]	C	D[c]	DR	DQ	DP(SB)
A1	B5	Bw4	Cw1	Dw1	DR1	DQw1(MT1)	DPw1
A2	B7	Bw6	Cw2	Dw2	DR2	DQw2(MB1)	DPw2
A3	B8		Cw3	Dw3	DR3	DQw3 (MT4,MB3)	DPw3
A9	B12		Cw4	Dw4	DR4	DQw4	DPw4
A10	B13		Cw5	Dw5	DR5	DQw5(w1)	DPw5
A11	B14		Cw6	Dw6	DRw6	DQw6(w1)	DPw6
Aw19	B15		Cw7	Dw7	DR7	DQw7(w3)	
A23(9)	B16		Cw8	Dw8	DRw8	DQw8(w3)	
A24(9)	B17		Cw9(w3)	Dw9	DRw9	DQw9(w3)	
A25(10)	B18		Cw10(w3)	Dw10	DRw10		
A26(10)	B21		Cw11	Dw11(w7)	DRw11(5)		
A28	Bw22			Dw12	DRw12(5)		
A29(w19)	B27			Dw13	DRw13(w6)		
A30(w19)	B35			Dw14	DRw14(w6)		
A31(w19)	B37			Dw15	DRw15(2)		
A32(w19)	B38(16)			Dw16	DRw16(2)		
Aw33(w19)	B39(16)			Dw17(w7)	DRw17(3)		
Aw34(10)	B40			Dw18(w6)	DRw18(3)		
Aw36	Bw41			Dw19(w6)			
Aw43	Bw42			Dw20	DRw52		
Aw66(10)	B44(12)			Dw21			
Aw68(28)	B45(12)			Dw22	DRw53		
Aw69(28)	Bw46	Dw23					
Aw74(w19, Th)	Bw47			Dw24			
	Bw48			Dw25			
	B49(21)			Dw26			
	Bw50(21)						
	B51(5)						
	Bw52(5)						
	Bw53						
	Bw54(w22)						
	Bw55(w22)						
	Bw56(w22)						
	Bw57(17)						
	Bw58(17)						
	Bw59						
	Bw60(40)						
	Bw61(40)						
	Bw62(15)						
	Bw63(15)						
	Bw64(14)						
	Bw65(14)						
	Bw67						
	Bw70						
	Bw71(w70)						
	Bw72(w70)						
	Bw73						
	Bw75,w76,w77(15)						

[a] "Broad" specificities (or previous designatins) are listed in parentheses after the appropriate "narrow" specificity (or "split") (e.g., A23(9) indicates that A23 is a subtype of A9).

[b] Bw4 and Bw6 are supertypic determinants, one of which is found on every HLA-B molecule.

[c] HLA-D antigens are defined by combined proliferative responses to HLA-DR, DQ, and DP region products: the failure to respond (proliferate) to particular homozygous typing cells (HTC) in *primary* mixed leukocyte culture defines the presence of that HLA-D type.

other individuals by exposure to repeated whole blood or platelet transfusions can usually only benefit from HLA-matched platelets.[27] Kidney disease (renal failure) patients can benefit if an organ transplant is possible from a sibling donor who has inherited the same combinations of HLA antigens, and cadaver donor kidney grafts are also successful when complete matching can be obtained for the serologically defined antigens, especially the DR antigens.[28,29] Bone marrow transplantation requires even more critical matching, since the transplanted immunocompetent cells not only can be rejected, but also can react against host tissue causing fatal graft vs. host disease.[30]

The polymorphism of the HLA system also, more incidentally, has lent itself to applications in relationship testing, particularly for purposes relating to immigration status and child support (paternity testing). Because the HLA system is so highly polymorphic, test results can be used as the basis for "inclusion" probability calculations,[31] as well as for determination of "exclusion". In an "inclusion" paternity case, for example, the likelihood that a nonexcluded man *is* actually the father of a given child can be calculated from the relative possibility that the child's particular HLA (and red cell) antigen combinatins could have been inherited from anyone else in the relevant population with access to the same mother.

THE HISTOCOMPATIBILITY SYSTEM IN DISEASE

Specific allodeterminants in the major histocompatibility complex have been associated with different diseases for three different kinds of reasons, as summarized in Table 2: those relating to the effects of linked genes, those relating to cell membrane properties of particular HLA molecules, and those relating to aberrant self- or foreign "antigen presenting" properties of particular class II epitopes. The latter two mechanisms usually involve multifactorial disease etiologies and can be studied in unrelated patients.

HLA Disease Associations Resulting from HLA-Linked Genetic Mutations

HLA-disease associations that result from mutations of HLA-linked genes, as determined by family studies, are the simplest to understand. From the current HLA gene map (see Figure 1), it is evident that several such diseases can be identified. These include deficiencies of the complement components C2 or C4 and deficiencies in the activity of the enzyme 21-hydroxylase (21-OH) (which is involved in the pathways of steroid metabolism through which its substrate, 17-hydroxyprogesterone (17-OH), is eventually converted to aldosterone or to cortisol). Deficiencies in the HLA-linked complement components are known to cause autoimmune diseases and renal failure,[32,33] while deficiency in 21-OH often causes severe masculinization of female neonates from intrauterine exposure to excess testosterone (formed from the excess 17-OH).[34] Differences in the properties of allelic variants of tumor necrosis factor (TNF) gene products may also affect the etiology of some diseases, notably systemic lupus erythematosus.[35] In addition, there is HLA linkage of the activity of some (currently unknown) gene with a product involved in iron metabolism which, when defective in both chromosomes, can cause hemochromatosis[36] and HLA linkage of a dominant gene that causes spinocerebellar ataxia in some (but not all) familial cases.[37] Location of this gene centromeric to the HLA complex was recently demonstrated by collaborative studies involving the author of this review.[38]

Mutations at the active 21-OH locus (21-OHB in Figure 1) are particularly common[39] and these have been associated with particular HLA antigens for at least two reasons. Firstly, local "founder" effects have been demonstrated wherein, for example, this author has shown that 21-OH-deficiency patients from different cities in Italy have been shown to have increased frequencies of different HLA types.[40] Secondly, some 21-OH mutations have been shown to involve relatively large DNA deletions[18] which could presumably result in stabilization of the chromosome because recombination is prevented when homologous DNA sequences cannot line up during meiosis. One such haplotype and some of the patients of diverse ethnic background in which it has been found[34,40,41] are illustrated in Table 3. Because of the lack of recombination events, the deficiency allele remains associated, in this case, with the particular HLA alleles HLA-Bw47 and HLA-DR7. As a result, these antigens *appear* to be associated "risk" factors for the disease (see Table 2). In reality, there are frequent 21-OH gene mutations[39] and it is unlikely that any particular HLA determinant per se confers any special risk for the occurence of such mutations or even for the clinical

Table 2
EXAMPLES OF WELL-KNOWN ASSOCIATIONS OF DISEASES WITH PARTICULAR HLA ANTIGENS

Disease	Representative population(s)	Associated HLA Antigen(s)	Postulated reason for association	"Relative risk"[a]	Ref.
Classical 21-hydroxylase deficiency	Caucasian	(A3,Cw6) *B47*, *DR7* (haplotype)	Linked gene; recombination-resistant haplotype	15.2	34
C2 deficiency	Caucasian	A25, B18, DR2 (haplotype)	Linked gene; recombination-resistant haplotype	Data not available	45
Hemochromatosis	Caucasian	A3	Linked gene; founder effect	8.2	47
Ankylosing spondylitis	Caucasian	B27	Route of infection or microbial cross-reactivity	87.4	47
	Japanese	B27		324.5	47
Psoriasis vulgaris	Caucasian	Cw6	Route of infection or microbial cross-reactivity	13.3	47
	Japanese	Cw6		24.9	47
Type I diabetes	Caucasian	DR3(DQw2); DR4(DQw8); DR3 + DR4	Aberrant immune response (IR) epitope(s)	3.3; 6.4; 33.1	53
Graves' disease	Caucasian	B8; DR(Dw)3	Aberrant (IR) epitope(s)	2.3; 3.7	47
Rheumatoid arthritis	Caucasian	DR4; "MC1"	Aberrant (IR) epitope(s)	7.0; 16.1	57
Lepromatous leprosy	Japanese	DR2; DQw1	Aberrant (IR) epitope(s)	3.6; 4.1	58
	Filipino	DR2; DQw1		7.7; 7+[b]	

[a] "Relative risk" (RR) represents the relative chance that a person with the particular HLA antigen (haplotype) will have the disease in comparison with a person without the particular antigen. The RR for these examples, which will be discussed in more detail later in the text, was in two cases calculated for this review from the published data.

[b] Since 100% of the patients were DQw1, the actual value of the RR cannot be calculated.

Table 3
THE DIVERSE DISTRIBUTION OF A Bw47/21-HYDROXYLASE DELETION HAPLOTYPE[a]

Patient I.D.	Ethnic origin	Relevant HLA haplotype									
		HLA-DR	C2	BF	C4A	21-OHA[b]	C4B[c]	21-OHB[c]	HLA-B	HLA-C	HLA-A
EM	German/Polish	7	C	F	1	p	0	0	w47	w6	3
RI	English/Irish	7	C	F	1	p	0	0	w47	w6	3
SH	French	7	ND	F	ND	p	ND	0	w47	w6	3
BO	German	7	C	F	1	p	0	0	w47	w6	25
ST	Lithuanian/French	7	B	F	1	p	0	0	w47	w6	3
LI	English	7	C	F	1	p	0	0	w47	w6	3
LA	English/Irish	ND	ND	F	1	p	0	0	w47	w6	3
TO	Italian	7	C	F	1	p	0	0	w47	w6	3
GR	Irish	7	C	F	1	p	0	0	w47	ND	24
A	Greek	7	C	F	1	p	0	0	w47	ND	11
BA	American black	7	ND	F	ND	p	ND	0	w47	w6	3

[a] Data reinterpreted, in some cases, from published[41,43,44] and unpublished tests performed by M. S. Pollack and G. J. O'Neill at the Memorial Sloan Kettering Cancer Center; ND, not done.
[b] p = pseudogene not expressed on any haplotype.[16,17]
[c] 0 = absence of detectable gene product; this large area of deleted DNA appears to contribute to the stability of this haplotype.

expression of the disease. The apparent associations with particular alleles are an artifact of the relative stability of chromosomes with particular HLA-DR and 21-OH allele combinations. For similar reasons, in this case involving a gene duplication, the B14, DR1 haplotype is statistically associated with cryptic or late - onset (mild) 21-OH deficiency.[34,43,44] The association of the A25, B18, DR2 haplotype with C2 deficiency[45] and the HLA-A3 antigen with hemochromatosis[46] may involve resistance to recombination or, more likely in these cases, a recent mutation ("founder" effect); these diseases are both ethnically and geographically limited in their distribution.

HLA Disease Associations (Multifactorial Diseases) Resulting from Membrane or Antigen Presenting Properties of HLA Molecules

HLA-disease associations resulting from the cell-surface membrane or self/foreign antigen presenting properties of HLA molecules usually involve diseases for which multiple additional etiological factors exist. These factors could include, for example, the necessity for exposure at the right age to a sufficient dose of a specific pathogen; factors relating to hormonal environment (e.g., female vs. male); factors relating to diet and climate, etc. For such diseases, multiple-case families may be rare or very difficult to analyze for linkage. Conventionally, a role for HLA determinants in such diseases was established by calculating the statistical significance of associations with particular antigens among unrelated patients. As in the examples of Table 2, data are often expressed as "relative risk"[47] which is calculated by the following formula:

$$\text{Relative risk} = a \times d/b \times c, \text{ with a, b, c, and d taken from:}$$

	Presence of antigen	
Presence of disease?	Antigen +	Antigen −
Patients +	a (+ +)	b (+ −)
Controls −	c (− +)	d (− −)

As noted above, there are at least two types of these multifactorial HLA-disease associations: those that appear to result because of specific interactions between microbes or their products and particular cell membrane HLA molecules and those which result from aberrant antigen presentation by different class II HLA epitopes.

HLA Disease Associations Resulting from Membrane Properties of Specific HLA Determinants

The HLA-B27-associated arthropathies and the Cw6-associated form of psoriasis are examples of HLA disease associations that appear to result from direct microbe/cell surface molecule interaction effects. In both these examples, the same HLA determinant is present in affected individuals from widely different ethnic and geographical regions.[47] For the HLA-B27-associated arthropathies, plasmids carried by some strains of bacteria (e.g., *Klebsiella, Shigella,* and *Yersinia*) have been postulated to code for factors modifying this specific HLA class I allele.[48] Antisera to one *Klebsiella* strain (K43) have been reported to selectively lyse lymphocytes from B27+ patients with AS, but not B27+ or B27− cells from healthy people. Moreover, 24-h culture filtrates of *Klebsiella* K43 contain a factor that modifies cells from healthy people who have B27 so that they will also be lysed by antibodies to K43. These results suggest that B27+ cells express a receptor for something produced by *Klebsiella* K43. Once modified by this bacterial product, patient target cells may be injured by antibodies specific for the bacterial antigens. However, alternative arguments, including cross-reactivity between a surface antigen on these bacteria and a HLA-B27 determinant,[49] have not been ruled out.

HLA Disease Associations Resulting from Antigen Presentation [Immune Response (IR Gene) Effects]

As reviewed by Gonwa et al.[50] and Parham et al.,[51] T-helper and T-inducer cells generally recognize foreign or modified self-antigens in association with particular epitopes of particular class II HLA molecules on antigen presenting cells, while suppressor and cytotoxic T-cells generally recognize antigens in association with class I HLA molecules. The recently analyzed three-dimensional structure of class I molecules (see Figure 2) suggests that antigens are actually bound within the cleft of the HLA molecule. The T-cell receptor thus interacts simultaneously with an HLA molecule and antigen, and the different polymorphic forms of the class I or class II histocompatibility molecules are said to "restrict" T-cell responses. The epitope that will function with antigen primed T-cells is determined during the initial immune selection. For example, if an individual who is homozygous for HLA-A3, B7, DR2, DQw1, and DPw4 is infected with influenza, infected cells will express influenza antigens and stimulate the expansion of selected T-cell helper, cytotoxic, and suppressor cell clones. The progeny of *particular* T-helper/inducer cell clones would only be restimulated to proliferate and express appropriate functional activity when they are exposed again to virus-infected cells expressing the appropriate, particular class II molecules (i.e., DR2, DQw1, *or* DP4) and the progeny of particular cytotoxic T-cells clones would only lyse virus-infected cells expressing particular appropriate class I antigens (i.e., HLA-A3 *or* HLA-B7). Individual deficiencies relating to class II antigen presentation, which is required both for T-helper/inducer cell proliferation and, therefore, for activation of class I restricted T-suppressor and T-cytotoxic activity, are often called "immune response" (IR) gene defects.

Autoimmune Diseases

The best characterized HLA-disease associations resulting from antigen presentation or IR gene defects include a number of organ-specific autoimmune diseases such as type I diabetes, Graves' disease, and rheumatoid arthritis (see Table 2). Particular class II HLA alleles (or *"trans"* molecules) may be important in the development of aberrant immune responses to cross-reactive antigens of the pathogen and host or to pathogen-modified autologous tissues because there is a relative lack or excess of T-cells with receptors for these particular class II epitope + antigen combinations or because only certain HLA molecules can bind particular antigens. Thus, the HLA region determinants may induce deficient T-suppressor responses or excess T-helper/inducer responses. Although HLA-associated immune response factors may also influence the initial viral infection, the pathogenetically important HLA related response appears to be directed against a determinant(s) created by virus-mediated tissue damage or cross-reactive viral determinants themselves. Several candidate pathogens, e.g., Coxsackie virus or rubella virus, for the initial tissue damage of type I diabetes mellitus, for example, have been proposed.[52] Incidentally, in type I diabetes mellitus, inheritance of either HLA-DR3 (DQw2) or HLA-DR4 (DQw3) is associated with increased risk, but the heterozygote HLA-DR3,4 is at significantly greater risk than individuals with either homozygous phenotype[53] (see Table 2). This particular disease *may* therefore involve a role for deficient suppressor or excess helper responses to antigen presentation by HLA-DQ molecules with novel determinants resulting from *trans*-complementation of gene products from each parental chromosome; since the DR α-chain is not polymorphic, there could be no novel *trans* DR determinants. More likely, however, more than one type of risk factor is involved. One risk factor has recently been shown to specifically involve DQ molecules in that all DR3, all DR4, and even all other DR haplotypes from affected patients have subtypes of DQw2, DQw3, or other DQ molecules in which there is *no* aspartic acid at position 57.[25] In contrast, DR haplotyes which are associated in population and family studies with resistance to type I diabetes all do have aspartic acid at position 57. This amino acid is presumed to be located at a critical site for binding antigens and its absence (or presence) could allow antigen binding that otherwise would not occur.

While in some cases of HLA-associated autoimmune diseases, HLA region genes appear to manifest their effects directly through T-cell-mediated damage, others appear to involve autoantibody production. The production of antibodies to the acetylcholine receptor in myasthenia gravis, to thyroglobulin in Hashimoto's thyroiditis, and to double-stranded DNA in systemic lupus erythematosus are examples of diseases in which a primary role for the relevant autoantibodies in pathogenesis has been documented.[54] In at least nine different autoimmune diseases in which both cell-mediated and antibody-mediated responses appear to be involved (type I diabetes, Graves' disease, celiac disease, Addison's disease, dermatitis herpetiformis, chronic active hepatitis, systemic lupus erythematosus, Sjogren's syndrome, and myasthenia gravis), it is interesting to note that Caucasian patients have a very high frequency of an HLA haplotype that includes determinants for HLA-B8 and HLA-DR3 (which have strong positive linkage disequilibrium). This haplotype is associated in healthy individuals with both hyperimmune antibody responses to conventional immunogens[55] and with suppressor cell defects;[56] i.e., there are deficiencies in natural levels of T-cells with IgG Fc receptors, which have been associated with suppression of some B-cell functions, and in suppressor cells induced by mitogens. Thus, in addition to disease-specific immune response factors, a more general HLA hyperimmune response pattern relating specifically to the HLA-B8,DR3 haplotype may play a role in autoimmune diseases.

One suggested sequence of events is as follows. Cross-reactive pathogens or modified cell surface antigens could directly stimulate immune responses by "forbidden clones" of helper/inducer T-cells that activate self-reactive cytotoxic T-cells or B-cells that form self-reactive antibodies, diverting the host-defense response to an autoaggressive mechanism. An insufficient number of precursor suppressor cells with receptors for the new antigen + self-class II epitope could instead, or in addition, indirectly allow excessive proliferation of these helper T-cells. Since regulatory interactions between suppressor and helper/inducer T-cell subpopulations depend upon the relative precursor frequency of each, genetic susceptibility may appear to be either recessive or dominant in multiple case families although the known gene products of the HLA complex are codominantly expressed on the cell surface. Graves' disease, for example, ordinarily appears to be recessively inherited in familial cases (deficient suppressor cells being compensated for in heterozygotes[52]), while diabetes shows both "recessive" and "dominant" inheritance characteristics.[53]

One could simplistically view this by assuming, as a model, that the basis for the association of HLA-DR4 with rheumatoid arthritis (see Table 2) is that many people *lack* T-suppressor cell receptors for the "DR4 + arthritis factor" (e.g., collagen) epitope bound to a DR molecule, but usually have sufficient numbers of receptors for other DR, "DRx + arthritis factor", epitopes. This would result in an apparent DR4 association with disease susceptibility even though the actual "defect" is at the level of the T-cell receptors. Alternatively, of course, DR4 molecules could more effectively present a self-peptide from the synovium to helper T-cells. Relatively weak HLA-DR, DQ, or DP associations undoubtedly result in some cases because of the heterogeneity of the serologically or cellularly defined marker determinants. Duquesnoy et al.,[57] for example, utilized both alloantisera and a cloned T-cell to identify a class II epitope called MC1 which is more highly associated with rheumatoid arthritis (RA) than DR4 (93.5% of RA patients have MC1 vs. 74.1% for patients with DR4). This epitope is also found on DR1 molecules and DR1 is indeed also associated with RA.

Infectious Diseases

Although there are many examples of HLA allele-autoimmune disease associations, there are only a few cases in which susceptibility to *infection* has been found to be HLA associated. Leprosy is one example of the influence of HLA on susceptibility to infection since both HLA-associations in unrelated patients (see Table 2) and linkage in multicase families have now been demonstrated.[58] Like other mycobacterial infections, an effective host defense to *Mycobacterium leprae* is related predominantly to effective T-cell-mediated immunity; patients with HLA-associated defective T-cell responses develop lepromatous leprosy, characterized by high antibody responses to *M. leprae* antigens, but specific skin test anergy. On the other hand, certain HLA alleles may be associated with relatively more effective cell-mediated responses, hence predisposing to the tuberculoid form of the disease. Haregewoin et al.[59] showed that *in vitro* unresponsiveness of T-cells from patients with lepromatous leprosy was reversed by addition of the T-cell growth hormone interleukin-2 to antigen-stimulated cultures, and Bloom and Mehra[60] directly implicated excessive suppressor T-cell activity in the unresponsiveness of at least some patients with lepromatous disease: removal of T8+ cells often resulted in high levels of antigen responsiveness. The relative importance of HLA gene products and other poorly defined pathogenetic factors (e.g., age of the patient at infection and size of the infectious inoculum) in inducing a state of effector T-cell tolerance remains unresolved, but HLA-associated host factors probably play a similar role in defining the characteristics of many diseases.

In most cases of infectious diseases, however, a clear relationship between HLA antigens and susceptibility to infectious diseases has not been established. For most infectious agents, the multiplicity of antigenic determinants would preclude the detection of the effects of antigen presentation (HLA class II immune response genes) by *individual* epitopes. Because of the antigenic complexity, particularly of bacteria, disease susceptibility may be influenced by the cumulative effects of different precursor frequencies and multiple, HLA-determined, T-suppressor, T-cytotoxic, and T-helper responses.

SUMMARY

HLA determinants are thus markers for diseases caused by or influenced by HLA-linked genes and influence the etiologies of some diseases by their membrane properties. They play an obvious role in the etiology of autoimmune diseases and undoubtedly also play a role in susceptibility to many infectious diseases although these are necessarily more obscure. To the extent that all of these kinds of diseases influence the survival of individuals, these immunogenetic aspects of the HLA determinants contribute to the role of the histocompatibility system in aging.

THE MAJOR HISTOCOMPATIBILITY COMPLEX IN AGING

As described below, both general changes in the functional activities of MHC gene products and differential effects of different polymorphic MHC alleles or linked genes contribute to MHC effects in aging.

Changes in the General Functional Role of MHC Gene Products with Age

Since immune system cell-cell interactions are the major physiological role of MHC gene products, general changes in the dynamics of these interactions with aging are an important component of MHC-aging effects. A great many studies of these changing interactions have been published and only a brief summary of some of the findings can be presented. These include, for example, the finding (from mouse studies) that there is a *loss* with aging in the number of suppressor T-cells which suppress the development of self-reactive cytotoxic T-cells with receptors for (modified) self-MHC determinants (thus increasing opportunities for autoimmune diseases).[61] Human studies have reported decreases with age in T4-helper/inducer cell numbers and mitogen responses.[62] The simultaneous loss in proliferative responses of T4-helper/inducer cells to autologous non-T-cells[63] also suggests a general mechanism for increasing opportunities for autoimmune diseases since suppressor

T-cells recognizing self-MHC determinants are the cells usually activated during that kind of autologous response.

The Role of Specific MHC Alleles in Aging

Individual difference in longevity per se and in the incidence of diseases of aging that relate to major polymorphic histocompatibility complex gene effects can result from at least three different mechanisms: the influence of parental HLA antigen sharing on fetal survival and postnatal immunological and hematological systems; the influence of difference HLA alleles on the incidence of autoimmune and infectious diseases; and the role of HLA-linked genes causing diseases that directly influence longevity and/or affecting other systems that influence aging such as free radical scavenger and DNA repair mechanisms.

Major Histocompatibility Complex Effects on Fetal Survival and Postnatal Immune and Hematological System Disorders

Since "longevity" is influenced even at the fetal stage by MHC genes, it is appropriate to review the increasing body of data relating to MHC influences on fetal health and survival in this section. Specifically, parental HLA antigen sharing has been assumed for several years to lead to an excess of spontaneous abortions. Although this has been difficult to test experimentally, a number of centers have succeeded in demonstrating that successful pregnancies can result when at least some chronically aborting women are immunized with their husband's leukocytes or pooled leukocytes from other donors.[64] Presumably, this immunization allows the production of suppressor T-cells and/or "blocking" antibodies which, in turn, leads to the so-called "immune protection" of the otherwise allogeneic fetus. The nature and route of immunization are critical and probably explain why some centers have been more successful than others.[95]

Data suggesting that an HLA-A locus linked gene(s) influences the formation of neural tube defects and spina bifida[65] are also relevant to this discussion since these affect fetal survival. Both of these conditions are associated with parental HLA-A antigen sharing, and the presumption is made that there is positive linkage disequilibrium between the actual genetic locus that causes the defect and specific HLA-A locus alleles. It is also of interest to note that HLA antigen sharing has been reported in the parents of patients with congenital immunodeficiencies, aplastic anemia, and acute leukemia.[66-68] The failure of the congenitally immunodeficient fetus to reject maternal cells appears to contribute to the clinical course of the immunodeficiency in some cases and such cells have been considered in other cases, in turn, to affect host stem cell populations and thereby lead to increased risks for leukemia or aplastic anemia in older children.

The Role of Specific MHC Alleles as Immune Response Genes in Aging

Since, as noted above, specific HLA alleles influence the incidence of autoimmune diseases and infectious diseases, these alleles per se are likely to influence longevity and the incidence of these types of diseases of aging. In addition, several different forms of cancer, notably both classical and acquired immunodeficiency syndrome (AIDS)-associated Kaposi's sarcoma,[69] mycosis fungoides,[70] chronic lymphocyte leukemia (CLL) and hairy cell leukemia,[71] renal cell carcinoma,[72] and testicular carcinoma[73,74] have been found in population studies to have significant positive associations with the specific HLA-DR antigens DR2 and, especially, DR5.

The specific HLA-DR antigen DR3 is negatively associated with malignancy in Kaposi's sarcoma,[69] hairy cell leukemia,[71] testicular carcinoma,[73] and basal cell carcinoma.[75] In two additional cases, Hodgkin's disease and prostate carcinoma,[76] a negative association with DR3 can be inferred from the data relating to B8 and B18 and the known strong positive linkage disequilibrium of B8 and B18 with DR3.[77] These data are particularly interesting because increases in the antigens B8 and B18 are actually found, but these are found only in retrospective studies and not in prospective studies.[78] The implication has been that B8 and B18 (and therefore DR3) are actually increased in the longer surviving patients because they offer a benefit in relation to survival (or resistance) to the cancer.

Since the HLA-DR antigen DR3 is associated with several autoimmune diseases, as described above, the association of HLA-DR3 with resistance to cancer suggests that the "hyperreactivity" of the HLA-DR3 haplotype may confer resistance to malignancy in general. Although few particular HLA antigens have been reproducibly associated with survival per se,[79] at least one study indicates that males, especially, with the B8 + DR3 haplotype have a general survival advantage,[80] with another study suggesting that women may lack this advantage[81] (possibly because of their relative susceptibility to autoimmune diseases). In addition, individuals with the A1,B8 haplotype (which is usually associated with DR3[77]) were found to lose significantly less capacity for allogeneic responses with age than other individuals.[82]

The Role of HLA Linked Genes Affecting Aging in Nonimmunological Ways

Several HLA linked genes, such as those involving complement deficiency, hemochromatosis, and spinocerebellar ataxia, all influence longevity and/or involve diseases of aging. C2 and C4 complement deficiency cause early death through immune complex disease and hemochromatosis through liver damage and/or diabetes. These diseases are often associated with particular HLA alleles through linkage disequilibrium (see Table 2). An unknown form of complement deficiency associated with the particular linkage disequilibrium haplotype B7, C4A3, appears to influence cognitive loss in Alzheimer's disease, possibly through deficient elimination of virus-infected cells.[83] Although spinocerebellar ataxia (SCA) is only distantly linked to HLA and there are no particular HLA-SCA associations, particular HLA antigens are markers for genetic risk for this disease of aging in individual families[84,85] and could be used for genetic counseling and even prenatal diagnosis. Moreover, it is highly likely that subclinical effects on aging of other variant alleles at the HLA-linked SCA locus will eventually be found. Other examples of HLA-linked gene effects on aging and diseases of aging will doubtless become apparent as molecular genetic analyses begin to better define the regions *between* the known HLA and complement genes (see Figure 1). Alleles of TNF are very likely candidates since this lymphokine has considerable effects on immunologic and inflammatory events and is already known to influence the HLA-associated disease SLE.[35,86]

Other HLA-linked genes that influence human aging could include genes regulating levels of free radical accumulation such as superoxide dismutase (SOD) and catalase. Walford and colleagues have shown that levels of repair capacity for UV-induced DNA damage segregate with alleles of the MHC in inbred mice and that induced levels of mixed function oxidases (MFO), which "detoxify" simple organic chemicals and are correlated with life span differences between species, are differentially influenced in mice by different MHC alleles.[87-89] However, the possible influence of different human MHC alleles or linked genes on *human* SOD and MFO systems remains to be determined.

Similarly, several studies have shown that particular MHC class I alleles influence the metastatic properties of tumor cells and affect intracellular responses to cell surface hormone/receptor interactions in mice.[90,91] Although many of these studies have likewise not yet been extended to humans, studies have shown the effects of specific MHC alleles on the formation of neoantigens by antibiotics and hormone/receptor interactions.[92,93] Endorphin binding, for example, is influenced by HLA class I type as, therefore, is the response to therapy in schizophrenia.[94]

Studies of nonimmunological interactions of MHC cell surface molecules with other receptors and substances are only in their preliminary phases. It can thus only be speculated that different polymorphic MHC gene products will ultimately be shown to have differential influences on individual variations in responses to hormones and drugs, etc. in relation to aging. The fundamental importance of MHC gene products in the immune response and the important effects of MHC linked genes, as noted above, are, however, already sufficient evidence for the significance in health, disease, and aging of these phylogenetically ancient cell-cell recognition molecules.

ACKNOWLEDGMENTS

The author wishes to thank Dr. Roy Walford, University of California, Los Angeles, for helpful discussions and Leesa Oliver and Lucy Stewart for expert typing and editorial assistance. The preparation of this review was supported in part by National Institutes of Health (NIH) Grant CA40552.

REFERENCES

1. **Hildemann, W. H., Linthieum, D. S., and Vann, D. C.**, Transplantation and immunoincompatibility reactions among reef-building corals, *Immunogenetics*, 2, 269, 1975.
2. **Scofield, V. L., Schlumpberger, J. M., West, L. A., and Weissman, I. L.**, Protochordate allorecognition is controlled by a MHC-like gene system, *Nature (London)*, 295, 499, 1982.
3. **Hildemann, W. H.**, Specific immunorecognition by histocompatibility markers: the original polymorphic system of immunoreactivity characteristic of all multicellular animals, *Immunogenetics*, 5, 193, 1977.
4. **Doan, C. A.**, The recognition of a biological differentiation in the whole blood cells with special reference to blood transfusion, *JAMA*, 86, 1593, 1926.
5. **Dausset, J. and Nenna, A.**, Presence d'une leuco-agglutinine dans le serum d'un cas d' agranulocytose chronique, *Soc. Biol. (Paris)*, 146, 1539, 1952.
6. **Dausset, J.**, Iso-leuco-anticorps, *Acta Haematol.*, 20, 156, 1958.
7. **Orr, H. T., Lopez de Castro, J. A., Parham, P., et al.**, Comparison of amino acid sequences of two human histocompatibility antigens, HLA-A2 and HLA-B7: location of putative alloantigenic sites, *Proc. Natl. Acad. Sci. U. S. A.*, 76, 4395, 1979.
8. **Parham, P.**, Histocompatibility typing - Mac is back in town, *Immunol. Today*, 9, 127, 1988.
9. **Kaufman, J. E., Auffray, C., Korman, A. J., Shackelford, D. A., and Strominger, J.**, The class II molecules of the human and murine major histocompatibility complex, *Cell*, 36, 1, 1984.
10. **Ko, H. S., Fu, S. M., Winchester, R. J., Yu, D. T. Y., and Kunkel, H. G.**, Ia determinants on stimulated human T lymphocytes, *J. Exp. Med.*, 150, 246, 1979.
11. **Maurer, D. H. and Pollack, M. S.**, The use of gamma interferon to increase HLA antigen expression on cultured amniotic cells used for the prenatal diagnosis of 21-hydroxylase deficiency, congenital adrenal hyperplasia, *Ann. N.Y. Acad. Sci.*, 458, 148, 1985.
12. **Pober, J. S., Gimbrone, M. A., Jr., Cotran, R. S., Reiss, C. S., Burakoff, S. J., Fiers, W., and Ault, K. A.**, Ia expression by vascular endothelium is inducible by activated T cells and by human interferon, *J. Exp. Med.*, 157, 1339, 1983.
13. **Gibbs, V. C., Wood, D. M., and Garovoy, M. R.**, The response of cultured human kidney capillary endothelium to immunologic stimuli, *Hum. Immunol.*, 13, 55, 1985.
14. **Giles, R. C. and Capra, J. D.**, Biochemistry of MHC class II molecules, *Tissue Antigens*, 25, 57, 1985.
15. **Bodmer, W. F., et al.** Nomenclature for factors of the HLA system, 1987, *Immunogenetics*, 28, 391, 1988.
16. **Bodmer, J. and Bodmer, W.**, Histocompatibility 1984, *Immunol. Today*, 5, 251, 1984.
17. **Carroll, M. C., Campbell, R. D., Bentley, D. R., and Porter, R. R.**, A molecular map of the human major histocompatibility complex class III region linking complement genes C4, C2 and factor B, *Nature (London)*, 307, 237, 1984.
18. **White, P. C., New, M. I., and Dupont, B.**, HLA-linked congenital adrenal hyperplasia results from a defective gene encoding a cytochrome P-450 specific for steroid 21-hydroxylation, *Proc. Natl. Acad. Sci. U. S. A.*, 81, 7505, 1984.

19. **Carroll, M. C., Campbell, R. D., and Porter, R. R.**, Mapping of steroid 21-hydroxylase genes adjacent to complement component C4 genes in HLA, the major histocompatibility complex in man, *Proc. Natl. Acad. Sci. U. S. A.*, 82, 521, 1985.
20. **Spies, T., Morton, C. C., Nedospasov, S. A., Fiers, W., Pious, D., and Strominger, J. L.**, Genes for the tumour necrosis factors alpha and beta are linked to the human major histocompatibility complex, *Proc. Natl. Acad. Sci. U. S. A.*, 83, 8699, 1986.
21. **Giles, R. C., DeMars, R., Chang, C. C., and Capra, J. D.**, Allelic polymorphism and transcomplementation of molecules encoded by the HLA-DQ subregion, *Proc. Natl. Acad. Sci. U. S. A.*, 82, 1776, 1985.
22. **Goyert, S. M., Shively, J. E., and Silver, J.**, Biochemical characterization of second family of human Ia molecules, HLA-DS, equivalent to murine I-A subregion molecules, *J. Exp. Med.*, 156, 550, 1982.
23. **Shaw, S., Johnson, A. H., and Shearer, G. M.**, Evidence for a new segregant series of B cell antigens that are encoded in the HLA-D region and that stimulate secondary allogeneic proliferative and cytotoxic responses, *J. Exp. Med.*, 152, 565, 1980.
24. **Angelini, G., Bugawan, T. L., Delfino, L., Erlich, H. A., and Ferrava, G. B.**, HLA-DP typing by DNA amplification and hybridization with specific oligonucleotides, *Hum. Immunol.*, 26, 169, 1989.
24a. **Bidwell, J.**, DNA-RFLP analysis and genotyping of HLA-DR and DQ antigens, *Immunol. Today*, 9, 18, 1988.
25. **Gorski, J., Niven, M. J., Sachs, J. A., Mach, B., Cassell, P. G., Festenstein, H., Awad J., and Hitman, G. A.**, HLA-DR alpha, -DX alpha, and DR beta III gene association studies in DR3 individuals, *Hum. Immunol.*, 20, 273, 1987.
26. **Todd, J. A., Bell, J. I., and McDevitt, H. O.**, HLA-DQ Beta gene contributes to susceptibility and resistance to insulin-dependent diabetes mellitus, *Nature (London)*, 329, 599, 1987.
27. **Yankee, R. A., Grumet, F. C., and Rogentine, G. N.**, Platelet transfusion therapy. The selection of compatible platelet donors for refractory patients by lymphocyte HLA typing, *N. Eng. J. Med.*, 281, 1208, 1969.
28. **Ting, A. and Morris, P. J.**, Matching for B cell antigen of the HLA-DR series in cadaver renal transplantation, *Lancet*, 1, 575, 1978.
29. **Thorsby, E., Moen, T., Solhein, B. G., et al.**, Influence of HLA matching in cadaveric renal transplantation: experience from one Scandiatransplant center, *Tissue Antigens*, 17, 83, 1981.
30. **Dupont, B., O'Reilly, R. J., Pollack, M. S., and Good, R. A.**, Histocompatibility testing for clinical bone marrow transplantation and prospects for identification of donors other than HLA genotypically identical siblings, in *Immunobiology of Bone Marrow Transplantation*, Thierfelder, S., Rodt, H., and Kolb, H. J., Eds., Springer-Verlag, Berlin, 1980, 121.
31. **Schacter, B. Z., Hsu, S. H., and Bias, W. B.**, HLA and other genetic markers in disputed paternity, *Transplant. Proc.*, 9, 233, 1977.
32. **Wolski, K. P., Schmid, F. R., and Mittal, K. K.**, Genetic linkage between the HLA system and deficit of the second component (C2) of complement, *Science*, 188, 1020, 1975.
33. **Pollack, M. S., Ochs, H. D., and Dupont, B.**, HLA typing of cultured amniotic cells for the prenatal diagnosis of complement C4 deficiency, *Clin. Genet.*, 18, 197, 1980.
34. **Dupont, B., Pollack, M. S., Levine, L. S., O'Neill, G. J., Hawkins, B. R., and New, M. I.**, Congenital adrenal hyperplasia, in *Histocompatibility Testing 1980*, Terasaki, P. I., Ed., Tissue Typing Laboratory, University of California, Los Angeles, 1980, 693.
35. **Jacob, C. O. and McDevitt, H. O.**, Tumour necrosis factor-alpha in murine autoimmune "lupus" nephritis, *Nature (London)*, 331, 356, 1988.
36. **Edwards, C., Cartwright, G., Skolnick, M., and Amos, B.**, Genetic mapping of the hemochromatosis locus on chromosome six, *Hum. Immunol.*, 1, 19, 1980.
37. **Suciu-Foca, N., Rohowsky, C., Godfrey, M., Khan, R., O'Neill, G., Starkman, S., and Johnson, W.**, HLA and spinocerebellar ataxia, in *Histocompatibility Testing 1980*, Terasaki, P. I., Ed., Tissue Typing Laboratory, University of California, Los Angeles, 1980, 943.
38. **Zoghbi, H. Y., Sandkuyl, L. A., Ott, J., Daiger, S. P., Pollack, M., O'Brien, W. E., and Beaudet, A. L.**, Assignment of autosomal dominant spinocerebellar ataxia centromeric to the HLA region on the short arm of chromosome 6 using multilocus linkage analysis, *Am. J. Hum. Genet.*, 44, 255, 1989.
39. **Speiser, P. W., Dupont, B., Rubinstein, P., Piazza, A., Kastelan, A., and New, M. I.**, High frequency of nonclassical steroid 21-hydroxylase deficiency, *Am. J. Hum. Genet.*, 37, 650, 1985.
40. **Pollack, M. S., New, M. I., O'Neill, G. J., et al.**, HLA antigens and HLA-linked genetic markers in Italian patients with classical 21-hydroxylase deficiency, *Hum. Genet.*, 58, 331, 1981.
41. **O'Neill, G. J., Dupont, B., Pollack, M. S., Levine, L. S., and New, M. I.**, Complement C4 allotypes in congenital adrenal hyperplasia due to 21-hydroxylase deficiency: further evidence for different allelic variants at the 21-hydroxylase locus, *Clin. Immunol. Immunopathol.*, 23, 312, 1982.

42. **Cobain, T. J., Stuckey, M. S., McCluskey, J., Wilton, A. N., Gedeon, A., Garlepp, M. J., Christiansen, F. T., and Dawkins, R. L.**, The co-existence of IgA deficiency and 21-hydroxylase deficiency marked by specific MHC supratypes, *Ann. N.Y. Acad. Sci.*, 458, 76, 1985.
43. **Levine, L. S., Dupont, B., Lorenzen, F., Pang, S., Pollack, M. S., Oberfield, S. E., Kohn, B., Lerner, A., Cacciari, E., Mantero, F., Cassio, A., Scaroni, C., Chiumelli, G., Rondanini, G. F., Gargantini, L., Giovannelli, G., Virdis, R., Bartolotta, E., Migliori, C., Pintor, C., Tato, L., Barboni, F., and New, M. I.**, Genetic and hormonal characterization of cryptic 21-hydroxylase deficiency, *J. Clin. Endocrinol. Metab.*, 53, 1193, 1981.
44. **Kohn, B., Levine, L. S., Pollack, M. S., Pang, S., Lorenzen, F., Levy, D., Lerner, A. J., Rondanini, G. F., Dupont, B., and New, M. I.**, Late-onset steroid 21-hydroxylase deficiency: a variant of classical congenital adrenal hyperplasia, *J. Clin. Endocrinol. Metab.*, 55, 817, 1982.
45. **Alper, C.A., Awdeh, Z. L., Raum, D. D., and Yunis, E. J.**, Extended major histocompatibility complex haplotypes in man: role of alleles analogous to murine t mutants, *Clin. Immunol. Immunopathol.*, 24, 276, 1982.
46. **Simon, M., LeMignon, L., Fauchet, R., Yaouanq, J., David, V., Edan, G., and Bourel, M.**, A study of 609 HLA haplotypes marking for the hemochromatosis gene: (1) mapping of the gene near the HLA-A locus and characters required to define a heterozygous population and (2) hypothesis concerning the underlying cause of hemochromatosis-HLA association, *Am. J. Hum. Genet.*, 41, 89, 1987.
47. **Ryder, L. P., Andersen, E., and Svejgaard, A., Eds.**, *HLA and Disease Registry*, Munksgaard, Copenhagen, 1979.
48. **Sullivan, J. S., Prendergast, J. K., and Geczy, A. F.**, The etiology of ankylosing spondylitis: does a plasmid trigger the disease in genetically susceptible individuals, *Hum. Immunol.*, 6, 185, 1983.
49. **Avakian, H., Welsh, J., Ebringer, A., and Entwistle, C. C.**, Ankylosing spondylitis, HLA-B27 and *Klebsiella* II cross-reactivity studies with human tissue typing sera, *Br. J. Exp. Pathol.*, 61, 92, 1980.
50. **Gonwa, T. A., Peterlin, B. M., and Stobo, J. D.**, Human Ir genes: structure and function, *Adv. Immunol.*, 34, 71, 1983.
51. **Parham, P., Coppin, H., Engelhard, V. H., Herman, A., Holmes, N., Holterman, M., and Ways, J. P.**, Biochemical approaches to understanding the structure and function of class I MHC (HLA-A,B,C) molecules, in *Lymphocyte Surface Antigens*, Heise, E. R., Ed., American Society for Histocampatibility and Immunology, 1984, 2.
52. **Sasazuki, T., Nishimura, Y., Muto, M., and Ohta, N.**, HLA-linked genes controlling immune response and disease susceptibility, *Immunol. Rev.*, 70, 51, 1983.
53. **Svejgaard, A., Platz, P., and Ryder, L. P.**, Insulin-dependent diabetes mellitus, in *Histocompatibility Testing 1980*, Terasaki, P. I., Ed., Tissue Typing Laboratory, University of California, Los Angeles, 1980, 638.
54. **Fauci, A. S., Lane, H. C., and Volkman, D. J.**, Activation and regulation of human immune responses: implications in normal and disease states, *Ann. Intern. Med.*, 99, 61, 1983.
55. **Reznikoff-Etievant, M. F., Muller, J. Y., and Patereau, C.**, An immune response gene linked to MHC in man, *Tissue Antigens*, 22, 312, 1983.
56. **Lawley, T. J., Hall, R. P., Fauci, A. S., Katz, S. I., Hamburger, M. I., and Frank, M. M.**, Defective Fc-receptor functions associated with the HLA-B8/DRw3 haplotype. Studies in patients with dermatitis herpetiformis and normal subjects, *N. Engl. J. Med.*, 304, 185, 1981.
57. **Duquesnoy, R. J., Marrari, M., Hackbarth, S., and Zeevi, A.**, Serological and cellular definition of a new HLA-DR associated determinant, MC1, and its association with rheumatoid arthritis, *Hum. Immunol.*, 10, 165, 1984.
58. **Pollack, M. S. and Rich, R. R.**, The HLA complex and the pathogenesis of infectious diseases, *J. Infect, Dis.*, 151, 1, 1985.
59. **Haregewoin, A., Mustafa, A. S., Helle, I., Waters, M. F. R., Leiker, D. L., and Godal, T.**, Reversal by interleukin-2 of the T cell unresponsiveness of lepromatous leprosy to *Mycobacterium leprae*, *Immunol. Rev.*, 80, 77, 1984.
60. **Bloom, B. R. and Mehra, V.**, Immunological unresponsiveness in leprosy, *Immunol. Rev.*, 80, 5, 1984.
61. **Gorczynski, R. M., Kennedy, M., and Macrae, S.**, Altered lymphocyte recognition repertoire during ageing. III. Changes in MHC restriction patterns in parental T lymphocytes and diminution in T suppressor function, *Immunology*, 52, 611, 1984.
62. **Thompson, J. S., Wekstein, D. R., Rhoades, J. L., Kirkpatrick, F., Brown, S. A., Roszman, T., Straus, R., and Tietz, N.**, The immune status of healthy centenarians, *J. Am. Geriatr. Soc.*, 32(4), 274, 1984.
63. **Moody, C. E., Innes, J. B., Staiano-Coico, L., Incefy, G. S., Thaler, H. T., and Weksler, M E.**, Lymphocyte transformation induced by autologous cells. XI. The effect of age on the autologous mixed lymphocyte reaction, *Immunology*, 44, 431, 1981.

64. **Beer, A. E., Quebbeman, J. F., Ayers, J. W. T., and Haines, R. T.**, Major histocompatibility complex antigens, maternal and paternal immune responses and chronic habitual abortions in humans, *Am. J. Obstet. Gynecol.*, 141, 987, 1981.
65. **Schacter, B., Weitkamp, L. R., and Johnson, W. E.**, Parental HLA compatibility, fetal wastage and neural tube defects: evidence for a T/t-like locus in humans, *Am. J. Hum. Genet.*, 36, 1082, 1984.
66. **Hansen, J. A., Good, R. A., and Dupont, B.**, HLA-D compatibility between parent and child: increased occurrence in severe combined immunodeficiency and other hematopoietic diseases, *Transplantation*, 23, 366, 1977.
67. **Werner-Favre, C. and Jeannet, M.**, HLA compatibility in couples with children suffering from acute leukemia or aplastic anemia, *Tissue Antigens*, 13, 307, 1979.
68. **Chan, K. W., Pollack, M. S., Braun, D., Jr., O'Reilly, R. J., and Dupont, B.**, Distribution of HLA genotypes in families of patients with acute leukemia, *Transplantation*, 33(6), 613, 1982.
69. **Pollack, M. S., Safai, B., and Dupont, B.**, HLA-DR5 and DR2 are susceptibility factors for acquired immunodeficiency syndrome with Kaposi's sarcoma in different ethnic subpopulations, *Dis. Mark.*, 1, 135, 1983.
70. **Safai, B., Myskowski, P. L., Dupont, B., and Pollack, M. S.**, Association of HLA-DR5 with mycosis fungoides, *J. Invest. Dermatol.*, 80, 395, 1983.
71. **Winchester, R., Toguchi, T., Szer, I., et al.**, Association of susceptibility to certain hematopoietic malignancies with the presence of Ia allodeterminants distinct from the DR series; utility of monoclonal antibody reagents, *Immunol. Rev.*, 70, 155, 1983.
72. **DeWolfe, W. C., Lange, P. H., Shepard, R., et al.**, Association of HLA and renal cell carcinoma, *Hum. Immunol.*, 1, 41, 1981.
73. **Aiginger, P., Schwarz, H. P., Kumits, R., et al.**, HLA-A,B,C and DR antigens and testicular cancer: a prospective study, *Proc. Am. Assoc. Cancer Res.*, 24, 190(Abstr.), 1983.
74. **Pollack, M. S., Vugrin, D., Hennessy, W., et al.**, HLA antigens in patients with germ cell cancer of the testis, *Cancer Res.*, 42, 2470, 1982.
75. **Myskowski, P. L., Pollack, M. S., Schor, E. S., et al.**, HLA associations in basal cell carcinoma, *Clin. Res.*, 31, 590A, 1983.
76. **Terasaki, P. I., Perdue, S. T., and Mickey, M. R.**, HLA frequencies in cancer: a second study, in *Genetics of Human Cancer*, Mulvihill, J. J., Miller, R. W., and Fraumeni, J. F., Jr., Eds., Raven Press, New York, 1977, 321.
77. **Baur, M. P. and Danilovs, J. A.**, Reference tables of two and three locus haplotype frequencies for HLA-A,B,C,DR,Bf, and GLO, in *Histocompatibility Testing 1980*, Terasaki, P. I., Ed., Tissue Typing Laboratory, University of California, Los Angeles, 1980, 994.
78. **Hors, J. and Dausset, J.**, HLA and susceptibility to Hodgkin's disease, *Immunology*, 70, 167, 1983.
79. **Hodge, S. E. and Walford, R. L.**, HLA distribution in aged normals, in *Histocompatibility Testing 1980*, Terasaki, P. I., Ed., Tissue Typing Laboratory, University of California, Los Angeles, 1980, 722.
80. **Proust, J., Moulias, R., Fumeron, F., Bekkhoucha, F., Busson, M., Schmid, M., and Hors, J.**, HLA and longevity, *Tissue Antigens*, 19, 168, 1982.
81. **Greenberg, L. J. and Yunis, E. J.**, Histocompatibility determinants, immune responsiveness and aging in man, *Fed. Proc.*, 37, 1258, 1978.
82. **Batory, G., Onody, C., Gyodi, E., Nemeskeri, J., and Petranyi, G. Gy.**, HLA and T-lymphocyte function in old age, *Hum. Immunol.*, 7, 187, 1983.
83. **Huletter, C. M. and Walford, R. L.**, HLA associations in Alzheimer's disease, in *Biological Substrates of Alzheimer's Disease*, Scheibel, A. and Wechsler, A., Eds., University of California, Los Angeles, 1985/86.
84. **Haines, J. L., Schut, L. J., Weitkamp, L. R., Thayer, M., and Anderson, V. E.**, Spinocerebellar ataxia in a large kindred: age at onset, reproduction, and genetic linkage studies, *Neurology*, 34, 1542, 1984.
85. **Zoghbi, H. Y., Pollack, M. S., Lyons, L. A., Ferrell, R. E., Daiger, S. P., and Beaudet, A. L.**, Spinocerebellar ataxia: variable age of onset and linkage to HLA in a large kindred, *Ann. Neurol.*, 23 580, 1988.
86. **Boswell, J. M., Yul, M. A., Burt, D. W., and Kelley, V. E.**, Increased tumor necrosis factor and IL-1 B gene expression in the kidneys of mice with lupus nephritis, *J. Immunol.*, 141, 3050, 1988.
87. **Hall, K. Y., Bergmann, K., and Walford, R. L.**, DNA-repair, H-2, and aging in NZB and CBA mice, *Tissue Antigens*, 17, 104, 1981.
88. **Gottesman, S. R. S., Hall, K. Y., and Walford, R. L.**, A thesis of genetic linkage of immune regulation and aging: the major histocompatibility complex as a supergene system, in *Developmental Immunology: Clinical Problems and Aging*, University of California, Los Angeles, 1982, 247.
89. **Walford, R. L. and Imamura, T.**, The major histocompatibility complex as an antibiosenescent system, in *Histocompatibility Testing 1984*, Albert, E. D., Ed., Springer-Verlag, Berlin, 1984, 644.

90. **Noah, I., Feldman, M., and Segal, S.,** Loss of the H-2.33 private specificity by 3LL tumor cells correlates with the tumor potential to metastasize across the H-2K region genetic barriers, *Exp. Clin. Immunogenet.*, 1, 170, 1984.
91. **Ivanyi, P.,** Some aspects of the H-2 system, the major histocompatibility system in the mouse, *Proc. R. Soc. (London) Ser. B*, 202, 117, 1978.
92. **Edidin, M.,** MHC antigens and non-immune functions, *Immunol. Today*, 4, 269, 1983.
93. **Schreiber, A. B., Schlessinger, J., and Edidin, M.,** Interaction between major histocompatibility complex antigens and epidermal growth factor receptors on human cells, *J. Cell. Biol.*, 98, 725, 1984.
94. **Claas, F. H. J., van Ree, R. M., Verhoeven, W. M. A., van der Poel, J. J., Verduyn, W., de Wied, D., and van Rood, J. J.,** The interaction between γ-type endorphins and HLA class I antigens, *Hum. Immunol.*, 15, 347, 1986
95. **McIntyre, J.,** personal communication.

INDEX

INDEX

A

Acetylcholine, 7, 220
Acetylcholine receptors, 26—27
Acetylcholinesterase, 85, 168
N-Acetylglucosamine, 166
Acquired immunodeficiency syndrome (AIDS), 222
ACTH; see Adrenocorticotropin
Actin, 42—43, 63, 165, 169
Action potential, 26
Actomyosin, 49
Addison's disease, 220
Adenosine diphosphate (ADP), 62
Adenosine monophosphate (AMP), 13, 62
Adenosine triphosphatase (ATPase), 26, 28
   calcium-activated, 48—49
   decrease in activity of, 63
   histochemistry of, 4, 28—32
   magnesium-dependent, 49
   myosin, 41—42, 45, 62—63
   ouabain-sensitive, 86
   in red cell membranes, 168
Adenosine triphosphate (ATP), 37, 113, 119
   levels of, 123—124
   in red blood cells, 85
S-Adenosyl-L-methionine, 127
Adenylate kinase, 85
Adhesion, 198
Adipocytes, 109
Adipose tissue, 109
ADP; see Adenosine diphosphate
Adrenaline, 13
Adrenocorticotropin (ACTH), 14
Agglutination, 87
Aging
   blood and, 89—90
   bone marrow and, 90—91
   granulocytes and, 89—90
   hematopoiesis and, 89
   HLA and, 222—224
   immune response genes in, 222—223
   lymphocytes and, 90
   major histocompatibility complex in, 221—224
   oxidation of red cell proteins during, 127—128
   platelets and, 90
   red blood cell protein modifications during, 126—128
   of red blood cells, 89, 122—125
AIDS; see Acquired immunodeficiency syndrome
Aldoses, 126
Alloimmunogenicity, 211
Amino acids, 11, 64; see also specific types
AMP; see Adenosine monophosphate
Androgens, 88, 89
Anemia, 88—89; see also specific types
   aplastic, 222
   sickle cell; see Sickle cell anemia
Ankyrin, 165, 169
Antibody responses, 203—204
Antigens, 86, 168, 170—171, 211; see also specific types
   cross-reactive, 219
   HLA and, 218—219
   presentation of, 198—199
   processing of, 198—199
Aplastic anemia, 222
Arthritis, 219—220
Ascorbic acid, 127
Asparagine, 166
Asparaginyl, 127
Aspartyl, 127
Aspirin, 11
ATP; see Adenosine triphosphate
ATPase; see Adenosine triphosphatase
Autoantibodies, 220
Autocrine factors, 7
Autoimmune diseases, 215, 219—223; see also specific types
5-Azacytidine, 144

B

Basal cell carcinoma, 222
Basal lamina, 14—15
Basophilic cells, 109
B-cells, 80, 90, 179, 181, 183
   cytotoxic, 220
   development of, 189—192, 201—202
   early development of, 189—192
   functional studies of responses of, 202—203
   HLA and, 213
   maturation of, 183—184, 189—192
   suppression of, 220
Beckwith-Wiedemann syndrome, 12
BFU; see Burst-forming units
Blood; see also specific components
   aging and, 89—90
   volume of, 87
Blood cells; see also specific types
   development of; see Hematopoiesis
   formation of; see Hematopoiesis
   red; see Red blood cells
   white, 78—80, 87—88, 211, 222
Blood clotting, 80, 82
Blood groups, 170—171
Blood islands, 104—106
Blood transfusions, 213—215
B-lymphocytes; see B-cells
Bone marrow, 102, 104
   aging and, 89—91
   B-cells in, 184
   differentiation in, 121
   lymphoid tissue and, 182—183
   in newborn, 86—88

postnatal development of, 109—110
reticulocyte release from, 116
structure of, 191
T-cells in, 184, 186
transplantation of, 214
Bone marrow-blood barrier, 109
Bone marrow hematopoiesis, 108—109
Bupivicaine, 64, 68
Burst-forming units (BFU), 121, 141, 170—171

C

Caesarean section, 83
Calcium, 26, 28, 42, 47—48
Calcium binding proteins, 49; see also specific types
Calcium pump, 48
Calsequestrin, 48—49
Capillary endothelial cells, 105
Carbamylcholine, 7
Carbohydrates, 166, 168, 170; see also specific types
Carbonic anhydrase, 85
Carboxyl methylated proteins, 127
Carboxyl methyltransferases, 127
Carcinoma, 222
Catalase, 85, 115
Catecholamines, 13, 28, 32; see also specific types
Celiac disease, 220
Cell death, 15
CFU; see Colony forming units
Chemoluminescence, 88
Cholesterol, 124, 165, 167
Cholinergic agonists, 7
Cholinergic antagonists, 6
Chronic active hepatitis, 220
Chronic lymphocyte leukemia (CLL), 222
CLL; see Chronic lymphocyte leukemia
Coagulation; see Blood clotting
Collagens, 14—15, 60
Colony forming units (CPU), 89, 121—122, 141, 170—171
Complement, 199—200, 215
Concanavalin A, 86
Congenital infection, 206—207
Connective tissue, 60, 62
Contractile proteins, 30—31, 37—43; see also specific types
Contractility of muscles, 25—32, 62
Creatine, 62
Cyclic AMP, 13, 62
Cytochrome-b reductase, 168
Cytosine arabinoside, 144
Cytoskeleton, 27

D

Degeneration of muscle, 15, 64—65
2-Deoxyglucose, 9
Deoxyhemoglobin A, 114
Dermatitis herpetiformis, 220
Dexamethasone, 13
Diabetes, 12, 219—220

Diet, 58
Digoxin receptor sites, 86
Diphosphoglycerate (DPG), 86, 114—115, 117, 119, 120
  binding of, 127
  levels of, 123—124
  oxygen affinity of, 127
Diphosphoglycerate (DPG) mutase, 122
Diphosphoglycerate (DPG) phosphatase, 119
Disopropyl fluorophosphate, 117
DNA
  double-stranded, 220
  RNA polymerase dependent on, 62
DPG; see Diphosphoglycerate

E

Embden-Meyerhof pathway, 113
Embryogenesis, 12
Endocytosis, 86, 169
Endomysium, 14
Endoplasmic reticulum (ER), 28
Endorphins, 223
Endothelial cells, 105
Endurance training, 63
Energetics, 49—52
Enolase, 85, 119
Enzymes, 123; see also specific types
  glycolytic, 168
  mitochondrial, 63
  nonglycolytic, 85
  oxidant stress and, 115
  oxidative system, 45
  red cell membranes and, 168
ER; see Endoplasmic reticulum
Erythroleukemic cells, 169
Erythroblastopenia, 123
Erythroblasts, 105, 136, 169
Erythrocytes; see Red blood cells
Erythroid cells, 77—78, 86—89, 105
  differentiation of, 121, 169—171
  primitive, 107
Erythropoiesis, 77—78, 88, 104, 107, 109
  in childhood, 88—89
  fetal, 78, 120, 143
  in infancy, 88—89
  major site of, 166
  metabolic changes during, 121—125
  stress, 143—144
Erythropoietin, 78, 87, 105, 108—109, 154
  in liver hemopoiesis, 108—109
  in newborn, 86—87
  in yolk sac erythropoiesis, 105
Euglobulin, 82
Evolution, 101—102

F

Fatigue, 26, 43
Fats, 45; see also specific types
Fatty acids, 11, 14, 113; see also specific types

Ferritin, 86
FGF; see Fibroblast growth factor
Fibrinogen, 80, 82
Fibrinolytic activity, 82
Fibroblast growth factor (FGF), 10—14
Fibroblasts, 8, 10
Folate, 89
Fungoides, 222

G

Galactokinase, 85
Galactose, 85
Galactose oxidase, 169
Gangliosides, 170
Gene deletion, 155—156
Genetic abnormalities of hemoglobin, 149; see also specific types
Genetics, 64, 149—150, 211—213
Globin chains, 77, 167
Glucocorticosteroids, 13; see also specific types
Glucose, 124, 127
Glucose-6-phosphate dehydrogenase, 85, 123—126
Glutamic-oxalacetic transaminase, 123—124
Glutathione, 85, 113, 115, 118, 127
Glutathione peroxidase, 85, 115, 120
Glutathione reductase, 85, 115
Glyceraldehyde, 85
Glycoconjugates, 168, 170
Glycogen, 5, 47, 61
Glycogen phosphorylase, 46
Glycolipids, 168, 170; see also specific types
Glycolysis, 113, 115, 118
Glycolytic enzymes, 168; see also specific types
Glycolytic pathway, 85
Glycopeptides, 169; see also specific types
Glycophorin, 165, 169—170
Glycoproteins, 168—171; see also specific types
Glycosylated protein, 126
Glycosylation, 126—127, 170
Granulocyte-macrophage precursor cells, 107
Granulocytes, 77, 80, 89—90
Granulopoiesis, 77, 107, 109
Graves' disease, 219—220
Growth factor receptors, 8
Growth factors, 8; see also specific types
    fibroblast (FGF), 10—14
    insulin-like (IGF), 8—10, 14
    platelet-derived (PDGF), 13—14
    transforming (TGF), 10—11, 14
Growth hormone, 10, 14, 58

H

Hairy cell leukemia, 222
Hashimoto's thyroiditis, 220
Hassals corpuscles, 204
HbC, 153—154, 158
HbE, 153—154, 158
HbS, 154
Hemangioblasts, 104—105

Hematocrits, 78, 82—83, 88—89
Hematocytoblasts, 77
Hematopoiesis, 78
  aging and, 89
  at birth, 82—85
  bone marrow, 108—109
  definitive, 105—107
  fetal, 77—78
  liver, 107—108
  migratory, 101
  ontogeny of, 101—110
    bone marrow and, 109—110
    embryonic development and, 102—109
    evolution and, 101—102
    fetal development and, 102—109
  origin of, 101
  primitive, 105—107
  settlement of, 101—102
  sites of, 102
  splenic, 108
  yolk sac, 104—105
Hemoblasts, 105
Hemochromatosis, 215, 218
Hemocytes, 101
Hemoglobin A, 86, 88, 105, 118, 121, 126—127
Hemoglobin F, 86, 88, 105, 115, 120
  hemoglobin switching and, 135—137
  HPFH and, 138, 142
  predominance of, 141
Hemoglobinopathies, 137—138, 143, 149, 151—154; see also specific types
Hemoglobins, 77—78, 82—83, 86, 88, 105, 135—144; see also specific types
  adult, 136—137
    cellular basis for production of, 141—142
    production of, 137—142
  aging and, 89
  binding of, 125
  damage to, 127
  denatured, 127
  disappearance of, 87
  fetal, 117, 120, 136—137
    cellular basis for production of, 141—142
    production of, 137—144
    reactivation of, 142—143
    therapeutic manipulation of production of, 143—144
  genetic abnormalities of, 149; see also specific types
  low-affinity, 121
  mean corpuscular, 84
  oxidization of, 115
  oxygen affinity of, 120—121
  production of, 135
    adult, 137—142
    cellular basis for, 141—142
    fetal, 137—144
    therapeutic manipulation of, 143—144
  subtypes of, 135
  switching of, 135—137
Hemolysis, 85

Hemopoiesis; see Hematopoiesis
Hemopoietic precursor cells, 105
Hemopoietin; see Erythropoietin
Hemostatis, 80—82
Hepatitis, 220
Hereditary persistence of fetal hemoglobin (HPFH), 138—140, 142, 153
Hexokinase, 115, 119, 123—124
Histochemical muscle fiber types, 61
Histochemistry
    ATPase, 4, 28—30, 31—32
    muscle, 43—45
    myosin, 4
Histocompatibility system, 211—224
    in aging, 221—224
    blood transfusions and, 213—215
    in disease, 215—221
    HLA system and, 211—215
    relationship testing and, 213—215
    T-cell-mediated, 221
    transplantation and, 213—215
Histogenesis, 3—5
HLA, 211—215, 223
    aging and, 222—224
    autoimmune diseases and, 220
    T-cells and, 221
HLA disease, 215—219
HLA-linked genes, 223—224
HLA-linked genetic mutations, 215—218
Hodgkin's disease, 222
Homing, 198
Hormones, 108; see also specific types
    growth, 10, 14, 58
    HLA disease and, 218
    interactions between, 14
    melanocyte-stimulating (MSH), 14
    myoblasts and, 7—14
    pituitary, 58
    thyroid, 12, 31—32
HPFH; see Hereditary persistence of fetal hemoglobin
Humoral immune system, 80
Hyaluronidase, 68
Hydrogen peroxide, 115
21-Hydroxylase, 213, 215
Hydroxyl radicals, 115
Hydroxyurea, 144
Hyperplasia, 57
Hyperthyroidism, 31—32
Hypertrophy of skeletal muscle, 57—64
Hypothyroidism, 31—32

I

IGF; see Insulin-like growth factor
Immunity, 197—208; see also specific types
    complement and, 199—200
    in congenital infection, 206—207
    humoral, 80
    nonspecific, 197—199
    in perinatal infection, 206—207
    specific, 200—207
    tolerance and, 205—207
Immunocompetence, 211
Immunodeficiencies, 222
Immunoglobulin, 204; see also specific types
Immunoglobulin A, 80, 203
Immunoglobulin D, 80
Immunoglobulin G, 80, 125, 203
Immunoglobulin genes, 201
Immunoglobulin M, 80
Immunohistochemistry of myosin, 4
Immunoregulation, 211
Indomethacin, 11
Infantile muscular atrophy, 30
Infectious diseases, 220—222; see also specific types
Inositol lipids, 62
Insulin, 9, 12, 58
    binding sites for, 86
    dexamethasone and, 13
    glucose and, 127
Insulin-like growth factor (IGF), 8—10, 14
Insulin receptors, 12, 171
Integral membrane proteins, 169—170; see also specific types
Interferon, 206
Internal membrane system, 28
Interstitial cells, 62
Iron, 89
L-Isoaspartyl, 127
Isoleucine, 64
Isoproterenol, 13
Isozymes, 31, 37, 40—42, 122

J

JCML; see Juvenile chronic myelogenous leukemia
Juvenile chronic myelogenous leukemia (JCML), 142—143

K

Kaposi's sarcoma, 222
Ketoamine, 126
Ketoses, 126
Klebsiella spp., 218

L

Lactate dehydrogenase, 63
Lactic acid dehydrogenase, 119
Lactoaminoglycans, 170
Lactogen, 10
Lactosaminoglycans, 166, 171
Laminin, 14—15
Lectins, 170; see also specific types
Lepromatous leprosy, 221
Leprosy, 220—221
Leucine, 64
Leukemia, 142, 222
Leukocytes; see White blood cells

Leukotrienes, 14
Lipids, 45; see also specific types
  inositol, 62
  in red blood cell membranes, 165
  in red blood cells, 167
Lipoproteins, 45; see also specific types
Liver hematopoiesis, 107—108
Lupus, 220
Lymph nodes, 181, 183
Lymphocytes, 78, 80, 87, 108, 181
  aging and, 90
  B-; see B-cells
  subclasses of, 179
  T-; see T-cells
Lymphocytosis, 87
Lymphoid cells, 107
Lymphoid tissues, 179—192
  B-cells and; see B-cells
  bone marrow and, 182—183
  liver and, 181
  ontogeny of, 179—183
  spleen and, 182
  T-cells and; see T-cells
  thymus and, 181—182
Lymphopoiesis, 80

M

Macrocytes, 135
Macrophages, 107, 197
Magnesium, 49
Major histocompatibility complex (MHC), 211—213, 221—224
Maternal antibody transfer, 203
MCV; see Mean cellular volume
Mean blood volume, 87
Mean cell diameter, 77
Mean cellular volume (MCV), 77, 84, 89
Mean corpuscular hemoglobin, 84
Megakaryocytes, 77, 80
Megakaryopoiesis, 107
Megaloblasts, 135
Melanocyte-stimulating hormone (MSH), 14
Membrane potential, 26
Membranes; see also specific types
  characteristics of, 26—27
  internal, 28
  phospholipids of, 113
  plasma, 165
  red cell, 86, 165—171
Mesenchyme, 105
Messenger RNA, 121, 156
Metabolism
  erythropoiesis and changes in, 121—125
  glutathione, 113
  oxidative, 45
  red blood cell; see under Red blood cells
  red cell precursor, 121—122
Methemoglobin, 113, 115, 117
Methemoglobinemia, 85

Methemoglobin reductase, 85
Methylation, 127
MHC; see Major histocompatibility complex; Myosin heavy chains
Microbial killing, 198
Microtubules, 88
Mitochondria, 121
Mitochondrial enzymes, 63; see also specific types
Monocytes, 80, 213
Morphogenesis of muscle, 27—28
Motoneurons, 8, 25, 29—30
MSA; see Multiple stimulating activity
MSH; see Melanocyte-stimulating hormone
Multifactorial diseases, 218—219; see also specific types
Multiple stimulating activity (MSA), 8
Muscle, 3—15; see also Myogenesis; specific types
  contractile proteins and, 30, 37—43
  contractility of, 25—32
  degeneration of, 15, 64—65
  energetics of, 49—52
  fast, 25
  fatigue in, 26, 43
  fiber production and, 15
  glycogen and, 47
  histochemistry of, 43—45
  histogenesis and, 3—5
  lipids and, 45
  morphogenesis of, 27—28
  myoblasts and; see Myoblasts
  myosin ATPase of newborn, 41—42
  oxidative system enzymes and, 45
  phosphorylase and, 46—47
  physiological development of, 25—27
  potentiation of, 26
  regeneration of, 6, 57, 65
  regulatory influences on, 31—32
  sarcoplasmic reticulum (SR) and, 28, 37, 42, 47—49, 52
  size of, 15
  skeletal; see Skeletal muscle
  slow, 25
  stretch of, 62
  tension in, 25, 62
Muscle fibers
  contractile proteins and, 37—43
  differentiation of, 25—32
  histochemical types of, 61
  loss of, 15
  production of, 15
  types of, 28—30
Muscular atrophy in infants, 30
Muscular dystrophy, 30
Mutations, 156—157, 215—218; see also specific types
Myasthenia gravis, 220
Mycobacterium leprae, 221
Mycosis fungoides, 222
Myelodysplastic disorders, 143
Myeloid cells, 78, 87, 107

Myoblasts, 3—5
  basal lamina and, 14—15
  catecholamines and, 13
  diversity of, 5—6
  fusion of into myotubes, 49
  hormonal influences on, 7—14
  hormones and, 14
  insulin and, 12
  L6, 9
  laboratory animals and, 5—6
  neural influences on, 6—7
  PDGF and, 13—14
  POMC and, 14
  proliferating, 8
  testosterone and, 13
  varieties of, 29
Myofibrils, 27—28, 37
  contractility and, 25—32
  postnatal muscle growth and, 57
Myogenesis, 4, 6—8; see also Muscle
  catecholamines and, 13
  contractility and, 25
  FGF and, 10
  leukotrienes and, 14
  PDGF and, 13
  prostaglandins and, 11
  secondary, 65
  timing of, 5
Myogenic precursors of muscles, 3
Myosin, 5, 28—30, 62
  actin binding properties of, 63
  developmental changes in, 37—41
  fast, 40—41
  loss of, 32
  slow, 40—41
  structure of, 37
  transition of, 31
Myosin ATPase, 41—43, 45, 62—63
Myosin heavy chains (MHC), 31—32, 37—38, 41, 68
Myosin immunohistochemistry, 4
Myotubes, 6, 8, 11
  ATPase and, 29
  basal lamina and, 14
  catecholamines and, 13
  degenerating, 15
  differentiation of, 27
  fast myosin and, 40
  fiber production and, 15
  formation of, 3, 27
  glucocorticosteroids and, 13
  hormones and, 14
  insulin and, 12
  multinucleated, 68
  myoblast fusion into, 49
  primary, 4—8, 37, 40
  regenerating, 68
  secondary, 4—6, 29—30, 40

N

NADH; see Nicotamide adenine dinucleotide
Natural killer cells, 199
NBT; see Nitroblue tetrazolium
NE; see Noradrenaline
Nesidioblastosis, 12
Neural influences on myoblasts, 6—7
Neutrophils, 87—88, 197
Nicotamide adenine dinucleotide (NADH), 45, 113
Nitroblue tetrazolium (NBT), 88
NK; see Natural killer cells
Nonenzymatic glucosylation, 126—127
Nonglycolytic enzymes, 85; see also specific types
Nonsense mutation, 156
Nonspecific immunity, 197—199
Noradrenaline, 13
Normoblasts, 121
Nucleated cells, 86, 105; see also specific types
Nucleoproteins, 45; see also specific types
Nucleoside phosphorylase, 120

O

Oligosaccharides, 166, 170; see also specific types
Ontogeny
  of hematopoiesis; see Hematopoiesis
  of lymphoid tissues, 179—183
  red cell membrane changes during, 166—168
Ornithine decarboxylase, 64
Orthochromatic normoblasts, 121
Orthophosphate, 85
Osmotic fragility, 89
Ouabain-sensitive ATPase, 86
Oxidant stress, 115
Oxidation, 126
Oxidative metabolism, 45
Oxidative phosphorylation, 116
Oxidative system enzymes, 45; see also specific types
Oxygen affinity, 120—121, 127
Oxygen delivery of red blood cells, 86
Oxyhemoglobin dissociation curve, 154

P

Paroxysmal nocturnal hemoglobinuria, 143
PDGF; see Platelet-derived growth factor
Pentose phosphate pathway, 113, 115
Peptides; see also specific types
Perinatal infection, 206—207
PFK; see Phosphofructokinase
Phagocytic cells, 197—198
Phagocytosis, 197—198
Phentolamine, 13
3-Phosphate dehydrogenase, 85
Phosphate pathway, 85, 113
Phosphatidylcholine, 165, 167
Phosphatidylethanolamine, 165
Phosphatidylinositol, 167
Phosphatidylserine, 165
Phosphoenolpyruvate, 126
Phosphofructokinase, 63, 85, 115, 118—120, 126
Phosphoglucomutase, 119
6-Phosphogluconate dehydrogenase, 85

Phosphoglucose isomerase, 85, 120, 126
Phosphoglycerate kinase, 85, 126
Phospholipids, 45, 113, 124, 165, 167; see also specific types
Phosphorylase, 46—47, 63
Phosphorylation, 116, 167
Pituitary hormones, 58
Plasma membranes, 165
Plasminogen, 82
Platelet-derived growth factor (PDGF), 13—14
Platelets, 78—80, 88, 90, 211
PMNs; see Polymorphonuclear neutrophils
Polyamines, 62; see also specific types
Polycythemia, 154
Polymorphism, 157
Polymorphonuclear neutrophils (PMNs), 87, 197
POMC; see Proopiomelanocortin
Postnatal muscle growth, 57
Potassium, 26, 86, 113
Potentiation of muscle, 26
Pre-B-cells, 201
Preleukemias, 143
Proerythroblasts, 169
Promoter mutations, 156—157
Proopiomelanocortin (POMC)-derived peptides, 14
Propranolol, 13
Prostaglandin E, 11—13
Prostaglandins, 11—12, 14, 27; see also specific types
Prostate carcinoma, 222
Proteins; see also specific types
   calcium binding, 49
   carboxyl methylated, 127
   contractile, 30—31, 37—43
   covalent modification of, 126—128
   degradation of, 61—62
   glycosylated, 126
   integral membrane, 169—170
   methylation of, 127
   oxidation of during aging, 127—128
   red cell, 126—128
   red cell membrane, 165, 167
   reticulocyte maturation and, 125—128
   synthesis of, 61—62
   thin filament, 42—43
Pyruvate kinase, 115, 120, 122—124, 126

R

Racemization, 126
Rapaport-Luebering shunt, 113—115
Red blood cells, 78, 84
   aging of, 89, 122—125
   circulating, 83
   count of, 82—83, 88
   diphosphoglycerate of, 86
   immature; see Reticulocytes
   lipids in, 167
   membranes of, 86, 165—171
   metabolism of, 85—86
      adult, 115—121
      fetal, 115—121
      newborn, 115—121
      pathways in, 113—115
   modification of proteins in during aging, 126—128
   nucleated, 84, 116
   number of, 77
   oxidation of proteins of during aging, 127—128
   oxygen delivery of, 86
   precursors of, 121—122
   in sickle cell anemia, 152
   size of, 77
   survival of, 87
   volume of, 87
Regeneration of muscle, 6, 57, 65
Regulatory influences on muscle, 31—32
Relationship testing, 213—215
Renal cell carcinoma, 222
Renal failure, 215
Repair of skeletal muscle, 64—65
Reticulocytes, 78, 84, 116
   characteristics of, 124, 128
   count of, 88, 118, 120, 123
   enucleation and, 169
   maturation of, 121—128
   protein changes during maturation of, 125—128
   transferrin and, 171
Reticuloendothelial system, 85, 121
Rheumatoid arthritis, 219—220
Ribosomes, 121
RNA, 121, 156
RNA polymerase, 62

S

Sarcoma, 222
Sarcomeres, 37
Sarcoplasmic reticulum (SR), 28, 37, 42, 47—49, 52
SDH; see Succinate dehydrogenase
Senescence, 121—128
Serine-linked tetrasaccharides, 166
Serum complement, 200
Sialation, 170
Sialoglycoprotein, 165, 169
Sickle cell anemia, 85, 135, 140, 142, 151—153
   cellular aspects of, 152
   clinical aspects of, 153
   molecular aspects of, 151—152
Sickle cell trait, 152—153
Sjogren's syndrome, 220
Skeletal muscle, 57—68
   adaptability of, 57
   biochemical adaptations in, 62—63
   cellular processes in, 68
   contractile properties of, 62
   degeneration of, 64—65
   embryology of, 3
   histochemical fiber types of, 61
   hypertrophy of, 57—64
   metabolic adaptations in, 62—63
   postnatal growth of, 57
   protein degradation and, 61—62

protein synthesis and, 61—62
regeneration of, 57, 65
repair of, 64—65
stretch of, 62
transplantation and, 65—68
work-induced growth of, 58—60
SM; see Somatomedins
SOD; see Superoxide dismutase
Sodium, 26
Sodium-potassium pump, 113
Somatomedin receptors, 12
Somatomedins, 8—10, 13, 62
Somites, 3
Specific immunity, 200—207
Spectrin, 165, 167, 169
Spermidine, 62, 64
Spermine, 62
Sphingomyelin, 165, 167
Spina bifida, 222
Splenic dysfunction, 85
Splenic hematopoiesis, 108
SR; see Sarcoplasmic reticulum
Stem cells, 106—107, 109, 121, 179
Stress; see also specific types
Stress erythropoiesis, 143—144
Stretch of muscles, 62
Succinate dehydrogenase (SDH), 45
Superoxide, 115
Superoxide dismutase (SOD), 115
Superoxide radicals, 115
Surface antigens, 168
Syncytiotrophoblasts, 171
Systemic lupus erythematosus, 220

T

T-cell-mediated immunity, 221
T-cell receptors, 219
T-cells, 80, 90, 179, 181—182
  cytotoxic, 219—221
  deficiencies in, 220
  development of, 184—189
  excess of, 219
  helper, 219, 221
  HLA and, 213
  inducer, 219
  lack of, 219
  maturation of, 183—189
  peripheral, 206
  receptor binding sites of, 211
  repertoire of, 204—205
  suppression of, 206
  suppressor, 219, 221—222
Tension in muscle, 25, 62
Termination codon mutation, 157
Testicular carcinoma, 222

Testosterone, 13
Tetrasaccharides, 166
TGF; see Transforming growth factor
Thalassemia, 135, 138, 140, 144, 149, 150, 154—157
  clinical aspects of, 157
  gene deletion and, 155—156
  molecular causes of, 155—157
Thalassemia syndromes, 89
Thalassemic hemoglobinopathies, 149, 157—158; see also specific types
Thiamphenicol, 121
Thin filament proteins, 42—43; see also specific types
Thymus, 181—182, 204
Thyroglobulin, 220
Thyroid hormones, 12, 31, 32
Thyroiditis, 220
Thyrosine, 32
Thyroxine, 12, 31, 109
T-lymphocytes; see T-cells
TNF; see Tumor necrosis factor
Tolerance, 205—207
Transferrin, 124, 171
Transferrin receptor, 171
Transforming growth factor (TGF), 10—11, 14
Transfusions, 213—215
Transplantation, 65—68, 213—215
Transverse tubular system, 28, 37, 42, 47—48
Tropomyosin, 42—43
Troponin, 42—43, 47—48
Troponin C, 42
Troponin I, 42
T-tubules; see Transverse tubular system
Tumor necrosis factor (TNF), 213
Typhoid fever, 64

V

Valine, 64
Vinblastine, 144
Vitamin B, 89
Vitamin E, 127

W

White blood cells, 78—80, 87—88, 211, 222
Work-induced growth, 58—60

Y

Yolk sac hematopoiesis, 104—105

Z

Zenker's waxy degeneration, 64, 127